EMERGENCE IN LANDSCAPE ARCHITECTURE

Emergence is the way in which complex systems and patterns arise out of a multiplicity of relatively simple interactions. Related to chaos theory and self-organizing systems, emergence highlights the ever changing and developing urban and natural world – and the need to work flexibly within this.

Just as an arborist must understand the development and functions of a tree in order to fully understand his or her actions in relation to it, so must a landscape architect understand the ecological functions of a specific terrain in order to appreciate how the landscape will respond to the conditions he or she establishes. This bottom-up approach is essential when dealing with the natural environment – and, as Barnett argues, the urban environment.

Covering critically the theory behind emergence in landscape architecture, the author also uses practical examples from international landscapes as a key tool in his mission to explain the basis of emergence and how it is critical to our understanding of both urban and natural systems.

Ideal for students taking courses in landscape architecture, landscape urbanism and architectural theory, this book provides a full discourse on the theory and practice behind emergence in landscape and features full color images.

Rod Barnett is Chair of the Master of Landscape Architecture course at Auburn University. He has published extensively on emergence theory in landscape architecture and taught design studios using techniques he has developed to explore emergent processes in the landscapes both of cities and rural areas. He has designed landscapes in New Zealand, Australia, the Pacific Islands and the United States.

EMERGENCE IN LANDSCAPE ARCHITECTURE

Rod Barnett

Routledge
Taylor & Francis Group
LONDON AND NEW YORK

First published 2013
by Routledge

4 Park Square, Milton Park, Abingdon, Oxon OX14 4RN

Simultaneously published in the USA and Canada
by Routledge

605 Third Avenue, New York, NY 10017

Routledge is an imprint of the Taylor & Francis Group, an informa business

British Library Cataloguing in Publication Data
A catalogue record for this book is available from the British Library

Library of Congress Cataloging in Publication Data
Barnett, Rod.
Emergence in landscape architecture/Rod Barnett.
pages cm
Includes bibliographical references and index.
1. Landscape architecture – Philosophy. 2. Emergence (Philosophy)
3. Landscape ecology. I. Title.
SB472.B355 2013
712 – dc23
2012028860

ISBN: 978-0-415-50655-7 (hbk)
ISBN: 978-0-415-50656-4 (pbk)

Typeset in Bembo and Stone Sans
by Florence Production Ltd, Stoodleigh, Devon, UK

CONTENTS

FIGURES

ACKNOWLEDGMENTS

This book is the result of twelve years of thinking, writing and teaching emergence in landscape architecture. Initially, I called the subject matter nonlinear landscape design, as the early experiments were mostly studio-based investigations of technique. Many, perhaps befuddled and certainly be-knighted, students were involved, and some of them probably wondered whether what we were doing had much relation to landscape architecture at all. To those undergraduate landscape architecture students at Unitec in Auckland who accompanied me into the new millennium in a blaze of nonlinearity I owe a huge debt of gratitude. As I learned more about natural processes my enthusiasm for emergence theory changed focus, away from design technique towards the possibility of an emergentist landscape architectural practice based on the properties of landscapes themselves. At this time a Master of Landscape Architecture was developed at Unitec, centered on research by design. A new wave of students helped me explore emergent processes in the urban context by developing their own projects around the investigation of open systems. I heartily thank these students as well, for their cross-referencing of nonlinear conditions with landscape urbanist notions of field, systems theory and indeterminacy in design.

A third generation of students, this time at Auburn University in Alabama, became guinea pigs – hostages to emergence – when I began to teach at that university, both in a seminar devoted to exploring emergence in landscape architecture and in the design thesis studio. The results of these investigations have been the most fruitful so far, and to the Masters students who devoted their time to my obsession I extend my gratitude and respect. I particularly thank my research assistants Kelly Homan, Qian Deng, Sudha Thapaliya, Pratisha Shakya, Long Zhou and Bonnie Szakel, who were instrumental in the preparation of this book. Students whose work I have used I acknowledge as better designers by far than myself, and these creative individuals are cited in the figure captions.

I have been lucky to have always had the support of colleagues, at Unitec New Zealand and at Auburn University. Two better groups of people could not be wished for. Lastly I thank my children Cassandra, Imogen, Sylvia and Boris, all of whom have suffered silently through my distractions. And then Jacqueline, my wife, without whom nothing is possible at all.

The following texts are included here with kind permission of the journals in which they originally appeared. Aspects of Landscapes of Noir first appeared as 'A Chance Encounter' in *Zx #4*, Art School Press, Manukau School of Visual Arts; and the section on the sacred groves of Ancient Greece was first published as 'Sacred Groves: Sacrifice and the Order of Nature in Ancient Greek Landscapes' in *Landscape Journal*, 26:2, University of Wisconsin Press.

01

INTRODUCTION

It might seem that a book on emergence in landscape architecture now would be enacting the end of a phase transition, reporting on a state of things that is moving to some new condition. After all, emergence in landscape architecture has been the next big thing for over fifteen years. The point, however, and the purpose of this book, in its explanation of emergence, is to suggest that in landscape architecture emergence has not yet emerged. It is not just that emergence in landscape architecture has been too closely tied to ecology – which it has – and therefore severely curtailed with respect to scope; nor that it has been completely entranced by problems associated with urbanism, which it has. The real reason for a book on emergence in landscape architecture now, is that there is no more useful conceptual tool for understanding the participation of humans in the nonhuman world they inhabit, and which landscape architects are continuing to construct. You will see immediately that the discussion is not about the social world exclusively – how humans live together – but that this moment of emergence in landscape architecture is an opportunity carefully to inquire as to the state of what used to be called subject and object, or human and world. We are suspicious of this distinction precisely because ideas associated with emergence (multiplicity, heterogeneity, difference, holism) have put into question the neat separation of a finite, isolated and singular subjectivity from an ontologically distinct condition that science studies, the physical world, a realm that has qualities that our privileged position enables us to gather into a category we call knowledge. The question "what is emergence in landscape architecture" therefore entails further questions about how humans know the world, how they interact with it, and how the world is thought to receive and respond to humanity. This is a question about the open character of the world.

> How is it that we have found inspiration from speculative philosophers when reflecting on the discovery by physics of its open character? The hypothesis that we would like to offer is the following: for these philosophers, it is . . . a matter of an experimental approach – not an experimentation on nature but on concepts and their articulations, an experimentation in the art of posing problems and of following the consequences with the most extreme rigor.
>
> (Stengers 1997: 55)

This book will lay out an experimental approach to concepts and their articulation in landscape architecture. Conducting this experiment will require the proposition that ecological thought is necessary in order to uncover the potential of emergence in landscape architecture, but not sufficient. To make up the shortfall it will recover from speculative philosophy the primacy of becoming, and connect this directly to an art of solving problems in landscape architecture. This art will be a matter of making. What this book makes, as part of that project, will be strong connections between encounter, disturbance and affect critical to the construction by landscape architects of the milieux in which human and nonhuman becomings occur. The landscape architectural domains that emergence theory might enrich, other than the ecological, include the historical, the cultural, the political – and the urbanistic which, while always found in these territories, will not be the only structuring condition and not the only terrain that is explored. This involves taking the idea of emergence that is scattered throughout a range of recent texts and projects, analyzing it and looking closely at its constituent parts; evaluating its pros and cons and identifying the challenges it sets; showing how it can be used to re-examine aspects of landscape architecture that have been overlooked. A certain openness is required. Held within the simple concept of *becoming* are a million ideas about the world and life, and accessing these ideas completely requires more than a straightforward exegesis. Emergence is everywhere. The most profound music, the great myth-systems, poems of triumph and loss, sprawling dynastic novels, insect metamorphosis, schools of sharks, prairie ecosystems, hurricanes born at sea, family traditions, drunken conversations, horizontal corporate structures, street life, economic boom and bust cycles, global trade relations, geopolitics, the dynamics of international airplane disasters. Urban landscapes. Our particular interests will be on the list. We will stick to them. But landscape architecture is not simply about the reformulation of urban terrain, or shaping new kinds of infrastructure, though these are important aspects of its mission. It is also the pattern of sunlight that radiates across a child's soft skin or the rain that falls through a fig tree to a curved concrete floor that has been shaped to receive it.

An elderly man holds a buttercup under his granddaughter's chin and says "It's glowing. That means you love butter." Landscape architecture also involves the way that thought crashes into consciousness like a wave on an angled quoin. Landscape architecture is not just the presentation of natural process, but the interaction of seemingly disparate realms of experience as with the convergence of wild animals and urban conditions in DeStefano's *Coyote at the Kitchen Door*, a book which explores the complex relationships between human and nonhuman life. Landscape architects create the sites of articulation of these relationships. This asks of the designer an overt contract with all the processes of life; it certainly does not require an effacement of the designer's intention, or a modest receding behind the abundance and mystery of nature. Instead it involves a critical engagement with the world through the adaptive re-use of styles, perspectives and languages of design as well as a sincere, motivated authorial involvement with material conditions, the articulation of human–human and human–nonhuman relations, even as these relations co-evolve through myriad instantiations of life and world. The book hopes in some small way to bring this idea of a collective evolution of the things of the world to life.

Cultural geographer John Wylie has noted that landscape is something we are *in*, and something we *look* at. Our experience of it has been thought to be both phenomenal and cognitive. This has required that we attend to the kinesthetic and sensorial, that is to say, the embodied nature of landscape encounter, as well as to the means by which landscape is represented to us by ourselves and by others. The nature of our embodied encounter with landscape, the story goes, is one both of construction and inhabitation. None of us are innocent of landscape construction – it is one of the great things we learn to do as we grow, historicizing and acculturating as we go. But both construction and inhabitation are one and the same thing. Cognition is a way of seeing, just as it is a way of feeling and thinking, and these are brought together in our encounter with landscape. It is with this encounter that the multiplicities of self and world are set free in, and by, the simple act of perception. The encounter is not mediated, it does not cross some theorized divide between mind and world, self and other, subject and object. It will be argued in this book that an emergentist approach to landscape architecture sees all the elements of the world as simple existents that are freely interpretable by any person, any animal, any thing, since all entities are understood as becomings whose ongoing purpose is a self-differentiation in which everything participates.

To construct, or interpret, landscape is to value some aspects and ignore others, to see what we want to see and all too often to not see, or not value, what is there. Many constructed landscapes have become naturalized – the farm landscapes of England and Wales for instance – and we resist the naturalization of others – railway yards, parking lots, canal systems. How we experience, value and consider landscapes depends primarily on how we respond to their representation across a range of discourses. David Matless argues, as have others, that "the power of landscape resides in it being simultaneously a site of economic, social, political and aesthetic value" (Matless 1998: 12). His book on landscape and Englishness includes countless photographs of people in landscapes, hiking, cycling, skiing, picnicking, harvesting, felling trees, building cottages. The book analyses these activities – showing how specific practices of landscape encounter have constructed a regime of English landscape cultures – and simultaneously participates in the ongoing construction of that regime. Matless clearly demonstrates the imbricated nature of the phenomenal and the representational. An emergentist landscape architecture does not distinguish between embodied practices and representational strategies. The reader will find in the pages of this current book an ongoing struggle to overcome the tendency for discussion to precipitate solely into one or the other.

Values live in words and dictate their use and therefore the formation of concepts. Writing in the overlapping zones of inquiry that landscape architecture inhabits requires a resilience to problems with words. There seems little point in using scare quotes for contested, ambiguous words such as nature and experience, or words that migrate from discipline to discipline taking on and shedding meanings as they go. Jargon will always be tricky – the specialist writer is tantalized by its seeming precision. Just as futile is the adoption of a skeptical tone toward conceptual debate. Rather than become distracted by these problems this book simply plows on. It is hoped that readers can supply their own distancing strategies and sift through the issues associated with the binaries that plague discussions which insist on separation when none exists. An example is the use of the collective pronouns us and them. Who is the we that is referred to in the last paragraph? Another example is the separation, just mentioned, between the representational and the sensorial. Kathryn Moore has argued convincingly that this division is based on the old, rationalist separation of thinking and sensing, the idea that the senses deliver a world to us that we then, through nonsensorial operations, attempt to know. It is a conceptual framework that has led to landscape discourse mobilizing ideas of sense of place and image of place as different things. The very notion of place is saturated with the rationalist devaluation of visuality (Moore 2012). Thinking of landscapes as emergent systems actually helps to avoid thinking of them in terms of images or places and adds another dimension to our understanding of how they work and how they work on us.

While this book does not go deeply into complex systems theory, as there are plenty of readable texts that illuminate this science, it necessarily draws on complexity, and uses some of its terms. Ultimately the book advances an argument that there is no need to split science and society, that science is necessary in the development of a theory of emergence in landscape architecture, but does not exhaust the potential of emergence as a theoretical basis for landscape architecture. Systems can be considered as performative, and this is a useful way to evaluate constructed landscapes: how well they perform when evaluated against set criteria. Certainly systems ecologists are interested in how biophysical landscapes function, and social ecologists are interested in how human environments fulfill human needs. However, it is important to take great care in the conceptual formation of landscapes as systems that have inputs, through-flows and outputs. If, as was mentioned earlier, landscape architecture is a way of expressing human–human and human–world relationships, and the great and complicated modes of interaction that these relationships call up, then it will be important – critical – to infuse performance-thinking with other types of heuristics. The word heuristic has its roots in the Greek *heuriskein*, to discover. Other pathways to discovery must be deployed. We live in a pluriverse rather than a universe,[1] a fact often noted by ecologists who argue that complex ecosystems must be observed through different frames, from multiple perspectives and scales, if they are to be understood. Each perspective on an ecosystem reveals only one facet of its complex and changing states (Kay 2008: 30). An example of this multi-perspectival thinking is seen in a series of books written from 1969 to 1974 in which philosopher Michel Serres invokes the figure of the Greek god Hermes to explain how it is necessary to move through different "regional epistemologies" gathering knowledge and understanding. Hermes is a complicated god. Known for his cunning and ingenuity, he is a protector of travelers and of boundaries and intersections. Mythology tells us that Hermes learned the art of telling the future by tracing pebbles on the ground. A geographer of space, or as Serres says, a weaver of space who crosses many spaces in his travels, Hermes calls attention to the myriad zones in which we simultaneously live; constantly on the move he guides the living (and the dead) across otherwise disparate spaces, connecting, disconnecting and reconnecting us with the different kinds of knowledge to be found in different places. Hermes the voyager explores without condition, and makes discoveries, about mathematics, music, weights and measures, relationships and scales: he is the god both of commerce and of theft, and therefore an expert on all types of exchange. "Mythical discourse undertakes a weaving together, a junction, a connection of places that are closed, isolated, inviolable, inaccessible, dangerous, or mortal – disconnected, in any case" (Harari and Bell in Serres 1982a: xxxiii).

This book is also a voyage of discovery that crosses many regions. In an effort not to be beguiled by the sirens of any one epistemology, it charts its itinerary across a range of conceptual materials. The passage is against the wind; the vessel tacks back and forth, weaving its course through turbulent waters. Ideas from Bruno Latour, and particularly Gilles Deleuze and Felix Guattari are crossed with current thinking in ecosystem science and cultural geography. From Deleuze and Guattari comes the important consideration that historical and social processes are material processes and that the world itself is a material reality. Nature is not a social construction, although the ongoing construction of nature is one of the great political activities of science and society. Instead the real processes, objects and relations that comprise the world, and which landscape architecture undertakes to design, organize and manage, are continually unfolding, producing further relations and making new connections. Further, this unfolding is not of a prior set of terms, or of a structure or substance with its points already laid out in a way that can be mapped; its movement is precisely productive of all the thoughts and concepts by means of which its possibilities may be understood. Emergence theory can help us understand the world not as an object or a process to be known, observed or represented, but as a dynamic, shifting assemblage in which humans, plants, animals, rivers, clouds, trees, suburbs, subways, heritage districts and all manner of other things are embedded, shifting and unfolding along with everything else. To work within this assemblage requires the development of a landscape architectural "logic of sensation," to use Deleuze's term. We are interested in the visceral, and the neural, reception of the world by the bodies that move through it; the enhancement and empowerment of those bodies through their encounter with the landscapes that we conjure out of our involvement with the world and all the beings that inhabit it. It is therefore an affective landscape architecture that emergence makes possible, one that develops a world held and inhabited unconditionally in common by – and this is Bruno Latour now – humans and nonhumans whose habits of living, whose matters of concern are respected by the collective they all form. A landscape architecture that sees the world as emergent and self-creative is one that will help to build a republic, Latour calls it, of humans and nonhumans that explore their common world, adjusting and improving it as they go (Latour 2004).

This weaving of the Deleuzian concepts expression, affect and sensation with Latour's vision of a republic of becomings shuttles into the rhythmic, open-ended thinking of what will be referred to here as the new ecologists, such as Nina-Marie Lister, Kristina Hill and James Kay and the recent works of cultural geographers like John Wylie, Hayden Lorimer and Mitch Rose. These writers, in ecology and cultural geography, have themselves been deeply influenced by Deleuze and Guattari just as much as they have by advances in the study and application of emergence theory to their disciplines. In recent ecology we find the problematization of the privileged scientific observer alongside a growing respect for the power of variable framing and scaling operations to overcome single perspective descriptions of phenomena. There is also an increasing emphasis on seeing the systems that are being observed as just as contingent and partial as the analyses that are made of them. Lister and Kay both describe their approaches as non-normal science, by which they mean it is neither reductive nor objective for, as Kay puts it, "there is not a true and correct perspective, just different perspectives that provide us with more or less insights" (Kay 2008: 32).

The recent twentieth anniversary of Peter Jackson's *Maps of Meaning,* a watershed publication in cultural geography, occasioned a series of reflections on the trajectory of that discipline in the journal *cultural geographies*. Contributors acknowledged the increasing role of expressionism in cultural geography studies and a corresponding weakening of interest in representation (although others, such as Wylie, look for a dialectic that brings both into an imbroglio of the visual and the phenomenological). Cresswell says his education in cultural geography "was very much about words like meaning, representation, iconography, text, and the image." In recent times, however, "[n]on-representational theory has turned our attention away from mediation . . . towards questions of emotions and affect, bodily knowledge and the fleeting and transient" (Cresswell 2010: 172). This should not be regarded as a disavowal of the social. Instead, the current book seeks to demonstrate that emergence theory can provide different ways to address issues of race, gender and poverty, and that territoriality and place, far from being replaced as subjects of study, can be invigorated by viewing them through the lens of emergence, and through the application of methods derived from an increased understanding of the emergent conditions of environmental and social meaning.

The book advances an argument that there is no single concept, no more productive condition that landscape architects of the future need to understand, than the cluster of ideas, practices and insights associated with the productivity and poetry of open systems design. In the last ten to fifteen years landscape architecture has become a leading design discipline. Texts such as *The Landscape Urbanism Reader* (Waldheim 2006a) and *Ecological Urbanism* (Mostafavi and Doherty 2010) situate the profession at the forefront of dynamic systems thinking with respect to urban planning and design. Professional firms like Field Operations and Stoss LU specifically evoke dynamic systems in their design processes, using analyses of complex adaptive natural and urban systems to generate large-scale urban interventions such as waterfronts, riverside parks, ecological corridors and brownfield developments. The expanding theoretical field of landscape urbanism provides a basis for these initiatives, gathering and focusing the new approaches to design in ways that suggest an emergent mode of practice. This practice derives its modus operandi from that cluster of ideas that can be loosely called emergence theory. But what is it doing with it? How are the challenges of emergence (and there are many) to be met?

Understanding how urban sustainability can be achieved through ecological and integrated systems design is of critical importance to landscape architecture students who are preparing to work with increasingly complex processes in the ongoing re-construction of the world. This requires a theoretical reformulation of the relationships between urban and natural processes that can guide and regulate landscape architectural design. Emergence theory provides just this link between cities and the natural systems that sustain them. Since the mid-1990s ideas to do with dynamic systems, initial conditions, morphogenesis, self-organization and field theory – key concepts of emergence – have become influential in landscape architecture, providing a bridge between urbanism and ecological design. But just how, and why, are these ideas changing contemporary landscape architecture discourse, and how is emergence theory informing critical practice? What exactly does emergence theory (and its related concepts, complexity, chaos theory, nonlinear dynamics) have to offer landscape architecture? Why is it important?

This book is not an instruction manual. Rather it is intended to explain what emergence theory is and where it comes from, how it relates to landscape architecture; how it has, or could, change landscape architecture. It purports to show why emergence is critical to our understanding both of natural systems and of urban systems. And since this blended or dissolved natural–social milieu is central to the ongoing development of landscape architecture as a design discipline, the book seeks to connect the rise of emergence theory in the cultures of science and society to its application in landscape architectural practice. It therefore investigates how it is being used in design, the problems associated with its use in design, and the challenges these problems are incubating. It is hoped that a deeper understanding of emergence theory can lead to new ways of thinking and contribute to enhanced practices. The book takes its theoretical mission further than description, explanation and critique, however. The central argument is that a thorough understanding of emergent processes in constructed landscapes is a key to the progress of landscape architecture as a leading environmental discipline. The text links its analysis of emergence to the types of interactions and encounters that landscapes can provide for humans and nonhumans as the world becomes more and more constructed.

It therefore provides examples of techniques and methodologies that can bed practice firmly in analysis of the complex open systems with which it is dealing. Instrumental considerations are balanced with readings of existing and imaginary landscapes that are intended to develop the subtle and nuanced sensibilities to human–world relations that landscape architects require for enhanced practice. The motivating force, the *élan vital* that drives the argument, is a conviction that the concept, or concepts, of emergence enable us to relate qualitatively different modes of existence to each other and through the connections thus established improve the lifeworlds of all. Emergence, inevitably, like Hermes, is about relations, connections and interactions. Central to this project, therefore, is the question how human subjectivity is to be understood in the light of the profoundly relational conceptual framework that an emergentist landscape architecture requires. How should we understand the place and role of subjectivity, or consciousness, or our human sense of selfhood, in terms of a theoretical agenda that immerses this traditional distinctiveness in the material flux of the world, and sees the self not as a singularity but as a multiplicity, an everchanging node in an evershifting sea of systems? Landscape architecture makes connections between humans and the world in which they live along with others: this is a question about the formation of the subject in the materiality of the world.

The book is organized into five main chapters. Chapter 02, Narratives, gives an overview of emergence theory as it has developed in recent physical science, in ecological science and in urban studies. Chapter 03 identifies nine Key Concepts in emergence theory and briefly discusses these. The following three chapters engage directly the lines of inquiry laid out above by means of engagements, in Chapter 04, Projects and Practices, with a range of design investigations of emergence, in Chapter 05, Encounters, through emergentist interpretations of landscape representations, and in Chapter 06 with a meditation on how emergence can help us re-think our approaches to the development of public space. The chapters are summarized as follows:

02 Narratives

This chapter briefly outlines the historical, philosophical and scientific background to emergence theory. Using the ancient dichotomy of the one and the many as a thematic frame, it explains how emergence has always featured as a condition with which landscape architecture has dealt. The non-mainstream thread of philosophical inquiry from Heraclitus (all is flux) through Spinoza (what can a body do?) to Bergson (form is a snapshot of time) is explored, as are developments in science from the atomist Lucretius through Einstein to Nobel-winning chemist Ilya Prigogine. The age-old figure of the gardener focuses these investigations. The gardener is an adaptive individual who understands the interactive multiplicities of weather, seasonality, growth and continually emergent spatial and formal conditions. She orchestrates an ongoing functional and aesthetic performance from a singular terrain. It is a performance that is never the same, and that never ends. The relevance of the gardener's light-footed, wily and adaptive responsiveness has become increasingly apparent in landscape discourse because her very openness to change bears on the conditions that set landscape architecture apart from other design discourses. Resilience, initial conditions, field theory, nonlinear dynamics, open systems theory and disturbance are concepts drawn from the behavior of natural systems and processes, and from the sciences that investigate these processes. The gardener knows intuitively that this cluster of concepts is extremely effective at provoking novelty – in ways of thinking and in ways of designing. Thinking and making new things are not accidental outcomes, however, for emergence is exactly a way of describing how novelty occurs in the garden – and in the world. Emergence theory is a speculative answer to the question how nature, continually and endlessly, creates novelty: new rhythms, new patterns, new processes; new species, new behaviors, new responses to new conditions. No wonder landscape architects have found emergence such a fertile source of new ideas! The chapter links the growth of natural philosophy to the development of cities and suggests that the undercurrent of emergence has always been a force for change through its periodic disruption of the status quo. This dialectic provides the background for an overview of emergence in current scientific thinking and enables a discussion of ecology's recent drift towards systems thinking and emergence theory.

Some readers may find it more useful to read Key Concepts before the more theoretical Narratives, since the Nine Guides to Emergence found in that chapter provide a range of landscape architectural examples that show how the discussion in Narratives both illuminates and is suffused with everyday making.

03 Key concepts

This chapter interweaves the ecological and the conceptual, cross-fertilizing the writings of ecologists like Lister and Kay with more philosophical approaches to ideas about order and chaos drawn from Deleuze and Guattari, Michel Serres and Manuel De Landa. Structured as a series of "guides to emergence," it introduces and explains a range of concepts that the author sees as underpinning emergence in landscape architectural projects and discourse. Each of the sections describes a set of ideas and provides examples drawn from urban situations and from projects undertaken by professional firms. While the guides aim for clarity of explication, and current and relevant exemplification, they are also provocative. They point to the difficulties of using emergence theory in practice, and discuss, for instance, the tendency for landscape architects to use it metaphorically rather than instrumentally. The chapter highlights the problem of bringing together top-down design intentionality with bottom-up landscape conditions, and investigates the challenges associated with the desire for order in situations that are essentially chaotic. The guides are interrelated and each one links back and forth to concepts discussed in other sections. These are:

1. Open systems
2. Situation
3. Meridian
4. Assemblage
5. Naturecultures
6. Field theory
7. Morphogenesis
8. Difference
9. Disturbance

The guides can be read in any order, as each one provides a different window on what is essentially the same material.

04 Projects and practices

An exploration of built and unbuilt projects that exemplify the concept of emergence, and of landscape architects who have developed programs for their practice that demonstrate a serious involvement in its protocols. The chapter examines recent landscape architecture practice, discussing projects that harness natural and social processes in the development of new ways of connecting diverse terrains and the people who live in them. One of the issues that comes into focus here is the seeming paradox that the term "ecological design" designates. Design – in landscape architecture – necessarily has a formal component that can be traced back to its root in *disegno*, while ecological emergence posits an inherent formlessness. Landscape architectural design has often represented the formlessness of the natural world by gathering and focusing its components into structures and patterns that purport to convey meaning through mediation, the resultant signifiers comprising a supplementary meaningfulness, an acculturation that develops an added value of signification, a layer inserted between human subjectivity and the objects and processes of unmediated brute fact. Emergentist design requires an investment in the embodied experience of this raw material, a seamlessness with it. Not so much a reformulation of the world as a rearrangement of it that avoids placing humans as subjects in an object realm.

05 Encounters

This chapter is the heart of the book. It consists of six sections that investigate in detail the potential for emergence theory to fundamentally alter the way that landscape architects think about their work. Using examples from historical landscapes, literary works, fine art and cinema, it discusses the intricate relations between human beings and the landscapes they inhabit, showing how landscape architecture can be considered as an art of encounter. The shifting, intimate relationships between the human and the nonhuman are explored, and the open nature of the world in which they co-exist is offered as a cause for emergence to be understood metaphysically as well as functionally and aesthetically. Human beings are in a constant state of productive practice: the production of a social assemblage of concrete, physical open-ended entities understood as compositions of forces in which other forces (including us) are always intervening.

These ideas, and many others, are explored through discussions of:

1. The medieval garden of love
2. The sacred groves of ancient Greece
3. The rococo landscape paintings of Watteau and Fragonard
4. The urban landscapes of film noir
5. The baroque gardens of *Last Year at Marienbad*
6. Exoticism and indigeneity in contemporary public space design

The point of the chapter is to show that landscape architecture can utilize the nexus of ideas associated with emergence to reformulate the relations between people and the places they encounter, both on a daily basis, and as they move around the world. In landscape architecture emergence provides a range of ways to rethink what landscape architects do, why they do it, how they do it and what they hope to achieve. Readers of this chapter may feel that the book has strayed away from landscape architecture. The contention, however, is that the structure of affect explored here is a unique disciplinary resolution of Hume's declaration that "The Reason is, and ought to be, a slave of the passions."

06 Propositions

This, the final part of the book, investigates a critical terrain in which emergence theory can provide new trajectories for design and discourse. Some issues will always require discussion and debate. As the world moves towards increasing personal mobility, as cities internationalize, as natural resources decline, and as spatial politics complexify, landscape architecture is participating in the construction of new conditions for life. How does understanding the city as a meshwork of open natural and constructed systems influence design and discussion of public space? Of place? The chapter suggests that the transition from static to dynamic models of the urban requires a shift from an equilibrium point of view to a disequilibrium perspective. From considering the social and the economic, as well as the physical and the ecological as closed or autonomous systems to a consideration of these crucial processes as strongly influenced by the flux of materials, populations and individuals who cross their borders. Public space becomes flexible, renegotiable, elastic. Infrastructure, which has delivered services to urban populations for 150 years by means of highly engineered structures such as pipelines and highways, loosens up, adapts, combines previously divergent systems and shrinks along with the shrinking world. Place making takes account of recent as well as traditional histories, pulls stories together from around the globe as well as from the specific locale. Place constructs cultures as much as cultures construct place. Place, like ecosystems, is always emergent, always making new things.

Emergence theory is not an environmental panacea. It is not a technique for designing landscapes. It is not a replacement theory for one that has just passed or is still passing, whatever that phantom might be. It is not represented by an intellectual bloc of like-minded thinkers and writers. The house has many rooms, and each room has windows and doors. Emergence may be considered as a window, or a door. After you have read this book, you may take your pick. Rest assured that at least one of the doors leads out of the house.

02

NARRATIVES

A genealogy of emergence

Narratives of origin

Emergence is becoming. It is process, change, evolution. Emergence *theory* attempts to describe *how* things and the interactive systems that comprise all things *can* change and develop. It seeks explanations for the continual creativity of natural systems, social systems, urban systems, that are always surging forward, overcoming disturbance, growing, redistributing their energy, adapting to new circumstances, propelling themselves into the future, becoming more and more complex. At the same time we must recognize that emergence is a concept we use to understand phenomena – it is not the phenomena themselves, and they do not behave according to the rules that we construct in order to explain them. They just do what they do, and we try to make sense of them. Such making sense will always take place according to the conceptual categories that we decide are important to us at the time. We cannot dispense with these categories, and new ones are always coming along. Thought, therefore, is emergent too. Sometimes, in order to understand the evolution of thought, it is useful to see how its origins are invented. In what follows, a connection is made between two narratives of advent, between stories of the beginnings of philosophy on the one hand and the origins of the city on the other. This small inquiry uncovers a speculative source for our study of emergence in the metaphor of one and many, a trope that has conditioned thinking about nature and change, and therefore about emergence, since what we are pleased to call the beginnings of history.

The one, according to Parmenides of Elea, active in the sixth century BCE, is being (Allen 1966: Fr. 2). The many, the phenomena, those things which change state, which become, which emerge and fade away, which "differentiate"; these, for Parmenides, are illusory. Since change and movement are sense phenomena it follows that the senses give us illusions. The one, apprehended intellectually, is the only real. This early metaphysical formulation, and versions of it, has underwritten natural philosophy and natural science since the time of the presocratic philosophers (Collingwood 1954; Russell 1961; Copleston 1962; Rorty 1979). It has also, therefore, underwritten the development of western conceptions of nature. We have, as has been noted regularly in environmental discourse, right here, the basis for a devaluation of the natural world. What is of interest in the context of landscape architecture is the ambivalent condition of sharing a universe in which to be human is to be divided between the real and the unreal, between the eternal and the transitory, between unity and multiplicity, and to be nonhuman is utterly to be lost.

The observation that things change is primordial. The transitory phenomena of the physical world are accounted for in the myths of every ethnic group. In fact, humans are so immersed in the phenomena of change – as agents of change and as changelings themselves – that it could be argued that almost everything we do is a response to the perception of plurality, division and difference. Some human activities, such as building shelter, may be attempts to refuse or defy change (by controlling its pace); others, such as gardening, engage with change and profit by it. Periodically, the sudden and unforgiving arrival of cataclysmic change throws whole populations into distress. Perhaps it is just because of the overwhelming and often violent condition of relentless process that Parmenides offered humanity the elegant and eternal solution of the one. The idea of the world as two systems (nature/culture; model/copy; mind/matter), one of which is privileged and the other diminished, is at the basis of the binaric configuration of knowledge and value. The philosophical dispute over being and becoming is rooted in the issue of the one and the many, and through this to what is understood as real, what is significant, what is valued and what is disregarded. Epistemologically and metaphysically, being has been privileged above becoming in western formulations both of what there is and of how we should live. It might be supposed that to consider emergence as a critical condition for landscape architecture, as this book does, is at the least to ignore these formulations, at the most to undo them completely. In reality the contrast is not so stark. Natural philosophy has privileged being, but the idea of becoming has always destabilized this emphasis, and many thinkers, from Heraclitus through Zeno to Ficino and Bruno to Einstein, Whitehead and Bergson, have considered becoming to be primary. It is therefore important to bear in mind that it is not suggested here that the philosophy of nature which has developed and evolved in the west since early times has somehow enshrined the one instead of the many, or vice versa, that they are somehow mutually exclusive. This book suggests, instead, that they have always both been present in most epistemological or metaphysical formulations representative of western philosophies of nature. While the counter-tradition to the Parmenidean one, thought to have been inaugurated by Heraclitus (fourth century BCE), who famously said that all is flux, is as old as the unitary tradition itself, there was a distinct resurgence in twentieth-century philosophy, aided and abetted by science. The advent of the theory of relativity, and the conceptual innovations it stimulated, caused a tidal wave in the hard sciences that rolled through all the various systems of knowledge formation and, finally, coursed through the design disciplines as well. Landscape architecture was taking advantage of this counter-tradition of flux before it was apparent in many other areas of life – its tenets and formulae have all the qualities required for a design practice that works both with structures and with processes. In many ways landscape architects (and their earlier incarnations as trades-people, builders and gardeners) have always dealt with emergent conditions, and have perforce considered how humans are intimately involved in these transformations. At some level we have always known that it is through a local and specific engagement with landscape conditions that landscape architecture confronts the dynamic of one and many, a figure which is often explicit in the very construction of the ideas of "local" and "specific."

Prior to Parmenides, Thales of Miletus (c.620 BCE) asked "What is nature?" Providing, perhaps for the first time, a rationalistic as opposed to a mythopoeic answer, Thales proposed that water is the underlying principle of nature. Nietschze emphasizes that this was the first time that all of reality was attributed to a natural element. Thales "begins to believe in nature, by believing at least in water" (Nietschze 1962: 42). The philosopher is attempting to address the phenomenon of emergence, asking, what is the one thing out of which all things are made? He does not seek so much to find the origins of plurality and difference as to explain them. In the seventh century BCE, for instance, Greek medics were divided over whether diseases (and their cures) were caused by gods and demons or whether they weren't, and were therefore natural – derived through material transformations from the operations of some underlying principle. The observation of the fact of change, of birth, growth, decay and death leads inevitably to this threshold, for "these wise men saw that, in spite of all the change and transition, there must be something permanent" (Copleston 1962: 36). Something primary, that persists. Something within, or intimately belonging to, the things of the world and which is the origin of their behavior. Thales is the source of the homogeneous primitive matter that local differentiation belies. In Thales, then, at the very beginning of philosophical and scientific discourse, we find the development of a number of concepts that have characterized the movement of this discourse:

- the distinction between natural and artificial
- the distinction between natural and supernatural
- local differentiation can be explained by recourse to an underlying principle
- this principle informs all matter (or substance)
- nature is intelligible
- the earth is alive
- nature is being.

These are the concepts, not all necessarily investigated by any one individual or group at any one time, that will guide environmental change for centuries and finally, in the nineteenth century, somewhat modified and a little the worse for wear, enter the new profession of landscape architecture as founding (if assumed and/or unconscious) tenets of the discipline. While Glacken argues that "the roots of modern attitudes toward nature are to be found in the Hellenistic age rather than in earlier periods" (Glacken 1967: 25), the emphasis here on the dialogical interplay of one and many pushes the origins of nature back to the presocratics simply because it is this figure, which they established, that sets out the field of natural philosophy. Aristotle acknowledges the primacy of the figure in the first books both of his *Physics* and *Metaphysics*. Implicit in the statements listed above is the assumption that the greatest knowledge of all is "the knowledge of the essence and core of all things" (as Nietzsche put it) and that this knowledge is attainable. As the list above attests, however, this world of nature had manifold aspects. It is not the plurality of appearances that is important here, but the multiplicity of accounts of the one that explains those appearances. The ancient philosophers of nature, Lloyd explains, had:

> their own distinct and divergent ideas, often in direct and explicit confrontation with their rivals. The concept (of nature) was forged in controversy . . .The paradox was that . . . the idea of nature was supposed to stand simply for what is there, for what can be taken for granted. Yet what that comprised was repeatedly contested, not just so far as the natural world went, but also as far as human nature is concerned.
>
> (Lloyd 1992: 21)

There was not just one concept of nature in early Greek thought, but several, and this diversity of agonistic concepts, so typical of natural philosophy, can be found always at critical moments in the history of western cultural life. These historic moments exemplify the interplay between the continuity of concepts of nature and the disruption of these concepts. They are moments at which certain doors closed on nature, and others opened: major bifurcations occurred. In each case, existing structures of explanation of nature had evolved to the point that, although they were able to account for more and more phenomena, they were not able to shift thinking or promote new ways of seeing. The gardens and texts reviewed in Chapter 05 are drawn from some of these periods of bifurcation. The sacred groves of ancient Greece, medieval gardens of love, seventeenth-century rococo landscapes and the urban landscapes of the modern period all exemplify historical eras when ideas of nature underwent transformation. They represent moments of critical instability – when, perhaps, the many came too close to the one. Above all, however, this short excursus into the history of philosophy serves to illustrate a special point. In the pages that follow it will be found that emergence theory encompasses a range of specific protocols or defining attributes. Suffice to say that these attributes, particular to emergence, are found throughout the period of Greek philosophy just discussed. The ancient period, prior to the rise of rationalism, was characterized by the unpredictable and open-ended realm of the gods, whose motivations were always indiscernible and indeterminate. As Ovid relates in his *Metamorphoses*, the lives of the gods were ruled by emergence. When Daphne changed into a laurel, Zeus into a bull, Minerva into an owl, these changes occurred at moments of critical instability in the lives of the people they affected. In the pluralistic world of the Greeks, the gods had the ability to disturb natural process at any time and enable new conditions to emerge thereby. The ontological structure of life permitted what was real to be blurred by what was appearance only, and humans could sacrifice to the gods in the knowledge that this ritual transgression of the order of things could change reality. In the sixth century BCE a bifurcation occurred that altered the course of western natures: rationalism and realism emerged at the threshold of history and the disjunctive synthesis of one and many was inaugurated. Nothing, as Manuel De Landa has observed, is more nonlinear than history itself.

Whatever theoretical paradigm has been employed in the understanding of nature – in the west – the presocratic discourse of one and many has inhabited it. For twenty-five centuries people of the European world were educated (formally and informally) in the history of antiquity. It was their mental habitus. The study of Greek and Latin, and knowledge of the classical world, its ideas, concepts and operations, *was* education. Even now, the literature, and art, the mythology and thought, and history of classical civilization flows into the mainstream of western education (and all its tributaries in the east and the new world). The presocratic philosophers represent, for western society, the entry into history. The formation of the key objects and concepts of this humanist discourse occurred at the same time as the development of the city-state. The same conditions made possible both the discourse of humanism as a rational exploration of the possibilities of human existence and the emergence of the polis as the condition of these possibilities. Thales, Heraclitus and Parmenides, and others like them accomplished their work within the framework of the city-state. The sixth century gave the west its epic framework. The pattern of Greek culture was still arranged according to myths, traditions and ancient ways of thinking; then, according to the humanist narrative itself, in the city-state came the step from *mythos* to *logos*, from mythic thinking to rational principles. It is in this moment that the liberal arts are invented and practiced in the cultural and political economy of the polis. Freeman sets out the connection:

> The first known electrical experiment was observed in Miletus; the first atomic theory in the town of Abdera in Thrace; the first statement of the transcending importance of numerical formulae for the understanding of matter came from Pythagoras, an islander from Samos, who migrated to Italy. The first geometric proofs demanding an assumption of moving lines and planes were evolved at Tarentum, Sparta's only colony, by its administrative chief magistrate Archytus. It was the small unit, the independent city-state, where everybody knew all that was going on, that produced such intellectual giants as Thucydides and Aristophanes, Heraclitus and Parmenides. If these conditions were not in part responsible, how is it that philosophy, science, political thought, and the best of the literary arts, all perish with the downfall of the city-state system in 322 BC, leaving us with the interesting but less profound and original work of men such as Epicurus and Menander?
>
> (Freeman 1950: 266)

Meanwhile, historian Fernand Braudel also describes the geographical richness of the Mediterranean region, as the field against which movements of thought and of people are to be descried and explained. It is on the fertile narrow coastal plains of Asia Minor and Southern Italy, he says, that the practice of "that urban vocation" began that was later to be called philosophy. The mountains were too harsh for the conditions which led to speculative thought; the hamlets and villages too small for the emergence of a way of seeing and inventing concepts that required urban conditions in order to flourish (Braudel 1995).This link between *locus* and *logos* is made concrete in the fact that the earliest thinkers did not pursue philosophy as a career. They had urban day jobs: Thales was an engineer, Anaximander a map-maker, Heraclitus a nobleman connected with the functioning of the state. If they moved around at all, they moved from city to city. Ionia, in archaic times, was the central part of the west coast of Asia Minor (now Turkey). It was settled in the eleventh century BCE by Ionians and colonized by the Greeks in the eighth century. Miletus, Thales' birthplace, was in Ionia (around 620 BCE), and both his pupil Anaximander, and Anaximenes, were born in Miletus too. Heraclitus came from Ephesus, another Greek colony on the coast of Asia Minor. Both Pythagoras and Parmenides were born in southern Italy, the former in Samos and the latter in Elea, both colonies of Greece. Miletus, Ephesus, Samos and Elea were all Greek city-states, four of scores of such settlements that spread around the Adriatic, the Tyrrhenian, the Aegean, the Ionian and the Black Sea during the period of the Greek expansion, from the eighth to the sixth century BCE.

Learning about life in common and thinking about natural processes: cities and concepts developed together. Freeman's description of the network of Greek colonies as a system in the passage quoted above hints of a connection between the development of the Greek city-state and the evolution of presocratic philosophy whose pattern of organization is, perhaps, not so much linear as interactive. A network of city-states proliferates centrifugally from a mainland Greece epicenter, incorporating both a hierarchical diagram of human–nature relations (god–man–dog) and a polycentric structure of geopolitical hubs and links. The hierarchical structure is present in the establishment of the acropolis at the heart of each city as the link to the gods and thus to nature. It finds its counterpart in the triumphant socialism of the agora. Perhaps it is the dwindling of the acropolis as a social/political force, and the corresponding rise of the agora, that best exemplifies the separation of nature from culture in the polis, and the subsequent redistribution of temples throughout the city (Wycherley 1949: 4). If the acropolis was a link back through kingly succession to the gods and ultimately nature, it was literally nature in the heart of the city. But the development of civil society and the hegemony of the agora challenged the power of the acropolis as a force of nature. Instead, temples developed at different sites within the urban field, as local intensities of nature/god (Scully 1979). Just as temples began to multiply and spread throughout the city-state, Greek colonies spread around the eastern Mediterranean, like nodes in an atomic network of independent entities. The motif of thematically connected but rigorously independent settlements characterizes nearly all the literature on the Greek city-state. The radical separation of the states from each other is often regarded as the downfall of Greek culture itself, this decline being mapped alongside the break-up of the states: Greek philosophy dies with Aristotle in exile at Chalcis. The spread of Greek cities, though, continued to develop, seemingly by drift, and to transmute from the clarity of the Greek model. The web of Greek cities around the Mediterranean is the product of a structure-generating, political algorithm of social and biological processes. Growing, evolving, the dynamic geographical unfolding of Greek culture generated a stable state that, when pushed far from equilibrium by a combination of internal (separatist) and external (Persian Army) forces, transformed into a much more complex system, the Hellenistic Age. The city-states drifted geographically outwards from the mainland of Greece in iterations that bear a certain similarity to their origins but avoid identity. The city-states are not clones, but rather non-identical repetitions. Their spread is a moving out from familiar formal territory to the unfamiliar ground of dissimilar invention that does not necessarily repeat the form of the polis, but often the program. As the idea of the polis spread through Asia Minor, individual cities adapted to local conditions, relying most often on the socio-political program of Greek citizenship as a principle of continuity, and sometimes simply using the Hippodamean grid to act as an ordering device and the signature of the reign of law. Greek thought, like the polis, seems also to have developed by means of drift and iteration. The highly speculative, forward-looking, opening-out of early presocratic thought demonstrates a rhizomatic quality similar to that of the expansion of the city-states.

Perhaps this is due to the origin of Greek philosophy in Greek myth. Many of the different versions of the Greek myths co-existed in pre-presocratic times in a mutually interrelated way. There were numerous versions, for instance, of the story of the Garden of the Hesperides. Its various instantiations were localized and particularized. While the advent of speculative thought begins the movement towards a dialectical approach to knowledge, finding its apotheosis in Plato, nevertheless the particular quality of presocratic thought is marked not so much by dialectical argument as by iterative drift. The strongest indicators of this structure of thinking are its fluidity and the overriding sense (present in Anaximander and Heraclitus in particular) of a network of gnomic utterances that relate to each other not logically but poetically, making little sense individually, but together forming a dramatic, active whole. While the individual philosophers may have wished their version of the cosmology to be taken as the best of equals, cosmological thinking fluidly contextualized, adapted and recreated itself throughout the matrix of Greek city-states as a new way of seeing. Despite the emergent character of presocratic thought, the model of the universe bequeathed to successive philosophers of nature by the influential Parmenides had a closed, concentric, centripetal structure.

Aristotle built on this model. He described a center, at which the earth was located, and successive spheres of heavenly bodies. Beyond the spheres there was heaven itself, the abode of God and the angels. Below the moon, everything changed all the time, beyond the moon all was eternal. The understanding of nature and natural things that accompanied this model was ambivalent but it was not vague. There was a clear distinction between the things that were simply there or given, and those that were made by humans. The stuff that was given was natural and the human-made was artificial. Likewise there was a distinction between the human and the supernatural. Natural objects and occurrences could be explained by recourse to an underlying principle that informs everything that is subject to transformation, growth and decay. This principle animates matter, and its lineaments can be discovered by human inquiry. It is the subject matter of natural philosophy. It ensouls the earth which is therefore alive. At the fall into time, at the beginning of history, humans become separate from things. Social and cultural production separates out from being. Is all that changes illusory, transitory, unstable? Or is becoming all there is? Crucially, human beings' role with respect to nature is ambiguated. Humans are a part of nature, but thought and speech are the central subject of identity. Nature can be known, but only by a human subject. These issues dominate thinking during the invention of the philosophy of nature, and were still critical at the time of the creation of the science of nature in the seventeenth century, the time of the scientific revolution. Throughout the history of the discourse of nature the theory and practice of urban and environmental design embodied this ancient set of ideas about humans and nature.

Becoming, multiplicity, materialism, transgression: the long genealogy of emergence begins not with philosophy, but with the turbulent, unpredictable realm of the gods. It was they who brought the dialectic of chaos and order to the earth. Their at once delightful and terrible devotion to the stochastic became discursive with the advent of Heraclitus, who astonishingly prefigured twentieth century scientific accounts of natural process with his aphorism that we cannot step into the same water twice, for new waters are forever flowing upon us. Aristotle's later theory of entelechy, or potencies, transferred emergence into an internal principle of growth that enables organisms to actualize qualities that are contained in a state of potential. He considered that the complete form of an animal, plant or indeed human being is already present in the individual, like a seed, and is directed towards its final form, or *telos*, by final causation. Finally, from the ancient stage, Epicurus (341–270 BCE) locates the source of free will in the random collision of atoms and chance that is at the heart of things, not in God, the one, or any supernatural being. According to this materialist philosopher, there is no underlying force, or demiurge, there is only the dance of atoms in the void. Four hundred years later, the neoplatonic Plotinus, who also ignored the Judeo-Christian tradition, returned to the notion of the one. He developed an account of God's continuing involvement in the world as an emanation or continual overflowing of His divine love for human beings. According to this theory the entire hierarchy of being emerges out of the one through a gradual process of becoming. God is so good that He spontaneously and unendingly pours forth His goodness, like the overflowing fountain that is to be found in so many medieval cloister gardens.

A thousand years after Plotinus a new a philosophy of universal becoming was developed by the German idealist Georg Wilhelm Hegel. His dialectic posits a flow of Spirit in which oppositions are overcome at each new level of emergence, presaging the disturbance-based propulsive becoming of an open system. Hegel's *Phenomenology of Spirit* (1807) influenced the early-twentieth-century French and English emergentists or, as they have been called "process philosophers," Henri Bergson and A.N. Whitehead, to whom philosophically inclined contemporary emergence theorists refer for a recent prehistory of their ideas. While there is not space to linger on the works of these two extraordinary thinkers, it is necessary to give a flavor of their wisdom, since it has been so influential, and so preternaturally accords with current emergence theory. Both Bergson (1859–1941) and Whitehead (1861–1947) wanted to develop a nonscientific metaphysics of time for, as Prigogine puts it, Newtonian physics "seemed to have no place for time" (Prigogine and Stengers 1984: 10). Both realized that science was a great endeavor and hugely successful, but they saw its emphasis on rational mechanics and reductivism as limiting. Bergson regards rational mechanics as the very essence of science, as he sees science as the product of a practical intelligence that develops the intellectual categories needed to achieve the domination of matter (Prigogine and Stengers 1984: 91). But science, for Bergson, has a radical limitation: it is incapable of understanding what he calls *duration,* which is apprehended through intuition rather than the operations of the intellect. Science, he says, reduces time to a sequence of instantaneous states linked by a deterministic law. But for him "(t)ime is invention or it is nothing at all" (Bergson [1911] 1998: 341). Here we see the role of change as a creative force and nature as a continual elaboration of novelty, a totality that is always creating itself in an open process of development *without any pre-established model.* Not only in Bergson is there an explanation of the energetic becoming of the world, but we also find a discussion of holism that is an insightful description of current emergence theory. In the following passage Bergson agrees with the principle of scientific mechanism and then goes on to explain that living systems are "part of the natural system that is the whole of the universe."

> That life is a kind of mechanism I cordially agree. But is it the mechanism of parts artificially isolated within the whole of the universe, or is it the mechanism of the real whole? The real whole might well be, we conceive, an indivisible continuity. The systems we cut out within it would, properly speaking, not then be parts at all; they would be partial views of the whole. And with these partial views put end to end, you will not make even a beginning of the reconstruction of the whole, any more than by multiplying photographs of an object in a thousand different aspects you will reproduce the object itself.
>
> (Bergson 1911/1998: 31)

Like Bergson, Whitehead sought to understand human experience as a process belonging to nature, and rejected the philosophical tradition that defined subjective experience in terms of consciousness and sense-perception. Instead, foreshadowing contemporary cultural geography's emphasis on embodied experience, he conceived all physical existence in terms of enjoyment, feeling, urge, appetite and desire. He invented the term "prehension" to account for a mode of activity in which the relationship between self and world is affective, the two flowing in and out of each other, causing differences in each other. The subject, he says, using an old Quaker formulation, has a concern for the object. "And the 'concern' at once places the object as a component in the experience of the subject, with an affective tone drawn from this object and directed towards it" (Whitehead 1961: 176). This account of the relationship between human and world is at the heart of Whitehead's philosophy. The creative evolution of nature (which includes humans) could never be conceived if it were defined as permanent, individual entities that maintain their identity throughout all changes and interactions. Ultimately Whitehead wanted to connect a philosophy of relation in which no element remains unchanged, with a philosophy of becoming which sees the world as radically innovative. In the process of its becoming, its genesis, its continual emergence, each thing that exists enhances the multiplicity of the world, since it adds to this multiplicity a new set of relations (Whitehead 1961: 20).

Despite this undercurrent of becoming, the classical notions of nature still found in twentieth-century landscape architecture polarized both theoretical and design discourses in that century and limited effective environmental design, often to formal considerations. Surprisingly many contemporary conceptions of nature, particularly in landscape discourse, are still derived largely from Greek philosophy with its division between the physical and the mental, itself based on a radical separation between the world of the human and that of nature. Landscape architecture has been struggling to overcome this dialectical image ever since its inception. How strange, then, that the answer has been at its feet all along, and delivered there by the natural sciences over the period of the whole twentieth century. A painstaking science-led reconstruction of nature has taken place during the last hundred years. This revision has resulted in an understanding of the world that simply ignores the nature/culture binary altogether – and yet still retains the realism required for an ethical dimension of life. A renewed interest in philosophical materialism provides the basis for the dissolution of the problem of the binary, where one term is always opposed to the other. Writers such as Deleuze and Guattari (1987), Serres (1992) and De Landa (1999a) have laid the groundwork for a conception of nature that does not separate it from the social and cultural. Instead *nomos* and *physis* are productions of historical and material processes derived from the same or similar generative processes. De Landa works against the attitudes that see history merely as the arena of texts, discourse, ideologies and metaphors, and traces the concrete movements and interplays of matter and energy through the domains of economics, biology and linguistics. He sees the source of all concrete forms in internal "morphogenetic capabilities" that lie within the flow of matter-energy itself. When understood as sharing both the diagrams and the processes of systemic self-organization, it does not matter whether the landscapes humans inhabit are natural or cultural; their valency is their energetic becoming (De Landa 1998, 1999a).

The actual term "emergence" was first used to define a philosophical concept in the nineteenth century, by George Henry Lewes, in his *Problems of Life and Mind* (1875). Through the twentieth century emergentism as a theoretical concept waxed and waned, despite the influential writings of Bergson and Whitehead. In the last two decades of that century its use as an explanatory device became pervasive due largely to new achievements in the life sciences, particularly the rise of the sciences of complexity, which include such subjects as chaos theory, nonlinear dynamics, network theory and self-organizing systems. The popularization of emergence as a way of understanding natural systems came about by means of a number of books written in the 1980s, 1990s and the early years of the twenty-first century. Prominent among these were Ilya Prigogine and Isabelle Stengers' *Order out of Chaos: Man's New Dialogue with Nature* (1984), Fritjof Capra's *The Web of Life* (1996), James Gleick's *Chaos: The Amazing Science of the Unpredictable* (1998), John Holland's *Emergence: From Chaos to Order* (1998) and

Steven Johnson's *Emergence: The Connected Lives of Ants, Cities and Software* (2002). In terms of the actual science, it was the research of the chemist Prigogine at the University of Texas at Austin, and the scientists gathered at the Santa Fe Institute working on complex systems, that brought emergence and related theories to fame.

Narratives of explanation

The relationship between theories of emergence and complex systems theory is important. If anything can be said to exhibit emergent behavior it is a system of some sort. It is therefore necessary to understand what is meant by the notion of system.[1] The first thing that has to be appreciated is that a system is a theoretical construct: a model or a metaphor. It is not a physical entity. It does not exist independently of the discourse in which it is evoked. The second consideration is that the concept of system involves time. Systems are dynamic: they move. In the 1930s, the theorist of biology Ludwig Bertalanffy introduced the idea of systems that change and evolve in accordance with information being received from their environments. This brings us to the third property of systems: their extent. Initially Bertalanffy thought that closed systems were quite simple. They could be said to have clearly defined boundaries, coherent internal dynamics and readily observable, quantifiable relations with their contexts. He thought it was possible for researchers to locate themselves outside these systems and develop generalizations about them. Nowadays all systems are seen to be radically complex. Their boundaries and categories are problematic, their levels and scales are not clearly separable, they continually restructure, and their components undergo continuous differentiation in relation both to one another and the system itself. Even a mouse in a box is an open system, or a marble in a boy's pocket, a stratum of pebbles embedded in a sedimentary rock. Moving from such objects to traffic on an interstate highway, clouds in the sky, sparrows on a branch, whirlpools in a river, euros in a foreign bank account, pages on Facebook, we can see that with these kinds of system – open systems – generalization and control are difficult and no privileged standpoint exists, because to describe the system is to frame it and cut it out of its embeddedness in other systems. The further point, therefore, is that ongoing assessment of the system requires engagement in the situation (Taylor 2005: 157). Often for a long time.

Open systems theory has been applied to many kinds of systems, from ecosystems to economic systems, from chemical maps to urban networks. Landscapes are extremely open systems, mostly complex, and always adaptive, always evolving, always, in fact, co-evolving. That is to say, they develop in relation to the organisms that interact with them, producing creatures, situations and assemblages that are exquisitely adapted to each other. It has to be said, redundantly, that complex systems are complicated. A branch of mathematics, called nonlinear dynamics, was developed in the 1950s and 1960s especially to deal with the processes that inform complex systems. Nonlinear mathematics provides ways of mapping everything from clouds to nervous systems. It is especially useful in describing patterns, configurations and organizations (Capra 1996: 83–117). Patterns cannot be measured and weighed: to understand a pattern a configuration of relationships must be mapped. The structures that emerge from such mapping are revealed as interactive webs or differential networks characterized by large numbers of highly mobile components that are linked to each other by various modes of communication. Not only are the components of these webs mobile, but the webs are too, since they are components of larger systems. An orchard-nursery-cool house-factory system is part of a farm-agribusiness-state economy system, and this is part of a national economy-global capital-trade surplus system.

Again, dynamical systems theory is not a theory of physical phenomena. It is a mathematical theory that is *applied* to a wide range of physical and nonphysical phenomena, a mathematics of relationships and patterns, of equations, of algebra and topology (Saunders 1997: 52–57). The contrast with the mathematics of Newtonian physics could not be greater. The transition from classical science to quantum mechanics and then dynamical theory was not just a change of emphasis. Both the relations between structures and the structures themselves were conceived differently. These latter were, for all intents and purposes, no longer discreet or absolute, but relational. The science of nonlinear dynamics was a response to the possibility of understanding how nonlinear events, such as eddies in rivers, clouds, weather patterns and population dynamics occur in the physical world. It soon incorporated social phenomena, with economists, sociologists and anthropologists applying nonlinear theory to their investigative programs. Once the theory got going it seemed as if nonlinear systems were everywhere. By the 1990s it was being applied to cities. Urban systems, too, are open-ended and unpredictable. It turned out that nonlinear systems theory had developed into a tool for devising new strategies in urban development. Why?

A nonlinear system is characterized by continual inputs of matter-energy that flush through it providing it with the energetic resources required for its existence. An important feature of nonlinear systems is that they require disturbance in order to grow and evolve. Perturbations in the system, or turbulence from its farthest reaches, cause the system to assimilate the disturbance and as a result achieve sometimes greater and sometimes simply different conditions of complexity and vigor. Nonlinear systems are open, and emergent.

The answer is: emergence. But how to define or characterize emergence is the subject of much debate. The best place to look for definitions is in the current controversies over the relationship between minds and bodies. Although introduced as a philosophical concept in the nineteenth century, by the 1920s emergence theory had migrated to the sciences where it eventually developed, across cognitive biology and other disciplines, into a powerful explanation for the mysterious connection between mind and body. The simplest definition of emergence is that it describes phenomena that arise from and depend on some more basic phenomena, yet are simultaneously autonomous from that base (Bedau and Humphreys 2008: 1). That, for many, is a working definition of mind. In the philosophy of mind, emergentism is the view that conscious mental states are not reducible to neurological states of the brain. Emergentists argue that conscious states make a contribution to the causal explanation of human behavior over and above any explanation that can be provided by neurological states alone. Consciousness is therefore seen to introduce into the world genuinely novel causal powers (Honderich 2005: 239). The mind–body problem is probably the most important area of science where emergence is taken seriously. Although the capture of emergence by the sciences is troubling and problematic, as it has limited its conceptual richness as well as its application, nevertheless it is important to consider its movement through the sciences as a provocative and controversial set of ideas. In the life sciences emergence has been considered to occur when components, or subsets of a system, combine to produce larger functional wholes, and new properties develop that are not present or evident at the level below (Odum 1997: 33). This could help to explain many of the developmental phases of a human being, or the transition from tadpole to frog. For reasons such as this, biologists, botanists, zoologists and ecologists are quite tender on emergence. In physics and chemistry, emergence has had a harder time, but is often advanced in explanation of what in the hard sciences are called phase transitions. A simple example is when water turns to ice. If they are cooled to 32°F water molecules that are quietly roiling in a state known as liquidity will suddenly cease their random tumbling and undergo a phase transition that locks them into "the orderly crystalline array" known as ice (Waldrop 1992: 82). Heat the water and the molecules will fly apart into vapor. In both cases the phase transition has meaning only at the level of the system – it cannot affect one molecule alone.

Perhaps, at this stage, to sense the breadth of the imaginative resources entailed in the notion of emergence we should consider the example of an orchard. We know that these horticultural landscapes are planted in grids whose dimensions are calculated on the basis of a range of efficiencies, including mowing, spraying, irrigation and harvesting. However, the landscape that emerges from the coefficience of practical husbandry, plants species, sunlight and water, and – usually – a monocultural understory, has for centuries also provided a simple but glorious habitable human space. Poems from Theocritus through Virgil to Robert Frost and Dylan Thomas attest to the use of orchards by people for many pleasures other than the tasting of the fruit the trees bring forth. Painters also have for centuries explored the affective potential of the orchard. The complex aesthetic phenomena of *Le dejeuner sur l'herbe* arise from the formal and spatial regularity of fruit trees in a grid, but are autonomous from their basis in the practical efficiency of the harvest.

Returning to science, while it seems quite simple and in some ways obvious, emergence theory nevertheless provokes careful consideration on the part of scientists and philosophers of science. Practically, and theoretically, science has been overtly reductionist since the time of Francis Bacon, who developed a new classification of the sciences in *The Advancement of Learning* (1605) and deeply influenced the classical physics of Isaac Newton. Even up until the time of Newton (1642–1727) the world was a vast living system: witches were burned, comets were portents, scientists were magicians. Newton's first law of motion eventually put a stop to all this. It effectively removed all traces of life from the laws of physics. For Newton the behavior of phenomena could ultimately be reduced to physical laws controlling the motions of inanimate matter. Reductionism, the term given to this extremely effective theoretical framework, has been very successful, reducing the more sophisticated to the less so for explanatory and instrumental purposes. Explaining the behavior of animals is – working down through "levels of reality" – a matter of explaining the behavior of atoms. Scientific emergence goes head on against reductionism. Running against the universalizing instinct of the sciences, it does not move from overarching rules to particular instances. Instead of starting from universal laws – social or natural – and taking local contingencies as so many weird particularities that should be either eliminated or protected, it starts from irreducible, incommensurable, unconnected localities which, often at a great price, end in provisionally commensurable connections (Latour 2005). It says that, because entirely new properties appear at each new level of complexity, entirely new laws are needed at each level. New laws, new concepts, new generalizations are necessary. We get greater explanatory power going "up" the levels rather than going "down." Crucially, new behavior emerges from the bottom up. A good example of this is the flocking behavior of birds. In 1986 computer scientist Craig Reynolds made a computer model of coordinated animal motion such as bird flocks and fish schools. This program, based on three-dimensional computational geometry of the sort normally used in computer animation or computer-aided design, shows that complex and fascinating behavior can emerge from collections of extremely simple components. He called it "Boids" (Reynolds 2012).[2] The flocking model consists of three simple steering behaviors which describe how an individual boid maneuvers based on the positions and velocities of its nearby flockmates. Instead of writing global, top-down specifications for how the flock should behave, or telling his boids to follow the lead of one boss boid, Reynolds used only three simple rules of local interaction. It is precisely this localism that allows his flock to adapt organically to changing conditions.

Although more and more scientists are reconsidering the role of emergence in the development of physical systems, there are many issues that draw discussion. Within the narrative of the sciences the study of emergence is said to be in its infancy, and the state of emergentist discourse is in considerable flux. As far as many researchers are concerned, a number of important questions still lack clear answers. While this book moves beyond the conceptualization of emergence in the physical sciences, there is much to be gained from a look at how emergence is constructed in these disciplines. The thinkers that are of special interest to landscape architects wishing to understand emergence owe a considerable debt to scientists and all invoke the physicists, biologists and chemists who proudly founded their resistance to reductionism upon it. In recent years the five key properties of irreducibility, supervenience, downward causation, unpredictability and novelty have been attributed to emergent systems and phenomena (Bedeau and Humphreys 2008; Daston and Lunbeck 2011). We will explore each in turn, noting in particular as we go how they interweave with each other to create a web of concepts from which no one property can be subtracted without affecting the whole.

Irreducibility

This is the anti-reductionist proposition that emergent properties are not reducible to the lower level phenomena from which they emerge. It implies that reality – the world, all that there is – is divided into a number of distinct levels or orders, a hierarchy of being. For instance, Odum has formulated what is now a quite famous classification system. It is predicated on a distinction between homeostatic and homeorrhetic states, and the thesis that different levels of biotic existence have different properties, which means that the effect of an outside disturbance may vary with the different levels. Emergence can therefore occur at different levels. Odum devised a chart to show what he calls "cybernetics in the organism–ecosystem hierarchy" (see sidebar).

Odum's hierarchy neatly divides the world into two realms (opposite to Aristotle's), the sub-individual level where homeostasis rules (and reductive scientists are happy) and one where emergence rules in the form of homeorrhetic processes. It's a sort of ecological dialectic that preserves balance but accounts for turbulence. In a development of this thesis, T.H.F. Allen and his colleagues have formulated a hierarchy theory that acknowledges the observer's role in the organization of complex systems for the purposes of study (Allen and Starr 1982; Allen

Waddington coined the term homeorrhesis in 1975, (from the Greek meaning "maintaining the flow") to denote the dynamic character of evolutionary and ecological process, as opposed to homeostasis which is used to describe physiological stability at the organism level.

Odum's hierarchy

ecosphere	HOMEORRHESIS Above the organism level systems are controlled by positive and negative feedback which maintains the system in a pulsing state.
regions	
landscapes	
ecosystems	
populations	
organs	HOMEOSTASIS From the organism level down, neural, hormonal and genetic mechanisms maintain a tight control on growth and bodily function. Feedback is considered by Odum not to occur at these levels, and that therefore emergence is not present.
tissues	
cells	
molecules	

(From Odum 1997: 71)

and Hoekstra 1992, Ahl and Allen 1996). This requires a further division into nested and non-nested hierarchies. While nested hierarchies involve levels that consist of, and contain, lower levels, non-nested hierarchies are more general in that the requirement of containment of lower levels is relaxed. Allen gives the following example. An army consists of a collection of soldiers and is made up of them: patrol, platoon, division, battalion and so on. Thus an army is a nested hierarchy. On the other hand, the general at the top of a military command does not consist of his soldiers and so the military command is a non-nested hierarchy with regard to the soldiers in the army. Pecking orders and food chains are also non-nested hierarchies. (Ahl and Allen 1996). As with Odum, in a nested hierarchy processes are present at each level of ascent and emergence can occur at each level.

Pickett et al. show how hierarchies of ecological existents are formulated exactly to explain causative influence from level to level. For instance, molecular dynamics influence cell behavior; the stimulus–response patterns of individual wolves influence the behavior of the pack; plant community seed dispersal mechanisms influence the composition of the ecosystem. Reductionist ecologists argue, contra Allen and Pickett, that nested hierarchies do not have emergent properties, and that processes at each level can be explained with reference to, and

Pickett has proposed the following nested hierarchy:

Biosphere
Ecosystem
Community
Population
Organism
Organ
Tissue
Cell
Subcellular structure
Molecule

(Pickett et al. 1994: 20)

are predictable from, processes at lower levels. Not only does Pickett disagree but goes one step further, asserting that even in non-nested hierarchies many higher-level events cannot be explained in terms of events at lower levels, and are therefore autonomous from their base. For instance, the personality of a general and the decisions he makes on the basis of his personality are not explicable in terms of the constituents of lower levels, and can therefore be said to be emergent (Keller and Golley 2000).

Supervenience

This describes the sense in which emergent phenomena, while having a distinct reality, nevertheless somehow "depend" on more fundamental existents. What is at issue here is this notion of dependence, which seems to slip from one meaning to another with the movement from one set of properties to another. Supervenience means that all the physical facts about the world, including biological, ecological and geological ones (those that concern landscape architects) are fully determined by, and can be explained by reference to, those entities and processes that are studied by physicists. Are all the biological facts about the world dependent on – determined by – the physical facts about the world? Once all the physical facts are fixed, is there room for the biological facts to vary? Put another way, does fixing all the physical facts simultaneously fix which objects are alive? Do biological properties therefore supervene on physical properties? Does animal behavior, for instance, supervene on more basic low-level properties such as mass, charge and spatio-temporal position, or on theorized forces and waves? Scientistic emergentists say supervenience is not a necessary relationship between two properties, or two modes of being. An important consideration here is whether or not the situation being observed or described is that of an individual, an ecosystem or an entire world. This raises the possibility of local and global supervenience, which recalls Odum's hierarchy that divides in the middle. Must the properties of a global system be supervenient upon those of a local system? To what extent is the world coffee trade supervenient, for instance, on labor relations in Brazil? Or global warming on the collective flatulence of cows in New Zealand? To advocate for some kind of supervenience relation is to require of emergentists that they explain the nature of the relationship between emergent phenomena and the other phenomena from which they arise. In order to avoid mystifying that relation emergentists must account for the appearance of novel properties or behavior in some material way. Does some kind of upward causation exist? If not, how does emergence come about? As we shall see, these questions have provoked answers of considerable importance to landscape architecture, particularly with respect to morphogenesis in design, and the role of feedback in this.

Downward causation

This term describes a relation between emergent phenomena and the circumstances from which they arise, that is the opposite of the supervenience relation. Also very important for landscape architecture, it is the question whether higher-level entities affect their lower-level constituents. (Note the invocation of levels in this discussion, and in the very phrase downward causation.) Also known as the problem of whole–part influence, the issue of downward causation introduces the crucial function of feedback, or autocatalysis. In autocatalysis, properties, events or objects serve as their own catalysts. Downward causation calls into question the reductionist position that all causation is ultimately to be analyzed in terms of microphysical causes, suggesting that higher-level entities can causally affect their lower level constituents. Some scientists argue that, as critical aspects of open systems, emergent features, processes and properties possess causal capacities that are not reducible to any of the causal capacities of the parts or to any of the relations between the parts (Clayton and Davies 2006: 8). In landscape architecture autocatalysis is almost taken for granted. It is however possible to accept other features of emergence theory without accepting downward causation, and many scientists sign up to this, the position known generally as weak emergence. The difference between weak and strong emergence can be understood as the difference between ontological emergence where new entities come into being as a result of feedback loops, and epistemological emergence where the appearance of whole–part influence is due to the limited state of our knowledge and will one day be explained through advances in microphysics.

Unpredictability

The discussion of downward causation introduces a condition of strong emergence (hereafter simply referred to as emergence) that is critical for landscape architecture. If developments occur in a system that are not the result of interactions at a more basic level, it follows that it is impossible to predict either the existence of those new features, or their behavior, on the basis of the properties and relations of the basic phenomena in the system. Even if all the facts about the basic phenomena are in, emergent phenomena will still be unpredictable. Unlike the Great Chain of Being, which is a precisely defined order determined by a supreme being, the hierarchies of levels proposed by emergentists are determined by the organizational complexity of the objects and their relations at each level. Since each level is assumed to contain at least one object or property that does not occur at a lower level these have novel causal powers that emerge from the organizational structure of material components. If the thing responsible for a system's, or organism's, behavior is not a thing at all then, it is argued, the condition that enables that organism's emergent behavior must be "an abstract control structure" (Waldrop 1992: 218). Prigogine coined the term self-organization to account for

The organization of phenomena into levels is primarily motivated by a need to classify. There are no levels "in nature," just as there are no systems, networks or hierarchies. These descriptors are cognitive tools that assist in the development of explanatory narratives. A classic example of the idea of levels of reality as an explanatory device is the ancient formulation of the Great Chain of Being, a religiously motivated sorting out of questions to do with the relationship between humans and nature. The historian of ideas Arthur Lovejoy shows how this chain stretches from the lowliest micro-organism to the Supreme Being in an unbroken series of links, each of which exists for the sake of the completeness of the series of forms (Lovejoy 1964). God made the Great Chain of Being, the narrative goes, in order to ensure that everything that could exist did exist, for to exist is better than not to exist, and furthermore, He ensured that what exists does so in its rightful place within an ordered universe. God is good; in His Infinite Wisdom he made the universe "in order that all possible forms of being might manifest themselves after their kind" (Lovejoy 1964: 50). By the time of the Enlightenment the metaphor of the ontological chain had reached its most sophisticated form. Every species was an end in itself, and yet linked to all others by means of the chain. Humanity, as the middle link in the chain, was part natural, part divine, the point of transition from the merely sentient animals to the intellectual forms of being called angels. An angel looks down upon a man, wrote Joseph Addison in the *Spectator* in 1714, as a man does upon a creature (cited in Lovejoy 1964: 191). The Great Chain of Being was highly influential as a conception of the natural universe. If a thing logically could exist, it had to achieve independent reality. This was the notion that sent explorers looking for chimera such as the Scythian lamb, half animal, half plant. The Great Chain of Being shows how it is important that such devices clarify rather than obfuscate. However, in ecology, as everywhere else, it is impossible to say whether the model "accurately" represents the phenomena it seeks to explain. All that can be asked is that the device operates consistently within the narrative code that is being used at any one time. When we say a process or a behavior is emergent it is reasonable to ask from what it is emerging. The answer, in scientific theory, is that the novel behavior is emerging from a lower level of phenomenal complexity. This is not a necessary requirement. Network theorists such as Callon and Latour repudiate the positing of hierarchies. The metaphor of the network adduces a trajectory of elements ("tokens") that circulate not from level to level but from pathway to pathway within a frame of reference that is the particular network's own. The network itself is emergent as are the aggregates from which it is composed.

systems in which seemingly goal-directed collective behavior occurs. Animal populations that behave collectively, without a leader, such as ant colonies and honey bees, are examples of self-organizing systems. When new and unpredictable patterns are realized in nonlinear systems – when they are pushed far-from-equilibirium – they reach a point of critical instability and undergo some kind of bifurcation or "jump" to a new condition of organization. Driven by autocatalysis, moving from level to level, open systems generate their own conditions of existence. As we saw with the case of Reynold's boids, the generative algorithms underlying

Far-from-equilibrium

Ecologists used to consider that ecosystems are always attempting to achieve some kind of balance; that when they are disturbed they return eventually to a state of equilibrium, and that this is good. After Prigogine and others' work on dissipative structures it became apparent that the opposite is true. Ecosystems are at their most vital and complex when they are pushed to far-from-equilibrium conditions by their interaction with other dynamic forces, such as new species entering them, periodic disturbance by fire or flood, or population instabilities. It is under such conditions that new state changes may occur. These are called phase transitions or bifurcations, and the ability of a system to generate new patterns of organization as a result of these Prigogine called self-organization. According to Prigogine, nonlinear systems not only maintain themselves in a stable state far-from-equilibrium, but may even evolve. When the flow of matter-energy through them increases, they may go through new instabilities and transform themselves into structures of increased complexity. Prigogine showed that, while dissipative structures receive their energy from outside, the instabilities and jumps to new forms of organization that characterize them are the result of fluctuations amplified by positive feedback loops. It is their ability to self-organize that gives ecosystems life. Equilibrium is death.

these systems can be expressed as a set of rules without regard to the material of which the system is composed. This is how bottom-up behavior happens.

Novelty

When a system has reached a critical level of complexity such that it cannot be described without recourse to a new conceptual or theoretical apparatus, then the system is said to incorporate emergent phenomena. This new conceptual apparatus is required to explain novel phenomena that cannot be explained within the terms or theory of the base phenomena. So far, this actually is an epistemological issue. But when irreducibilty, unexplainability and novelty are taken not just at face value – are not just conceptual – but as a reflection of a deeper ontological interpretation of emergence the new entities are regarded as genuinely novel features of the world. Emergence is not therefore defined only relative to a theory, or a level of analysis. In order to understand why not, we now consider what ontological categories physical and social scientists think of as emergent. On one hand these can be described abstractly, as properties, substances, processes, phenomena, patterns, rules, even laws – the terms we have been using so far. On the other hand they can be described more specifically in a way that invokes different forms of life. Aspects of the world that can be considered emergent may include, for instance, consciousness, life itself, human social organizations, nonhuman social organizations, physical aggregations and biological conditions (Bedau and Humphreys 2008: 4). These are open-ended phenomena – they evolve. Change comes about when they are disturbed and they have to respond. The response is the production of something new. A novel material condition, a new structure, a new pattern of organization.

Narratives of nonlinearity

Nonlinear ecology

The five defining aspects of emergence outlined above have contributed to the development of new thinking in ecology and a new paradigm is emerging as a result. For instance, it has been realized that non-physical informational properties (conceptual and emotional data as much as physical and electronic) have causal effects in the material world of rivers and roads, ecological systems and transit networks. But let's turn this conception around. Another way to consider how landscape change occurs is by asking what landscapes do, instead of thinking of them as passive fields that are affected by external processes. Conceived as *agents* landscapes may be considered to evolve by capturing, storing, analyzing and reprocessing different types of material *and* social information. The nonmaterial features may include socially constructed elements such as interpersonal space, cultural values and the price of food at concession stands. The material elements may involve such things as the passage of genetic information through generations of insects, the rotting of leaves on the forest floor, adaptive mammals like weasels and rats entering the habitats of insects and birds. Landscapes process these activities; they are active (but not intentional). In the same way trees photosynthesize, landscapes are continually creating new conditions for life. Emergentist scientists, mathematicians and philosophers, for instance professor of mathematics George Ellis and philosopher Philip Clayton, argue that many levels of reality emerge in the world, and all the objects and processes that occur at these different levels have their own types of existence (Ellis 2006: 84–85; Clayton 2004: 62, 148). Emergence takes place within, and partly enables, these multiple forms of existence. This is particularly true of landscapes, where many explanations are simultaneously applicable.

However, the last two decades of theory construction in ecology have been characterized by debates about the extent to which landscapes in general can be considered to comprise emergent phenomena. While the spectrum of positions has been wide, stretching from radical reductionism at one end to extreme holism at the other, it seems that the balance has moved towards an emergentist view of landscapes. The new fields that have opened up in ecology as a result of this transition have developed a keen interest in the role of the body as a material condition of existence, and a complexification of relationship, agency and engagement has occurred. This transition has been reflected in other disciplines that bear upon thinking in landscape architecture, such as cultural geography. Over the last ten years, geographer John Wylie writes, "varied notions of performance, practice, materiality and embodied agency have increasingly come to the fore in human geographical research" (Wylie 2007: 163). There has been a substantive shift from studies of representations of landscape, nature, identity, space, place, the body and so on, to studies investigating the performativity of these tropes (Wylie 2007: 163). In Chapter 04, the implications for landscape architectural discourse of this turn in cultural geography will be explored. Meanwhile, it is time for a closer look at how, in ecology, ecosystems and biological systems have been increasingly understood as complex and adaptive, instantiating – and problematizing – the concepts of irreducibility, supervenience, indeterminacy, autocatalysis and ontological novelty. The ecological turn, like the one in cultural geography, has been towards relational topologies: connectivities, reciprocities and interactions. Relations are seen as primary, the fundamental existents.[3] While they are regarded as actually creating the world rather than occurring in an already-given topography of space and time, however, in ecology relations are accounted for by reference to the data of scientific observation, and the world they create is a physical thing.

This section focuses on systems-based ecology. Most, but by no means all, ecologists nowadays regard the entities of ecological investigation as dynamic amalgamations of biotic and abiotic components. This does not necessarily mean that these ecologists reject reductionism in favor of a holistic approach that privileges the whole system over the interactions of its component parts. Instead, for many, a systems-based approach to the building of ecological understanding includes both the analytical methodologies of, for instance, population ecologists, and the synthesis of ecosystems ecology. As Ulanowicz writes, the contrast between mechanistic, or steady state, descriptions of nature and thermodynamic, or flow, descriptions (which both seem to describe what goes on) means that "some connections must exist between state descriptions and flow narratives." It is highly plausible, he says, that ecologists are creating pluralistic descriptions of the same phenomena. "Which description one prefers turns upon one's own predilections and what information is at hand" (Ulanowicz 1997: 110). Some reductionist ecologists demur: "Continued use of the term emergent property . . . can serve only to direct limited time, attention and resources from the real focus of ecological inquiry." That was written in 1981 (Edson et al.), and things have moved on. It gives a flavor of the debate that characterized twentieth-century ecology, however – a debate that pits holological (or holist) ecologists against merological ecologists. Merological ecologists, typically trained in evolutionary biology and focused on genetic replication, tended until recently not to see organization above the level of the individual, and did not pay much attention to the physical environment. Consequently they argued that significant properties do not emerge on a systemic level (Keller and Golley 2000: 175). Holological ecologists, on the other hand, bed their theoretical framing of emergence in the macro-realm of bodies interacting in space, and consider this interaction as the key subject of their science. It is the very interanimation of specific material processes, they say, that constitute the space–time coordinates in which physical events actually take place.

Systems thinking is not new to ecology, but it has been overshadowed by reductionist approaches and it was not until the 1980s that researchers such as Odum, Golley and Salt worked out a set of holistic ideas that related to the overall structure and function of ecological communities. They introduced the idea of self-organization into ecology and began to think of ecosystems as unique wholes with emergent properties while still, particularly in the case of Odum, trying to reconcile emergentism with reductionist objectives (Odum 1983; Golley 1993). This is where E.P. Odum's chain of being, discussed above, comes in. He analyzed ecosystems in terms of constituent subsystems through the detailed study of smaller and smaller components even as he examined them for emergent properties on a gestalt level, in the sense of trying to understand large components as functional wholes (Keller and Golley 2000: 177). Others, such as Levins and Lewontin, rejected the reduction/holism debate altogether in favor of a "dialectical ecology" that saw whole systems as comprising parts that were continually redefined and recreated in the process of their interaction (Levins and Lewontin 2000). This nuanced, interactionist holism was a movement in the direction of systems thinking which ascribes the emergence of the material realm itself to chains of contingent, cumulative events in the course of the history of life on the one hand, and mutually adaptive, self-organizing interactions occurring here and now on the other (Haila and Taylor 2001: 96–98). No wonder there is debate: this is a provocation to the hard sciences such as physics and chemistry, which avow the primacy of the electron and the quark. A kind of "postclassical" holism arose, deeply affected by advances in the sociology of science, in which scientific objectivity and evidence are considered as comprehensible only if understood as socially and historically located. This put the ecologist within the system being studied (as implied by Ulanowicz), and regarded the observations of the ecologist as constitutive of the information being harvested. Ecological knowledge became situated. The ecological theater, as Haila and Taylor put it, is not so much the ordered stage of interaction that classical ecology constructed, as a complex of interactive elements that includes the social and the discursive within its operations (Haila and Taylor 2001: 96–98).

Once ecologies were understood to incorporate both physical and social factors, the door was open to the idea of landscapes themselves as heterogeneous webs of socio-environmental events: contingent, mutually adaptive, self-organizing. This made landscape ecology much more complex, and inevitably the domain of more than one discipline. In 1992 Levens wrote:

> Our task as ecologists therefore includes the study of the dynamics of the combined human-natural system in which the rate of reproduction of fish is a co-parameter with the coefficient of bureaucratic inertia, and the variability of rainfall interacts with the variability of prices. Any recommendation for a regulatory scheme must take into account the responses not only of the natural system but also of the human participants with their partly compatible and partly opposing goals.
>
> (Levens quoted in Haila and Taylor 2001: 93–102)

An example of socio-ecological research that incorporates a rich investment in complex cross-disciplinary interaction and human participation is Peter Taylor's "heterogeneous constructionism." In a case of the provision of aid to livestock herders in West Africa through the US Agency for International Development in the 1970s, Taylor describes the complex network of relations between scientists, agencies, members of Congress, technocrats, aid workers and the nomadic pastoralists whose economies depend on ecosystem health in the semi-arid landscapes of Sahel. Taylor argues that these relationships constituted a "heterogeneous web" in which no one resource or domain of action stood alone, but instead together formed a framework that linked components in multiple points of engagement. Causality and agency were therefore distributed rather than localized, as it is not the components themselves but their relations that constitute the causes (Taylor 2005: 129–133). Similarly, ecologists in the 1970s developed an approach called Adaptive Environmental Management (AEM), which promoted the use of multiple models and their ongoing revision in recognition that any ecological situation is a moving target, continually restructuring and always embedded in other situations (Holling 1978). The imbrication of human structures of thinking and perceiving in the development of scientific knowledge is not new. Much Anglo-American philosophy of the twentieth century was concerned with the issue of objective knowledge. Quine showed that a human – and by extension a scientific – language that is not already theory-laden is impossible, and that therefore all statements about the physical world of things are influenced by how we see the world. This put an end to the possibility of a scientific knowledge constructed independently of consciousness.

Reflecting the transition from static to dynamic models, early ecologists who once saw ecosystems as homeostatic now see them as homeorrhetic. Odum, for instance, stated in 1971 that "Homeostasis . . . is the term generally applied to the tendency for biological systems to resist change and to remain in a state of equilibrium" (Odum 1971: 34). Twenty-six years later he wrote: "Organization and function at the ecosystem level are [not] tightly regulated [and display] more pulsing and chaotic behaviour . . . in other words they exhibit homeorrhesis as opposed to homeostasis" (Odum 1997: 72). Pulliam and Johnson explain the paradigm shift in ecology in terms of two major changes: a shift from the equilibrium point of view to a disequilibrium point of view, and a shift from considering ecosystems as closed or autonomous systems to a consideration of ecosystems as open and strongly influenced by the flux of material and individuals across their borders (Pulliam and Johnson 2002: 51). Keller and Golley describe this movement – from essentialism to probabilism – succinctly: "Ecology has been dominated by the entity (Being) approach at the expense of the process (Becoming) approach" (Keller and Golley 2000: 31).

The models of classical physics, thermodynamics and even quantum mechanics had excluded time, reducing it to an abstract entity. Research into the laws of nature was associated with a reversible, determinate description of time in which no distinction was made between past and future. The whole point of the Carnot cycle (proposed in 1824) was to describe an ideal condition in which the conversion of heat into work was reversible, so that the motor could return to its initial mechanical and thermal state, with perfect temporal symmetry, and the cycle of work could begin again (Pulselli and Tiezzi 2009: 6). In nineteenth-century thermodynamics time did not live in the world. It was the chemist Ilya Prigogine, whose influential work was conducted in the 1950s and 1960s, who made the breakthrough to a scientific description of emergence which understood time not as reversible but as a propulsive, forward moving construction that conditions life. Prigogine's theory of dissipative structures has been immensely influential. A dissipative structure is a nonequilibrium system that operates on "a multiplicity of levels that are all connected, none of which may have a claim to preeminence" (Prigogine and Stengers 1984: 300). In other words, it is nonlinear.

The points for landscape architecture to draw from Prigogine's work on dissipative structures are that:

- The interaction of the system with the outside is the condition of life for the system.
- Increasing the matter-energy flows through the system increases its vitality and health.
- The fact that such systems are in continuous relations with external forces means that they are undergoing continual disturbance.
- Disturbance is in many instances necessary for the system's development.

It is in systems ecology that emergence theory has had most impact in terms of ecological science itself. One of the major areas of biological organization in which systems ecology has deployed the protocols of nonlinear dynamics is the area of biodiversity depletion. The most direct cause of biodiversity loss is habitat loss. This occurs any time a piece of land is converted from its current state to some other land use or land cover (Collinge 2009: 3). It is usually the result of population growth requiring greater and greater tracts of nonurban landscapes, and technological change often in the form of new farming practices such as wind farms and massive biofuel agribusinesses. Another significant factor in the loss of biodiversity through habitat destruction is the fragmentation of habitats by transit infrastructure projects. Fragmentation breaks wholes into smaller pieces. While it is not possible to have habitat fragmentation without habitat loss, it is possible to have loss without fragmentation (Collinge 2009: 5). These distinctions are important because the landscape architectural response to them is calibrated on the specific mechanisms responsible for shifts in biodiversity in relation to landscape change. Landscape architects (among others) are instrumental in the development of strategies for reversing the trends of biodiversity losses, whether by creating new kinds of housing subdivisions that have high ecological performance values, by designing wildlife bridges that enable fragmented habitat to reconnect, or by multi-tiered approaches to infrastructure and technology installations, bundling infrastructure, and designing wetland wind farms. These and many other new perspectives on the problem of biodiversity conservation are using the systems-based approach to ecology that has been on the rise since the 1990s. "The traditional perspective of biodiversity," Lister wrote in 1998, "is being fundamentally challenged by the emergence of a post-normal, systems-based science . . . the hallmarks of which are uncertainty and complexity" (Lister 1998: 124). This new science is seen to be providing new tools for biodiversity conservation planning, in the form of operational and management procedures that draw from general systems theory, complex systems theory, chaos theory and hierarchy theory, among other disciplinary fields which seek to study and perhaps harness emergence (Waltner-Toews, Kay and Lister: 2008: ix). These fields, notably, are all scientific.

Systems ecologists can cite literally thousands of instances where natural processes are best described using emergence theory. Ecosystems exhibit a "circular logic" where cause and effect require nonlinear explanations using morphogenetic causal loops. In wetlands, for instance, form – or structure – is determined by, and determines, its own development. The emergent properties of wetlands, Kay argues, cannot be deduced by studying more intensely their individual elements in isolation. For instance, a classical experiment designed to measure evapotranspiration in a wetland separates the plants into pots and the water into pans and measures the evapotranspiration from each of these sets of entities. However, the result is a higher value than when evapotranspiration is measured from plants in an actual wetland. There are a number of possible reasons for this, including the increased humidity that results from the transpiration in situ, thus decreasing the evaporation from the open water. "The nonlinear causality in the loops typical of such systems makes distinguishing causal order impossible" (Kay 2008: 4).

Lister provides another compelling case for the explanatory power of emergence in the understanding of ecosystem behavior. She argues that, apart from its energetic resources, a biodiverse ecosystem is generally thought of in terms of its ecological structure and its function. The work of Allen and Hoekstra (1992) and O'Neill et al. (1986), on ecological hierarchies, provides a frame for the definition of structure and function. Hierarchy is an important feature of Lister's case, as a levels-based ranking structure recognizes the scale-dependency of many ecological phenomena. The functionality of different levels is best observed through the use of a scale appropriate to that level. When representing the nesting patterns of herons in urban waterways, we must use a scale that permits all the relevant behavior and conditions to be linked up and taken into account. The requirement for different analytical tools for making measurements at each level, each scale, indicates that there is no single convenient method of measuring biodiversity. This is Lister's point about the relationship between structure and function. But, as mentioned earlier, measurement is also a function of the observer of the system. Kay has outlined how the observer introduces variability with respect to interpretation by making decisions about boundary conditions (what is considered to be the whole system?), context (what externalities affect the system?), type (what frame provides the descriptive language and criteria?) and scale (Kay 2008: 19–25). Lister argues that conventional definitions of biodiversity do not recognize how open systems operate, and how humans are involved in the construction of these systems through their elaboration of research agendas and research frameworks that influence what they see and how it is understood. The traditional structural perspective is that "biodiversity is the totality of genes, species and ecosystems in a region." The 1992 Global Biodiversity Strategy (GBS) considers that:

> Genetic diversity is the variability of the genetic code within species and amongst populations.
>
> Species diversity is the variety of genetically similar organisms present in a given spatial unit.
>
> Ecosystem (or landscape) diversity, the most complex scale, is the variety of large-scale ecosystems or distinct landscape patterns in a given region.

Lister's point is that definitions of biodiversity that are limited to either one perspective (e.g. structure) or one level of the hierarchy (e.g. species) are problematic "when they fail to recognize explicitly the scale and observer-dependency of the diversity concept" (Lister 1998: 126). Prigogine had already argued that emergence "does not suppose any fundamental mode of description" and that "each level of description is implied by another and implies the other" (Prigogine and Stengers 1984: 300). Therefore Lister proposes a broadened definition of biodiversity that incorporates the influence of scale and context, and the qualities of uncertainty and complexity. This new pluralistic definition includes the social and cultural conditions through which biodiversity, ultimately, is constructed and understood.

The overall heuristic framework of systems ecology permits a reconsideration of many ecological concepts. It is worth mentioning one more that Lister discusses. This is the role of so-called keystone species in the new conception of ecosystems as shifting steady-state mosaics. Because ecosystems are constantly moving through multiple states at different speeds, it will be the case that species play different roles in ecosystem functioning at different times and in different ways. While ecologists have sought to identify keystone species that play critical roles in ecosystem health and vigor, Erlich and Erlich (1981) suggest that all species play important roles, depending on the state of the mosaic. Instead of identifying a few keystone species and marginalizing "redundant" species to the sidelines as intermediaries which make no difference, it is thought by systems ecologists that asking which species play which role is asking the wrong question. It is more likely that species play many roles although at different times and in different contexts. This contention can be related back to one of the five scientistic aspects of emergence theory, the irreducibility requirement, where it is seen that rules established at one level of a hierarchy cannot be transferred meaningfully to another scale or context to make generalized statements about the whole system. The interesting upshot is that the biodiversity of a whole ecosystem cannot be reduced to a simple threshold. Instead, being a context-, state-, scale- and observer-dependent condition, the best that landscape architects can do with respect to biodiversity is develop protective and conservation measures based on these insights. This will necessarily entail processes of design decision-making and political judgments that, like the ecosystems under consideration, are adaptive, flexible and participatory. "(M)anagement," Norton says, "must have as a central goal the protection of the systems' creativity," its ability to develop novel structures and new modes of behavior

(Norton 1992: 37). Ecosystem management – and enhancement – must, therefore work with the organizational unpredictability of the systems it administers. This is made possible partially by mapping, monitoring and adjusting the system from within. Since the system is ever mobile, this work is never finished. And since it is emergent, and responsive to fluctuations and disturbance, its future states are unpredictable.

Many ecologists are concerned with reconciling the predictive role of the sciences with the relativity and uncertainty of systems ecology. It is exactly because ecology is increasingly asked to offer a leading role in identifying and solving environmental problems that Oswald Schmitz says there is a "huge premium" on predictability. He calls for the resolution of complexity in ways that enable general predictions to be made about how ecosystems will function in response to myriad anthropogenic and nonhuman-caused disturbances. Broadly speaking, normative science ecologists use two kinds of empirical method to derive understanding of ecosystems. The first is experimentation, which uncovers causal relationships, and the second is macro-analysis, which is believed to offer predictive insights. The trouble is that these two approaches lead to quite different understandings of the relationships between a system and its parts (Schmitz 2010: 2). If different plant functions are revealed at different locations within a single plant community, which one should be used when making predictions about ecosystem response to, say, loss of diversity? Neither experimental nor meta-analytical approaches necessarily produce predictive insights that can be applied to ecosystem management, Schmitz concludes, unless they explicitly confront the issue of contingency. Norton, among others, has suggested that we must change the kinds of questions we ask, and the kinds of values to which we link ecosystem management, in order to deal adequately with complex adaptive systems (Norton 1992). A major reason for this is that in a complex system there is no fundamental, enduring strata or base to which system behavior can be reduced. System behavior at the macro level is most likely produced spontaneously at that level. Hence the indeterminacy, the *contingency*.

To account for contingency, Schmitz posits the existence of "fundamental mechanisms that are evident among all systems regardless of spatial extent and temporal scale." He argues that contingency arises when "the nature and strength of ecosystem functioning in different locations are different realizations of the same underlying process" (Schmitz 2010: 4). He claims to overcome this problem by identifying a unifying field that accounts for the variable relation, and postulates a set of rules, like Ted Reynolds' for his boids, that are not a script, but a set of limit conditions. These parameters enable the ongoing creative improvisation of all the ecological actors in his theater of ecologies. In this theater evolutionary ecological acts play themselves out differently in different ecological contexts, even with identical sets of actors. This implies that contingency is an emergent property of the ecosystem. To explain contingency requires understanding the improvisatory rules. For Schmitz, this means identifying fundamental mechanisms that are evident among all systems, regardless of spatial extent and temporal scale. He clearly aligns with the ontological materialism of strong emergence, which arises, he says, when "a new set of causal principles become active in a complex system" (Schmitz 2010: 5). However, he also wants to entertain a certain level of reductionism, believing that through a synthesis of emergence and reductionism he can develop a causal connection between the biotic (lower level) components of an ecosystem and its biophysical (upper level) components. It is just the necessity for this connection that autocatalysis obviates.

Schmitz's ecological framework is predicated on the postulation of an ontological difference between humans and nonhumans. But systems ecologists seem to be moving toward a more radical description of the world, such as Latour's empirical or "practical metaphysics" which argues that there are multiple, contradictory realities inhabited by different agents whose actions cannot be explained in terms of some external structure (Latour 2005). According to this view, known as actor-network theory, to work within the system as systems ecologists suggest, is to treat the human and nonhuman agents that comprise the system as what actor-network theorists call actants, components that are described in the same ontological terms. The rationale for this is that any differences between actants are generated by the relations within the system itself, and should not be presupposed. Given the synergies that exist between urban and natural systems, this is a powerful proposition that has serious implications for the design of urban environments. It is considered in more detail in Chapter 03, Key Concepts. Now, however, it is necessary to turn to the discipline of urban ecology, a research area that has expanded rapidly in recent years. How does urban ecology bring together innovations in urban systems and ecosystems research, how does it incorporate emergence theory and how does this impact landscape architecture? Not, it has to be said, much.

Recent urban ecology texts attribute its rise to an increased interest in "how an improved understanding of the ecology of urban areas may be vital for . . . human health and wellbeing" (Gaston 2010: xi), and the fact that most people now live in urban regions, a trend that will continue. Note the limitation of the rationale to human affairs.[4] Gaston defines urban ecology as the scientific study of the interactions between organisms and their environment within urban systems, with particular emphasis on the flows of energy and materials through ecosystems. Urban ecology can be viewed from two related perspectives (Grimm et al. 2008; Wu 2008). The first addresses "ecology *in* urban areas," echoing "traditional ecology" in that it typically investigates ecological patterns and processes in urban areas in much the same way that they might be examined in other environments (Gaston 2010: 3). The second is a more "systems-oriented approach." This conceptual shift is registered as "ecology *of* urban areas," a concern for how urban areas function as "aggregated wholes," the interaction between ecological and social systems (Niemelä et al. 2011). Neither approach has precedence, and the combination of studies *in* urban areas with the study *of* urban areas "is likely to bring about the most rapid advances in understanding, particularly because of the inherent complexity of urban systems" (Gaston 2010: 4). According to Wu a major goal of urban ecology is to understand the relationship between the spatio-temporal patterns of urbanization and ecological processes (Wu 2008). Few urban ecologists so far foreground the theoretical frames of complex adaptive systems, emergence or nonlinear dynamics in their discussions of the ecologies of entire cities, however, and few are concerned with collaboration with landscape architects or urban designers. This is not to say that urban ecologists do not collaborate with landscape architects, or have nothing to say on the topic of the design of cities. In fact there is increasing interest in collaboration. For instance urban ecologist Colin Meurk and landscape architect Simon Swaffield together argue for cities as complex landscapes. They advise landscape architects on design directions that exploit biodiversity opportunities through carefully planned plant-based landscape configurations (Meurk and Swaffield 2007). Again, Felson and Pickett (2005) suggest that linking plant ecology with urban design (architecture, landscape architecture, civil engineering and urban planning) can help to integrate research and understanding of plants into the structure of cities and, significantly, to make use of urban design projects as ecological research tools. Plant ecology, they argue, "is increasingly engaging urban ecosystems as integrated natural-human systems, in which human agency is part of the complex of feedbacks" (Felson and Pickett 2005: 1). Their credo:

> (E)cological design illustrates the linkage of plant ecological research with the ongoing transformation of urban systems by urban designers and civil society. Quality of life, human health, public appreciation of ecological processes in cities, and scientific understanding can all be enhanced by participating in a cycle of ecological urban design.
> (Felson and Pickett 2005: 1)

A very distinctive departure from the plant-based approach to urban landscape architecture just cited is to be found in Lister's "Insurgent Ecologies: (Re)Claiming Ground in Landscape and Urbanism" (2010). A conservation biodiversity specialist, rather than an urban ecologist, Lister discerns a two-way street between the practices of ecology and urbanism. On the one hand "the ground plane of the contemporary metropolitan region has reshaped the paradigm of ecology," and on the other "(a)ll of our ecologies – multiple, complex, and insurgent – collectively inform the design of our urban and urbanizing landscapes" (Lister 2010: 536). She sees a confluence between ecology and the design of cities, a "synchronicity" between "new understandings in ecology" and "a new breed of urbanism." Both are using protocols and practices derived from complexity theory; urban systems and ecosystems equally derive their nonequilibrial force from emergence. Lister's article is a manifesto and a call to arms through the promotion of a methodology she calls "adaptive design." This is a whole-systems approach to the design and management of social ecologies that "draws on current ecological science." That is to say, it is based on the work of herself and her colleagues which, while significant and timely, has not yet found its way into the ecological mainstream. Lister has absorbed theoretical studies on urbanism that see the city as a living field of forces and flows and has discovered a congruence between the landscape urbanist conception of urban design and her own systems-ecology approach. Contemporary urbanism, Lister says, "encompasses the notions of form, function, field and flows across and between the dynamic layers" of the city, weaving together aspects of culture and nature. Systems ecology and landscape urbanism speak the same theoretical language. Let's call it the language of ecological urbanism. It is therefore no surprise that Lister's article is to be found in the book *Ecological Urbanism* edited by Mohsen Mostafavi and Gareth Doherty of Harvard University's Graduate School of Design. This book claims an advance on the discourse around landscape urbanism and more specifically links urban landscape architecture with the wide range of disciplines now involved in city thinking.

For one of the limitations of landscape urbanism to date has been a reluctance to locate this field of practices within a broad theoretical framework. Most (not all) writers tend to adopt an outside-the-frame observer's position and, rather than developing a position from within, distance themselves from a definite statement of philosophy. While this gives the impression of permitting landscape urbanism to adaptively self-organize and avoid orthodoxy, it also has the drawback of encouraging it to float free from any form of signifier and therefore remain uncommitted to political and social change. But flexibility is no stranger to finitude, and in his foreword to *Ecological Urbanism* (2010) Mostafavi counters this tendency and constructs a theoretical framework for ecological urbanism. Following Felix Guattari's concept of ecosophy, he positions the agenda for ecological urbanism within an understanding of the urban as a location for three ecological registers: environment, social relations and human subjectivity. Mostafavi fashions a broad conceptualization of ecology as a model for the engagement of "new subjectivities and collectives," in the provision of greater opportunities for "social and spatial democracy." While Lister sees the power of ecology as a metaphor for strategizing human–nonhuman relationships, speculating on the role of humans in world-creation, and representing natural/social interactions, she does not (yet) discern, as Mostafavi perhaps does, the problematic that all this cuddling up of ecology and urbanism engenders. There are three primary problems (which have already been identified with respect to science in general by proponents of actor-network theory such as Latour 2005, and Daston and Lunbeck 2011):

1. The world is divided into subject and the object, and these are separate entities.
2. Ecology assumes unreflectively the role of representer of nature to human society.
3. Ecology takes over the primary function of constructing human–nonhuman relations.

In other words, the engagement of ecosystems ecology with the philosophy of emergence is not yet fully achieved. Moreover, urban ecology as a discipline distinct from, but potentially related to, ecosystem ecology is still a long way from such an engagement. Mostafavi's ecological urbanism has moved toward the development of a philosophical basis. While not based in complex systems research, ecological urbanism (considered as a field of endeavor) has readily connected points of convergence with emergence theory. From the ecology side, only Lister (1998) and Cook (2000) have connected emergence explicitly to landscape architecture. In the meantime we should reflect on the extent to which landscape architecture is being co-opted by ecologism and wonder if the potential that emergence theory offers landscape architecture is not being gobbled up by ecologists who see biota as the only self-organizational material of interest in landscape design. Is there not a possibility that emergence reveals something more profound the more it departs from ecologism? What is left of emergence if we delete ecology? Despite the title of his book Mostafavi's introduction hints that emergence offers more than a new way to do ecological design (unless the "ecological" in ecological design is expanded to such a degree that it is no longer ecological). Encounters, the fifth chapter in this book, gives an account of emergence that simply leaves ecology behind.

Nonlinear urbanism

Once open systems theory became established in the physical sciences, differential, nonlinear structures were discovered everywhere. In western Europe alone, more than fifty centers for nonlinear processes were founded between 1985 and 1995 (Prigogine 1996: 70). By the 1990s open systems theory was being applied to cities. Chaos theory (the popular term for nonlinear dynamical theory) developed into a tool for devising new strategies in urban development when it was found that urban systems, too, are open-ended and unpredictable. Writers such as Kwinter (1992, 1994, 1998) Salingaros (1998, 2000), and De Landa (1998; 1999c) began to explain urban processes in terms of dissipative systems. De Landa, in particular, staked out the urbanistic territory of Deleuze and Guattari's rhizomatic process. High-profile practitioners like Allen (1997), Corner (1999a and 1999b) and Koolhaas (1995) drew explicitly on the work of scientists and science popularizers such as Capra (1996) and Gleick (1998) to develop urban landscape design strategies that took account of the unpredictable and open-ended character of urban systems, and used it to generate design proposals that were time-based and

Dissipative structures and cities

In *The End of Certainty* Prigogine cites the city as the simplest example of a dissipative structure that can be evoked by analogy. "A city is different from the countryside that surrounds it," he says. "(T)he roots of its individuation lie in the relations it entertains with the adjacent countryside: if the countryside was eliminated, the city would disappear" (Prigogine 1996: 62). Urban function is inseparable from urban structure, he says, and together they express the interaction of the town with its environment. Prigogine's point is that metropolitan regions are physical systems in contact with a huge range of sources and sinks. Excess production flowing from one set of urban systems (manufacturing and hospitality industries for instance) to other systems such as agribusiness and recycling industries, enable ordered structures to be formed, maintaining them in time; in the end most urban flows are dissipated as wastes and other pollutants to external sinks. When biosystem scientists Pulselli and Tiezzi describe urban systems as nonlinear, they base this description in the openness of urban systems. This openness is characterized by the input of low entropy resources such as coal and oil and high entropy output such as transit systems and industrial estates (Pulselli and Tiezzi 2009: 21). The more open the system the lower the entropy it produces. Updating Prigogine's point and applying it to the large contemporary city that is part of a vast region, they see the multiplicity of urban processes as having wide ranges of action and a global dimension. The vast energy flows going in and out of cities enable them to build what Prigogine calls "ordered forms" and maintain them in time. If these energetic processes ceased in any city, the entire urban system would be abandoned and fall to ruin (Pulselli and Tiezzi 2009: 21). In entropy terms, this means that a city needs continually to reduce its internal entropy by self-organizing, by making structures, creating and transmitting information, building social assets, economies and cultures. Flows of resources feed the nonlinear city. They are metabolized as in an ecosystem, as if the city were a set of organisms; they are spent constantly to conserve the city's structure in time. The greater the system, the greater the size of the flows involved in its processes; the greater the system, the more urgent the need to absorb resources (low entropy inputs) and to get rid of entropic wastes (high entropy output). Once the dissipative nature of urban economic processes and dependence on natural sources and sinks are recognized, the problems and demands for running a city or a large regional system may be reformulated in a new light (Pulselli and Tiezzi 2009: 23).

adaptive. Despite their emphasis on structures of organization, and transformation over time, the earlier, ground-breaking work both of Ian McHarg and Christopher Alexander, who discerned, respectively, landscape and urban patterns, lacked a sense of the propulsive character of change and the recognition of the ability of systems to self-organize without recourse to external agency. As Batty notes, early systems theories of the city "were structured around parts adding up to wholes in terms of an equilibrium which could not cope with any form of change" (Batty 2000: 167–168). A city may be thought of as a pattern, yes, but "a pattern in time" (Holland 1998).

Have cities not always been nonlinear? Despite the old concept of the city as a cultural condition that separates itself from nature, even the walled urban landscapes of ancient and medieval cities were dynamic, complex and adaptive. The Greek polis itself was permeated by rural systems of trade, supply and agricultural endeavor. Even the religious year was driven by the rhythms of the rural landscape. The whole network of Greek colonies had an intimate connection to agrarian production through the seasonal supply of pine nuts, cypresses, grains, herbs, olives, goat's milk and sheep's wool that provided urban merchants with their living. It was through the various activities associated with agriculture, horticulture and pastoralism and, importantly, the festivals that celebrated the productive cycle of the year, that the urban system received and reticulated the matter-energy it required to evolve. It was not until the seventeenth century, with its emphasis on an urban order achieved through visual and spatial technologies such as the aerial map (often emblematized by Giambattista Nolli's 1748 map of Rome), that the baroque characterization of the urban condition as an arrangement of buildings and other structural elements (sculpture, fountain, column) across a relatively unproblematic floor developed. The great urban projects of the nineteenth century – Haussmann's makeover of Paris, Vienna's Ringstrasse, Nash's plan for West London – introduced the celebrated planning ideals of modernism. Mid-twentieth-century urban planners such as Lynch (1960), Cullen (1961) and Bacon (1967) sought to provide theorems for urban design based on empirical, spatial and visual categories designed to increase order and organization in the urban system. Specific systemic conditions were often neutralized as cities were zoned and spatial conditions formulated in ways that decreased the interactivity of social, natural and physical systems. Planners tended to see the city as a potentially ordered condition that was prey to turbulence and disorder. Through better planning theory and practice, urban order could be maximized and disorder minimized. Ironically, given the intimacy between polis and plantation, this approach stems ultimately from classical philosophy of nature and the ancient Greek antinomies can still be found at the basis of much current urban discourse. As discussed earlier, presocratic philosophy can be seen as a dramatization of the struggle between order and chaos. The natural/artificial binary developed by Greek philosophers led to an extremely influential conceptual model of the city as an artificial realm from which nature is divided or separated. Even in Plato's *Republic* we find an urban theory that walls the city off from natural processes, conceiving the latter by turns as primitive and disorderly, and in the *Phaedrus* Socrates remarks that he prefers the social realm of the polis to the natural world of the agrarian countryside (Plato 1952a). The city originates with a cleansing ritual, Michel Serres provocatively concludes, and nature is that which is swept away (Serres 1982f: 176). It is an ancient and beguiling thesis that has led to the time-honored response, such as with Olmsted Sr., that the mission of landscape architecture is to restore nature to the city (Olmsted Jr. 1928: 248). The current emphasis on bringing nature "back" into the city, common among landscape architecture students, relies on a conception of the city as not-natural. Under conditions in which nature is divided from urban life, it is permitted to "re-enter" the city only as a framed, acculturated and attenuated other.

But now that nature is no longer stable, predictable or knowable, now that it is chaotic, there is little to distinguish the natural and the artificial at all. This conception establishes a new framework for the investigation of nature by design. It tallies with the sense that society too is chaotic, and with the conceptualization of the contemporary city as an emergent phenomenon unfolding through multiple time frames. Since neither nature nor cities are in a state of equilibrium, both natural and social structures emerge and converge in a complex process that involves changes from one state to another. Just like the ecosystems studied by conservation biologists, cities are unstable, far-from-equilibrium, and can be directed by different attractors into new patterns of organization, new conditions of social–natural interaction. The study of organization is crucial to urban nature systems, because systemic properties arise from a configuration of interrelationships. Their patterns share a common property: whenever we look at natural systems we look at networks. Whenever we look at cities, we look at networks (not forgetting that networks – webs, matrices, "meshworks" – are not the only structure-generating patterns, and that systems are usually amalgams of both meshworks and hierarchies). The emphasis on patterns of organization rather than solely on structures has led to cross-disciplinary developments in network theory. Networks, or connected systems, across a variety of situations, exhibit similar behavior patterns and structural organization (Barabasi 2002). "Natural" and "artificial" structures such as internet search engines, terrorist organizations, living cells, ecosystems, urban transit systems and cities themselves have a common organizational resource that determines structure and pattern. This common resource can be diagrammed. A diagram of the organizational patterns of a city will show the matter-energy flows that constitute that city (De Landa 1999c: 121). Since the 1990s, the possibility of dynamically diagramming the city has engaged a number of urbanists and designers in Europe and America.[5] Drawing often on Deleuze and Guattari, these writers and designers promote a way of seeing the world for which the natural/artificial distinction is simply not relevant.[6]

If urban systems are designed in such a way as to harness the conversion of certain types of matter and energy to other types through managed dissipation, they can be used to contribute greatly to the energetic resources of cities. The question is how landscape architectural design can work with the metropolitan field when it is conceived as a "complex, infinitely entailed, dynamic system or fluid manifold?" (Kwinter 2001: 24). How can landscape design connect with matter-energy networks at specific and local points, assisting the various systems with which it engages in their transformations? Can it, for instance, activate specific behaviors or new patterns of connectivity by generating instabilities in the system? Will the system respond, as nonlinear theory implies, by transmitting a "perturbation" from level to level, dissipating it through different sets, until a singularity emerges, a kind of difference produced by transformations at particular points along particular flows? Will this singularity combine, or be combined, with other flows to produce other differences at other scales and other levels? If what we call "things" are energized by morphological processes that occur nowhere else but in themselves, how does the designer get involved? Will it turn out to be, as Waldheim seems to imply in his review of "problematized authorship in landscape architecture," that designing with landscape and urban emergence is a matter of action at a distance? These are the questions that are explored in Chapter 03, Key Concepts, and some answers are proposed in Chapter 04, Projects and Practices.

Before we leave this overview of nonlinear urbanism we should touch on one last area of interest to discussion: how urban planning has incorporated emergence. Branches of mathematics, urban geography and urban planning have converged in a proto-discipline that might be called urban complexity. Writers and researchers such as Salingaros (2000) and Batty (2005) have merged complexity theory with planning theories of urban form, space and growth. Coming from mathematics (geometry and algebra) these planning approaches to urban self-organization have a tendency to be formal and metrical, even as they utilize the insights of emergence theory. Batty states that "(E)mergence is key: it suggests no overall control, a limit on predictability, the notion that where we start matters – history matters – and that what ultimately comes about is dependent on the path we take – path dependence" (2005). There are problems, however, with urban complexity research. The histories, for instance, that Batty is interested in tend to be those of formal patterns and fractal geometries rather than those of the lived materialities of the city. And while Salingaros does apply his "theory of complex interacting systems" to urban design in an effort to develop what he calls "urban coherence," his use of complexity theory is wedded on the one hand to the spatial discourse of urban design and on the other to the place-based doctrines of new urbanism. Both of these discourses, so far, are inimical to emergence. When space and place are considered as environments within which emergence occurs, rather than as conditions produced by emergent systems catalyzed through downward causation and characterized by unpredictability, novel forms and structures cannot be brought into play. Resilient and adaptive urban conditions cannot evolve in already-given spaces or places forged by typological predilections. Salingaros has much of interest to say about "people–people and people–object interactions" that "provide the primary motivation for mankind to erect buildings and cities" (Salingaros 2000: 3), but his analysis, like those of Batty and the Centre for Advanced Spatial Analysis, is difficult to realize in terms of design practice (even though they both want this). Their principle-based generalizations are difficult to reconcile with the specificity and singularity of the intensive urban distinctions to which landscape architects connect as designers.

Both Salingaros and Batty are interested in examining spatial configurations of urban regions and predicting urban futures on the basis of these. The work of the Centre for Advanced Spatial Analysis, dedicated to a "science of cities," mobilizes digital technologies to explore geography, space and the built environment. It uses emergence theory to develop proposals for urban planning and design by means of agent-based modeling and geosimulation. However, as landscape architect Nicolay Popov notes, their simulations rely to a large extent on demographic and economic data inputs processed through Geographic Information Systems software, as well as complicated (rather than simple) set rules. The outputs of their models are often symbolic, topological or geometrical at low resolutions, and confined to large scales, 1km squared being the smallest. These simulations cannot have an immediate influence on landscape architects, for whom productive dynamic modeling needs to make use of finer scales and simple, entirely local rules (Popov 2011). Popov, echoing Waldheim, emphasizes that computational systems modeling implies a change in the role of the author. But for him, the landscape architect becomes "a system designer, who defines space, form and morphology as a complex system based on feedback" (Popov 2011). This is an intoxicating thought for designers who see the future of cities in terms of interrelated regional geographies and broad horizontal webs of architecture, infrastructure and designed public space. As Salingaros has found, if you pull back far enough, the infinitely complex web of systems seems to become intelligible, manageable, plan-able. The ever extending list of urban problems, documented regularly in doorstop-sized books stuffed with information, is based on the idea that statistics, graphs, maps and tables can usefully be generalized and digested. Thinking in terms of systems does allow us to use the new representational tools of urban landscape architecture to direct the flows of matter-energy that might otherwise overwhelm us. The challenges of urbanism are too urgent, too intense, too costly, too harmful to consider in any other way. However there is a problem: the further you pull back from the city, the further you pull back from the landscape and the species that inhabit it. You pull back from the simple, entirely local rules. You pull back from the middle ground across which we all move on a daily basis.

In 2007 American landscape architect Chris Reed asked how landscape urbanism can help "research, frame, design and implement expansive new public works and civic infrastructures"

(Reed 2007: 227). The same year Australian landscape architect Richard Weller noted that the discipline still needed to acquire a "genuine reconnection with aesthetics, poetics, philosophy and representation" (Weller 2006/2007: 70). Both landscape architects have gone on to answer the first question through the development of projects that, as Reed says, "allow for future inputs and adaptations, perhaps even resiliency to changing conditions, while simultaneously allowing for transformations of form, program and role within the city" (Reed 2007: 227). The second issue is more difficult. Reframed for this book as a question it becomes simply, how can emergence theory enable the development of new approaches to landscape design? In order to contextualize this question, Chapter 03, Key Concepts, attempts to make connections between concepts derived from, or inherent to, emergence theory, and practices of landscape architecture. Chapter 04, Projects and Practices, attempts to show how some of these concepts can be used to help answer Weller's question.

Nonlinear landscapes: the theory

Any landscape, for instance a street, a park or a woodland, has developed from particular historical circumstances. It has occurred in a certain geographical situation as the result of specific, locatable events available to it. Everything it has become and is becoming can be referred to these events, which should be thought of as imminent and natural to it: by-laws regarding the width of sidewalks, long-term erosion or periodic tornado damage, for instance. Despite the fact that a landscape has no explicitly determinable future state, there are restrictions to the number of ways that it can develop or change. These degrees of its evolutionary freedom can be enumerated and mapped, just as the parts of a bicycle can. The wheels of a bicycle can move in a circular but not a transverse motion, for instance. The degrees of freedom of any landscape are going to be much greater in number than those of a bicycle, but they can be prioritized and understood. For example, a street may currently be dominated by vehicles but it could become pedestrianized. Once this has occurred by means of an analysis, the state of the landscape may be captured and represented – well, poorly or indifferently as the case may be. We have now modeled the landscape and have a sense of its state space. If we can model it in time we may be able to register its potential changes of state. In effect we are capturing a process and can even, through this modeling, assay the long-term tendencies of the system. We will notice that a large number of the landscape's components seem to form spontaneously and in doing so make evident the constraints that cause their development to happen in specific ways. But a landscape is not a machine, such as a bicycle, a washing machine or a hydroelectric turbine. It does not operate in a linear fashion or reach a steady state, an equilibrium. A washing machine, when washing, is pulled into a topological form (or set of possibilities) by its long-term tendencies, which are known as "point attractors" because they guide the system to a single, regular condition that, when in action, does not change. The washing machine will continue to wash until the cycle stops. During that period it repeats a single action. The attractor defines the state of possibilities of the system.

A landscape, however, is nonlinear. It consists of a set of singularities all of which attract different components of the system (plant communities, bird habitat, food chains) along different trajectories into different states. By this means different structures emerge, a progressive differentiation occurs. In different places different sets of transformations are developing and, interestingly, even as these transformations reach out to the world certain aspects of the group that defines them remain the same, or invariant (topology was invented to describe how some things stay the same while others change). The woodland is still a woodland. Aspects of the group (a tree–bird–insect ecology, say) respond to the events that occur to them in different ways. The degree of transformation caused by an event is called the group's symmetry. This is not an intrinsic property of the group since it is always only ever relative to the transformations that are occurring within the group. In landscapes we get groups within groups, so that a particular process can cause another one to change radically by becoming susceptible to transformation. When this happens the landscape or part of it could be said to be undergoing a symmetry-breaking transition. A steep incline in an area of an urban park, having eroded slowly for years, finally slips completely away revealing an old underground passage that everyone knew was somewhere "around here" but not its exact location. Such an event is called a phase transition, or bifurcation. In the laundry a phase transition might occur if the washing machine's stop-valve malfunctions and the water in it begins to bubble on to the floor. Phase transitions switch a physical system such as a landscape from one state to another. In landscapes they are relatively common. There is always a critical point at which a bifurcation will occur. When a bifurcation results from some fluctuation in or outside of the system, a change of state ensues. This may, through downward causation, or feedback, initiate changes to the group in which the bifurcation occurred, unleashing a sequence of instabilities or a cascade of bifurcations. This is autocatalysis. When it happens the whole group may shift to a different state altogether (measured as a slight or a major transformation) which is itself not turbulent since it has absorbed the turbulence and refashioned. The sequence described here leads a system to have a particular morphology and therefore a morphogenetic pathway. The system is always on the pathway, which may diverge at any time. We have just described a multiplicity, and multiplicities give form to process, not to a final product. In reality a landscape is a continuum of multiplicities continually self-differentiating. Such a landscape has extensive and intensive physical properties, the former being metrical and therefore divisible into equal parts, and the latter indivisible without involving a change in kind. When the ambient temperature of a vineyard (an intensive property) rises, the amount of sugar in a grape (an extensive property) gets greater. Generally, as intensive properties differentiate, physical (material, spatial) properties change. But they change in unpredictable ways. Viticulture can understand the weather systems, and vintners can select the grape variety according to climate and soil, but the price of wine will go up and down because the grape–climate-landscape system itself is essentially nonlinear. It involves risk, without which there would be no investment, as investment thrives on risk. As we move away from equilibrium we move towards the unique, the specific, the local; towards richness, variety and life.

To recap: a landscape system is a multiplicity. It is defined by distributions of tendencies in a process; and by a series of critical transitions that can take several such distributions embedded in them and unfold them. This process forms a real dimension of landscapes in which transformations in intensive properties create all sorts of different kinds of metric space and the discontinuous structures that inhabit it. Just how the landscape evolves will depend on the initial conditions plus the parameters or constraints we have called its degrees of freedom. Within these parameters any of a range of future states, achieved by a range of trajectories, is possible. The trajectories, once activated, are determined by their phase transitions on the basis of interactions with events combined as singularities. In most cases a whole population of trajectories displays certain regularities because the history of a landscape system is shaped by global regularities as much as by local fluctuations. It is exactly because a landscape has many different attractors that it breaks the link between necessity and determinism, enabling it to select between different destinies and making later phases of its ongoing transformation a combination of chance and determinism. Once we realize that a nonlinear system such as has been described is promiscuous, in the sense that it can be attracted to different singularities for different reasons, and that there are as many attractors as there are fish in the sea, we can see that its organization is contingent and more or less arbitrary, depending on the degrees of freedom it enjoys. A bifurcation can lead a single system into a distribution that no one could have foreseen. That is, which alternative obtains is decided by chance fluctuations in the environment.

03
KEY CONCEPTS

One of the more sustained inquiries regarding the potential of emergentist landscape architecture to connect with human experience is Connolly's project on landscape affect. Deeply influenced by the writings of Deleuze, Connolly places affect at the very heart of human encounter with landscape, seeing the openness of open systems not as a physical or functional condition but as a way to "inventively follow" nature's own self-organizational practices. "Landscape architecture," Connolly asserts, "has always been loosely about openness, even if landscape architects . . . have been weak at affirming such openness." Self-organization in natural systems "re-orients and clarifies . . . practices of openness" that in landscape architecture are misconstrued or tentative. Connolly invokes Deleuze's redefinition of affects which, for Deleuze, are not feelings or affections but transitory thoughts or things that occur prior to an idea or perception, the changes, or variations, that occur when bodies collide, or come into contact (Colman 2010: 11). The notion of affect puts the human encounter with landscape in a central position in human life. An affect is physically and temporally produced through the individual's encounter with other bodies and processes (animal, mineral, vegetable) both by chance and by organization. It consists of a variety of landscape-related factors that include geography, biology, meteorology, astronomy, ecology and culture. For Deleuze affect is intimately related with experience, which is not a subjective condition, but an empirical "place" that is connected organizationally with other milieux. The whole nature of the human–nonhuman relationship is affective. In attempting to bring this insight into landscape architecture Connolly turns to the language of emergence. He sees landscape as a "heterogeneous continuum" that is continually re-sorting itself according to the multitude of systems that comprise it. Human beings are plugged into these systems naturally and unconsciously. Landscape architects forget this and partition the landscape continuum into discrete disconnected functionings that can be programmed, instead of opening up the "connectibilties and trajectories" that provoke and produce affect, that most exquisite of human sensations. Connolly's formulation both provides a background for landscape architecture's unbounded fascination with instrumentality, and identifies the problematic that this fascination causes. Landscape architects, according to Connolly, are blind to the productivity of the landscape, and find its capacity, through self-organization, to create singular situations completely "indiscernible" (Connolly 2006: 214).

Connolly's critique opens on to a consideration of emergence that, in the end, this book is all about. How can emergence theory extend the practices of landscape architecture into the realms of the imaginary that Corner, Weller, Mostafavi and so many others regard as critical to the creation of human and nonhuman environments? What will make landscape architecture more than "an art of instrumentality?" This chapter explores a number of key aspects of emergence theory in an attempt to develop them into conceptual frames for design inquiry. The circulation of ideas derived from emergence theory has increased in landscape discourse because these ideas bear on the very conditions that set landscape architecture apart from other design discourses. Resilience, initial conditions, field theory, nonlinear dynamics, open systems theory and disturbance are concepts used to describe the functioning of natural systems and processes. This cluster of concepts has proved to be extremely effective at provoking novelty – in ways of thinking and ways of designing. Thinking and making new things is not an accidental outcome, however, for emergence is exactly a way of describing how new things occur in the world. Emergence theory is a speculative answer to the question how natural systems, continually and endlessly, create novelty: new rhythms, new patterns, new processes; new species, new behaviors, new responses to new conditions. This chapter incites an engagement with the speculative possibilities of emergence by introducing and explaining a range of concepts that underpin emergence in landscape architectural projects and discourse. It describes each set of ideas and provides examples drawn from recent speculative projects, proposals for a range of urban situations and from projects undertaken by professional firms. While it aims for clarity of explication, and current and relevant exemplification, it is also provocative. It points to the difficulties of using emergence theory in practice, and discusses, for instance, the tendency for landscape architects to use it metaphorically rather than instrumentally. It highlights the problem of bringing together top-down design intentionality with bottom-up landscape conditions, and investigates the challenges associated with the desire for order in situations that are essentially chaotic. A relatively new branch of literary studies known as ecocriticism provides an orientation. Ecocriticism explores the literary borderlands where human and nonhuman meet. Literary texts such as poems, novels and plays are presented as sites where contemporary environmental concerns with how humans relate to, construct and inhabit the world coincide with how these things are articulated. "Green readings" of Roman poems, medieval epics and Renaissance plays remind us that humans are always interacting with other organisms in particular ways (bodies encountering bodies) and that at particular times these interactions are forged by ideas of becoming that are far different from the division of the world into human and other that we have inherited from the Enlightenment (Rudd 2007; Siewers 2009). Ecocriticism suggests how, as with literature, landscape architecture needs to push ecology beyond the application of scientific principles to human environments into a consideration of how scientific models might play out in the popular imagination. In this way it could provide an important example of how facts and values, rather than being separate, are deeply intertwined. The nine sections in Key Concepts can be read in any order: the concepts overlap, iterate and reiterate, telling the same story from different perspectives. The whole chapter is intended to provide a bridge from the discussions about emergence in the sciences and ecology to the essays in Encounters that explore issues to do with embodiment and representation.

Open systems

What does not kill me will make me stronger. It will cause me to adjust. I will improve. I will survive the catastrophe and go on. Not just endure; enhanced, ennobled by my experience I will meet the world with greater resilience, more useful tools, a wilier intelligence, an increased capacity for dealing with the slings and arrows that are thrown my way. I will emerge from my small struggle a better person. Myths and legends, local stories, family anecdotes, movies, romances and crime novels are rooted in the device of overcoming, of the hero (me, you, her) breaking a conflict situation and through this catharsis becoming something new or better. The narrative of self-improvement is based on the trope of renewal after disaster – because of disaster. The epic poem, such as the *Odyssey* or the *Iliad*, requires disaster to enable the hero to create a new condition; for the old order to be swept away and a new order ushered in. We see it in films and novels too: any Clint Eastwood movie, for instance *Gran Torino* or *The Unforgiven*; Jane Austen's novels, where the heroine and hero alike become improved through hardship and the asymptotal nearness/farness of the loved one. It is the same in real life: war cleanses and purifies, the rat eats her young and the rat community flourishes, the frightened young officer is wounded and, spurred on, leads her company to victory. And who is this subject for whom the figure of emergence is so apt, so peculiarly descriptive? This subject is, of course, a system, for it is systems that are emergent in this world. The subject is a set of systems, nested hierarchically, level after level of system from macro to micro, from the lofty heights of nonphysical consciousness to the microscopic, almost nonphysical but so material systems of submolecular events in terms of which some say consciousness itself can be explained. And, blending with this vertical systemic subjectivity is a horizontal field of systems overlapping, drifting through each other, passing into other systems, social, geographical and cultural. Vertical systems and horizontal systems comprise the subject in contemporary life. When we meet a friend or hold a flower to our nose one set of systems confronts and mingles with another. And all these systems are open . . .

It was noted in Chapter 01 that "system" is a metaphor, a theoretical construct that enables us to understand (perceive, describe, explain, interpret) observed behavior in a language that is formulated especially for the job. One of the most important features of the construct when applied to people is the sense it gives of human entanglement in other systems, the complete and unavoidable imbrication – embeddedness – of landscape and body. The sense, as Merleau-Ponty put it, of the human body "being caught in the fabric of the world" (Merleau-Ponty 1969: 256). The idea of system reinforces a phenomenology of landscape–human relations as indissoluble and of the human agent as a becoming that occurs always inside the intertwining landscape systems in which it is immersed. The supposition that there is a mind (or a subject) over here and a landscape (an object) over there, mute, physical, external, is no longer tenable. Any discussion of open systems will entail the decisive concepts of emergence: time, disturbance, instability, equilibrium, inclusiveness, resilience, self-organization, adaptivity. All of these concepts, developed or adapted to explicate certain kinds of phenomenal behavior, are characteristic of open systems, merging the open systems of human existence with those so often considered separate, of landscape.

The study of open systems, and the development of this language, began in earnest in the twentieth century after nineteenth-century innovations in thermodynamics and related disciplines. Classical, Newtonian physics had based its account of natural phenomena on the analysis of the behavior of closed systems. In a closed system a kind of experimental purity is possible. System events and circumstances can be controlled and repeated, as they are not influenced by anything but the experimenter's interventions. Observations of the system's behavior are therefore reliable and determinate. However, the observation of closed systems is now considered an artificial and misleading way of finding out about nature, including human nature. Natural processes are always interacting with each other and are highly susceptible to change. A landscape, therefore, is a system that changes and evolves according to information continually being received from an environment that includes itself. Once we have open systems, we can speak of self-organization, of which autocatalysis is a crucial characteristic. Readers will remember that autocatalysis, or positive feedback, occurs when the output, or result, affects the input of the system, thus altering its operation. Put another way, information generated can influence the generation of further information. An autocatalytic process is one, therefore, that catalyzes or accelerates itself. Classical physics did not have the tools for finding this out. Newton could predict the moon's orbit from the laws of gravity, but did not have the equations to describe the nonlinear feedback produced if another moon is introduced into the system, when orbits become chaotic and linear prediction impossible. For the first time in history, the study of feedback loops enabled researchers to distinguish between the pattern of organization of a system and its physical structure.

As noted earlier, an open system is part of its environment. While this means that its future behavior is not able to be determined, it can nevertheless attain a structure and maintain it in far-from-equilibrium conditions. This undermines the traditional view that systems must be examined as if they were isolated from their environment. Instead they are seen as embedded, or nested within an interactive cascade of mobile networks. For this reason scale is critical in the consideration of landscape systems. In landscape studies space is dynamic, flexible, relative. It is not an absolute, not a container for form. Within a watershed or a city we can observe spatial changes at different scales. As a whole, a city does not seem to change much in a week, but at the level of the street the change in one week can be radical: buildings vanish, families move on, crushing accidents occur. What is most crucial is that we can see different relationships depending on the scale at which we look. At the scale of Oceania, we can observe economic flows between Pacific islands, and this causes us to understand these islands in certain ways, as part of a global system. At the scale of the bay we see different aspects of these economic flows, and understand their relationship to landscape conditions: harvest, management and husbandry; weather patterns, kinship influences and so on. In fact, scale itself is a framing device that separates and isolates systems that are in reality connected.

The flow of energy in open systems allows them spontaneously to self-organize by developing novel structures and new modes of behavior. Self-organizing systems are therefore said to be creative, and their bifurcations the source of diversification and innovation (Prigogine 1996: 70). They are adaptive, and actually require conditions of instability and disturbance in order to evolve to new levels of order and vigor. Matter-energy, and its flows and coagulations, is the basic subject-matter of landscape architecture. The regulation of these flows and coagulations as a design strategy is developed by Zaera-Polo, a designer with Foreign Office Architects. For FOA, design is seen as a way to "permit organizationally complex landscapes to emerge through the production of topographies artificially generated by a mediated integration of rigorously modeled orders" (Zaera-Polo 2005: 23). Open systems are complex. Their parts are so numerous that there is no way a causal relationship between them can be established. Instead, their components are connected by networks of feedback loops operating at different levels, different scales and different rhythms. Landscapes work like this. This is why it is not necessarily useful to think of them in terms of types. As complex adaptive systems their higher order patterns are the result of their continual interactions within and without, of their ongoing openness and responsiveness to quite specific, changing conditions. We should think of landscapes as generated by, and therefore imminent to, the specific conditions they comprise. How we remodel landscapes is a question how we gather and direct these forces and, indeed, this is just what gardening is: the production of difference through the husbandry of natural processes.

In Chapter 02 cities were described as dissipative structures. As such, they are in contact with various sources and sinks. Flows of matter-energy from sources to sinks enable ordered structures to be formed, maintaining them in time. In the end these flows are dissipated as wastes and other pollutants to external sinks. If a landscape is static, like a nineteenth-century park, it becomes a sink not a source. It requires maintenance in the form of physical and economic input. One of the objectives for the landscape architect is to make sources not sinks. The project is to encourage the development of open systems, and to place human beings into these systems in such a way as to enhance the exchange that occurs between the multiplicity that is human personhood and landscape multiplicity. The operations of landscape architectural making – the poetics of landscape architecture – are entirely adequate to this project. Garden making, city making, the creation of parks great and small, have always required of landscape architects a special sensitivity to the openness of becoming. An immersion in the unique processes of human–nonhuman interaction that only landscapes can afford, and a careful rehabilitation of the operations that are specific to landscape making – a commitment to the intimate association of humans and the world that makes them human – can invigorate the discipline and re-orient its approach to the complicated terrains that confront it. This will require a revision of expectations on the part both of landscape professionals, and the constituencies with whom they work, to realize the designed open systems of the twenty-first century. An open system will evolve into something that has not been envisaged. Any design investigation must take into account the responses not only of the natural system but also of the human participants with their partly compatible and partly opposing goals. The system is not simply the terrain and its affordances. It includes subsystems: the weather, climate cycles, municipalities, funding organizations, real estate values, stock market reports. By means of the transmission and exchange relations it enters into with these the system selects its own trajectory. While this trajectory may not suit everybody, everybody has been involved in it. The system does not just take, it gives.

Koolhaas and Mau's Downsview Park Tree City is emblematic. Now widely assumed to have failed, Tree City was an early example of designed landscape emergence, the winning entry in a competition to repurpose an abandoned airfield as an urban park in Toronto. Its nonlinear strategy was as open as is possible to future entailments, and evolved accordingly. This always meant that the "rapidly changing and constantly shifting urban field," as Waldheim put it in his 2001 review of the competition, could move the system in unforeseeable directions, as a result of the downward causation that typifies emergence. After the three phases that established the green infrastructure of the park, the scheme left the remaining terrain "to be returned to its native ecological state or left fallow for an indeterminate future" (Waldheim 2001: 85). Only by 2006 would pathways and programmed activities be introduced. Therefore managing public expectations, as Smith says in an article written after a visit to the park in 2006, was a critical part of the project that was somehow overlooked, or under-realized, and the project has indeed shifted (Smith 2007: 37). New commercial and residential developments are now under construction, working reassuringly towards a final state, rather than the awkward and unintelligible (because slow and informal) natural and cultural succession phases that would never reach an achieved condition. It might be, as Smith states, that the problem was a lack of clear communication strategies that might explain the basis of the proposal and what stage it was at in regular and appealing transmissions. It might be the lack of interim detailing to provide formal, legible visual, physical statements about progress. It might be, as Berger suggested in a 2001 review, that an inattentiveness to the specifics of topography and hydrology meant that the envisaged self-organizing plant ecologies were never going to flourish and the whole thing just sit there looking like an underfunded, half-hearted arboretum project (Berger 2001: 131).

But this is too simple. An urban system comprises, among other things, sets of social systems and ecosystems. The success or failure of Tree City cannot be judged in terms of one or the other. It is particularly because humans are part of the Tree City system that it could branch off into any one of a number of different organizational states. There is no "correct" or "preferred state." Like the three moon problem that Newton could not explain with classical dynamics, the park that evolves rather than remaining static will elude description unless we attend to the frame by means of which the description is developed. This requirement is as fundamental to the establishment of the system in the first place, as it is to future analyses of it. Both the designers and the readers of any self-organizing urban landscape must necessarily identify the important components of the envisaged landscape, their interrelationships, a sense of the system as a whole, and the system's relationship with its context. And these aspects of the landscape must be understood as undergoing continuous change. Ecologist J.J. Kay has explained how a system description is always developed from the perspective of an observer. Since a system is not "out there," but "inside us," he says, any situation can be described in a number of ways. No description is complete or correct, and there is always the question which description should be used. Kay emphasizes that when we look at a system we see its structure but not the processes that inform it (Kay 2008: 17). The urban processes that Tree City selected as its conditions of evolution are complex and difficult to pin down. Additionally, the state that an urban landscape occupies at any one time is as much a matter of its history as its current context. As self-organizing regimes, designed landscapes will develop according to what has happened in the past. The Downsview Park trajectory began in 1994 when the Canadian federal government announced the site as an urban park, and its future states were portended by the establishment of Parc Downsview Park Inc. as mandated to create an "urban recreational greenspace" in 1996. These initial conditions formulated the DNA that would actualize the park by defining its possibilities. Political, social, financial and cultural perturbations could push the system from one regime into a very different one. Tree City tells us a lot about the difficulties of strategizing open urban systems.

As Kay's insistence on the appropriateness of system descriptions implies, a big picture is made up of many little pictures. Returning to ecocritical literature for insight, we find Robert Watson, in his "Ecology of Self in a Midsummer Night's Dream," noticing that Shakespeare's characters are "motivated by uncontrollable forces within and without." They occupy worlds that are reliant on each other, that even merge. The play's interspecies couplings and metamorphoses remind us that there is more going on than "cool reason comprehends," and that an environment does not necessarily refer to clouds and trees (R.N. Watson 2011: 13 *passim*). Setting aside cool reason, then, let it be stated that an open system is a multiplicity. As a multiplicity a landscape offers itself to composition with other multiplicities, such as ourselves, in ways that urge us to become involved. A landscape, in this sense – as multiplicity – is not defined by simple materiality, or the space it occupies, or its organic structure, and nor are we. When we become involved with a landscape it is the relations between its parts that strike us, and the actions and reactions these provoke. So it is not simply tree, but how the tree develops its distinctions through placement, orientation, distance ("external" milieux) and sunlight filtration, scent, color gradients, interaction with other plants and materials, the spatiality it creates, the ephemerality it assumes, the involvement it enables other bodies to develop with each other under its influence. A tree is a body, I am a body, and a body is more than the physical suggestion of my frame. Above everything my body has a capacity for being affected by other bodies, and when there are landscape systems that consist of magnitudes and dimensions that are in continual change – like me – I can open up to them, enter their elements and relations, their porosity, intransigence, elisions and divisions, and become affected by their powers. I bring the entire field of my lived experience to this encounter: past encounters, significant events, historical determinations, concepts, memories, capacities, habits, persuasions, predilections and powers.

As an open system the landscape simultaneously lays out its own semiotic, material and social flows. It does this, as I do, in ways that elude the separation of the situation into world, representation and subjectivity. A landscape has no outside, for its connectivity to other multiplicities is always complete. Any figuration of a multiplicity is, in the manner of a systems description, at once a subjectification, a projection and an artificial partitioning. Perhaps it is more apt to say that a landscape is only and always an ecotone, an edge, a continuous immanent spatiality, a set of borderlines that continually present themselves as elements to be dissolved, disassembled and reassembled through my contact with them. This compelling admixture of body and body, produced by congeries of landscape difference (height, color, light, moisture gradient, adjacency, etc.) is not, then, an interpretation or an experiential condition, even though it is phenomenal, but an individual realization of the concrete constituents of time and place. When we encounter landscape multiplicities, we are always encountering them in time and place.

The human subject-system that is doing this is itself a moving location of affects, not a thing. The kind of individuation we find in a person is more akin to the variable differentiation of something like a day, a season, a life, a climate, a wind, a fog, a swarm, a pack. A landscape is like this too. It is not some kind of backdrop or background that situates subjects or holds things to the ground. It is the entire assemblage incorporating the situation and its aggregates. In my capacity as an effect, an event, a component of the assemblage that is arranging around me, I am inseparable from the hour, the season, the air, the street, the "weeds" in the cracks of the paving. I am always in composition with the landscape I am connecting to. The wetland I discover at the end of the street is as much a part of the walking human–boardwalk–heron assemblage as the designer who initiated it and the collectivity whose neighborhood it skirts.

Situation

Philosopher of science Isabelle Stengers advances the need to uncouple "the analytical approach" from "the scientific view of the world." Analysis in her view is a question of asking the right questions rather than distinguishing between objective truth and arbitrary decision (Stengers 1997: 4–5). As a scientist Stengers is permitting us to tease analysis away from reductionism, and to accept the advantages of an analytical method in the light of the possibility that it may directly contradict the generalization of reductionism. For landscape architects wishing to found their analysis of a complex landscape situation on something more useful than the reductive measures of normative science or on intuition or freeform interpretation, Stengers' work conjures up an interesting and relevant possibility. According to *Webster's Dictionary*, analysis is the process of breaking a complex topic or substance into smaller parts in order to gain a better understanding of it. The word comes from the Greek *analusis*, "a breaking up" which brings together the root words "and" meaning "up" and "lyein" meaning "loosening." It implies the separation of a whole into its component parts, or the "examination of a complex, its elements and their relations." Analysis, which can be conducted conceptually, linguistically, manually or with a device, is a crucial part of landscape design. The question what to analyze comes before the question how we analyze, and yet these are two sides of the same coin.

Of special significance are the relationships between people and the dynamic conditions that they create and inhabit. Close observation of behavior is necessary. What are the specific assemblages that comprise these human–nonhuman interactions? How do we find this out?

In landscape architecture we have to take account of the immersion of humans and nonhumans in the environments that support and maintain them, and determine how these environments are continually transformed on the basis of this immersive condition. The actant in any system, say, a neighborhood, that is far-from-equilibrium, represent what Stengers calls "fluctuations." That is to say they are responsible for the uncontrolled variability of individual behaviors and may play a role in changing what she calls "the macroscopic regime" of the situation. These fluctuations emerge from a background noise that may or may not be important. The designer needs to know with respect to the noise what is significant and what is not. The people in a community may be particularly sensitized to a political discourse that affects the potential futures of the community. Whether this discourse will have an effect is largely decided through the collective regime of activity, which ultimately decides what is insignificant noise and what must be taken into account. We do not know *a priori* what a neighborhood community can do, and when it is far-from-equilibrium, which is most of the time, we cannot tell once and for all the difference between what must be taken into account and what we can ignore. Furthermore, the collective learns, and changes as a result. It is a moving target. Its emergent effects continually alter the composition of the situation. Other actants in the system, such as the community organizations that provide the funding for landscape projects, and the school system that wishes to be involved, as well as the polluted creek that runs through the community park, are all players in the game. Kristina Hill draws attention to the variability of readings of nonhuman species existing within situations. She gives the example that mowing a meadow in spring, when juvenile mice are dispersing may, in the absence of tall grasses, make these young animals more visible to hawks. "On the other hand an insect that lays its eggs in the meadow may find its population maintained if the meadow is not mown until summer, after its eggs have hatched and the insects have developed wings to fly away" (Hill 2005: 147). What we have is a complex system. It is not possible, or at least it is dangerous, to isolate these players in order to deal with them. Kay calls this sub-optimization. "When one part of a system is optimized in isolation, another part will be moved farther from its optimum in order to accommodate the change. Generally when a system is optimal, its components are themselves run in a suboptimal way" (Kay 2008: 19). And Stengers warns: "isolation is a dangerous game, and those who believe they can purify their objects in fact intervene actively in the significance of the object they observe" (Stengers 1997: 16).

These issues have been raised in order to demonstrate that the analysis of existing conditions within an open system, a system that is continually emergent on the basis of internal and external interactions, requires a special set of tools and procedures. The study of morphogenesis shows that discernible structural and organizational patterns underlie these interactions. When we understand these underlying blueprints we can start to seed processes with catalysts for the development of new patterns and forms. Such understanding involves a letting go of understanding in general and an acceptance of the kind of framed description mentioned in Open Systems. The French philosopher Alain Badiou describes a condition – the situation – that enables us to refine the way we look at terrain. A situation, he says, is:

> an ordinary multiple, a multiple that is obviously infinite because all situations in reality are infinite. It can be a historical, political, artistic, or mathematic situation; it can even be a subjective situation. I take situation in an exceptionally open sense, and to capture that openness I say it's a multiplicity.
>
> (Badiou 2000: 64n6)

While anything can be a situation – a dream, a farm, a divorce, a stock prediction, a death in the family, a supermarket, a street, a bad hair day, a crime scene – it is the usefulness of the concept for landscape architectural analysis that will occupy us here. Situations come in any modality; they can be necessary, contingent, actual, possible, virtual, potential (Feltham and Clemens 2003: 10). We have seen in the exploration of open systems that a landscape situation is a set of multiples, all of which are composed of further multiples, and all of which can belong to different situations. There is no unified totality that encompasses situations. Their relations are always adaptive and dynamic. Two types of landscape situation were mentioned in the above list: farm and street. Clearly, humans are involved, but the kind of human involvement that Badiou's notion entails differs radically from that which appears in most landscape architectural formulations of site. We can see readily how a street is a situation in Badiou's sense. It is a multiple that participates in other systems through the obvious medium of its role in transportation networks that overlap and interact. A street participates in less obvious systems too, such as the vertical system of a high-rise building. The social meshwork of the street sometimes gives way to a stratified diagram that structures sociability in the high-rise office tower, however: more than one urbanist has commented on the reduction in ethnic and – once upon a time – gender diversity the higher you go in a building. The situation of the street really is, as Badiou describes it, an infinity. In terms of structure the compositional elements of the street are countless: physical, ecological, vegetational, architectural and material. Then there is its intensive character: comings and goings, speeds and slownesses, temperature gradients, proximities, affordances, not to mention its ability to create human worlds out of the bringing together of bodies and subjectivities that the street so wonderfully makes possible.

Badiou argues that a situation is a field of murky possibilities in which the human subject may encounter something special, a critical event that disrupts the situation, or transforms it, opens it up to something Badiou calls "truth." This event is turbulent and unpredictable, and is of crucial significance to the subject, who encounters it as a difficult condition. Think of this. The situation is now more than an externality. You are implicated in it: it is yours. Every situation belongs to, is part of, a human subject's reality. This is perhaps the most useful aspect of the notion. Human beings are always acting in response to, or in complicity with, a chance event that disrupts their situation, which makes them belong to the situation. When we understand how embedded humans are in situations, and how multiple these situations are – how infinite – we can begin to see how important are the modes of analysis we bring to bear on them. What we need to discover within the situations that we break into smaller parts are the transformational potentials of our own involvement in those situations as designers. The decisions we make as landscape architects have transformational consequences for the situation. Our interventions make a difference to the lived experience of the situation for those whose lives are entangled in it, and in whose lives the situation is itself entangled. It may well be landscape architecture's emphasis on lived experience that makes the urban field so real and present to its inhabitants. We create the city by discovering both the subjective and the physical material of the urban field and making these the means by which the urban is recovered from its dormancy. Physicality and subjectivity are critical to our experience of experience. This is the message of the situation. What urban landscape design can give us is the utter complicity of the situation in the construction of the subject's lived experience. What is gathered in analysis, how it is gathered and collated, and how it is interpreted, depends on what it is being gathered for, and how it is to be used. Since, in landscape architecture, all data that is used in the design process involves a crucial set of relationships between humans and landscapes, it is important that this relational condition is taken into account in the practice of collection and interpretation. Landscape analysis is often considered to be a combination of the following analytical methods.

Site reconnaissance

Reconnaissance (literally: recognition) is a military term denoting exploration conducted in order to gain information. It is usually undertaken by scouts or military intelligence soldiers by means of direct observation. Reconnaissance is part of combat intelligence, and contributes to and is managed by, the government-level cycle of intelligence management. It may be compared to counter-intelligence and surveillance, which are the passive gathering of data. Sometimes reconnaissance is used by geologists to refer to "the examination or survey of the general geological characteristics of a region." The term is also used in computer networking for the exploration or enumeration of network infrastructure.

Site forensics

Forensics refers to the application of a broad spectrum of sciences to answer questions of interest to a legal system. It encompasses the accepted scientific methodology and norms under which the facts regarding an event, or an artifact, or some other physical item (such as a corpse) are ascertained as being the case. In this regard the concept is related to the notion of authentication, whereby an interest outside of a legal form exists in determining whether an object is what it purports to be, or is alleged as being. Forensics comes from the Latin, where it means "of or before the forum." In Roman times accuser and accused would put their arguments before a group of public individuals in the forum. The individual with the best argument and delivery would win. This would usually be the person with the sharpest forensic skills. Some forensic techniques (i.e. "scientific methods") have been found to be questionable. For instance, comparative bullet lead analysis, forensic dentistry (later invalidated by DNA evidence) and fingerprinting. With respect to the latter, a *New York Times* article recently noted that no one has proved even the basic assumption that everyone's fingerprint is unique.

Site surveying

This is the technique or science of accurately determining the terrestrial or three-dimensional position of points and the distances and angles between them, and delineating these on a two-dimensional surface. Often used to establish maps and boundaries, it deploys elements of geometry, trigonometry, engineering, physics and law. The theory and practice of surveying can be widened to include the determination of cultural details above, on or beneath the surface of the earth. Information is gathered through observation, measurements in the field, questionnaires, research of legal documents and data analysis. This information is examined as to its condition, situation or value. In ordinary parlance the term is widened to include ideas of inspection, appraisal and scrutiny. Accuracy is one of the key components of a survey, as surveys are used to develop architectural and engineering projects on the one hand, and to settle land disputes on the other.

Analysis, then, combines intelligence gathering (reconnaissance) with accurate measurement (surveying) to present a compelling argument (forensics). It is just this approach that *the situation* renders inconceivable. What is faithful to the situation cannot be observed, measured or argued for. How can an indivisible condition that reaches through the very boundaries that such analysis erects become accountable to that analysis? This condition *can* be captured, to a certain extent, by imaginative drawing, writing, videography and photography, and it is these creative and constructive performativities that must be embraced and used to augment landscape architectural analysis. In this way, unhooking the analytical approach from the scientific view of the world, using drawing and writing and photography and videography as a our means of inquiry, we may enter into the situation and determine what may or may not become useful in our design. Its utility is not discovered but invented. What is loosened up is not parts from a whole but events from a situation. The next Key Concept, Meridian, advances this argument.

Meridian

Using a scale is drawing the borderline of a multiplicity – it gives it an edge. Take an estuary and its creatures. Individual species occupy this amorphous borderline in such a way that other species are drawn to its moving edge condition. When they come in, the original population makes an adjustment. Sometimes it is only individuals that realign, moving slightly in or out of the group, even taking members with them; in which case a part of the population redefines the border randomly as it were, re-establishing the group and its place within the estuarial field. These adjustments may occur hourly, daily, seasonally, as a result of tidal fluctuations, long- or short-term weather patterns and the migration of species. (A street corner in an urban core may be similarly described.) Although this self-ordering assures a kind of stability, even as the multiplicity moves and changes, swells and diminishes, reacting to outside circumstances, it is impossible to define its extent. But scaling implies extent, size. And it is supposed to tell us not only how many components there are in a specific location, but what kinds of relations hold between them. Examining a system that is always becoming something else therefore requires more than one scale. As Nowicki notes, "the condition of being different or having differences does not provide a yardstick with which to measure diversity" (quoted in Lister 2008: 86). To avoid any sense of limit or line cast across a gradient of shifting qualities, different scales must be used simultaneously. When we do this, processes and relationships come in and out of focus, and it becomes clearer which of these we are interested in with respect to the frame we are developing. Scale, then, is observer-dependent, just as the identification of a specific level for analysis is observer-dependent. For biologists examining the estuary described above, the middle scale – that of the species whose movements were observed – is the most useful. It is in the middle that the individuals in which they are interested tend to be most obviously embedded. Observing the interactions between populations, however, requires a different scale. In either case, the relationships that the specific elements enter with each other and other individuals, including habitat structures and passing biota, "there is a staggering degree of uncertainty" (Lister 1997: 126). But the middle scale will always be the most tangible and is often the most common. This is the scale of the macro-organism, of the beast, the bird and the human being – of the holological ecologist. And yet, "It's not so easy to see things from the middle, rather than looking down on them from above or up at them from below, or from left to right or right to left: try it, you'll see that everything changes" (Deleuze and Guattari 1987: 23). Serres uses the figure of a surfer catching a wave to make the same point. The surfer waits until the right moment and then, capturing the matter-energy of the wave rides it until it dissipates in a thunderous crash on the shore and becomes undertow, pulling sand and pebbles back into the sea. The ocean is a vast movement system, but the surfer catches the wave in the middle, as it is peaking, before it has broken (Serres 1995).

A situation, whether it be primarily ecological or primarily social, will always have components of both, and in each case the potential for using middle-scale indicators of situation extent is fraught with difficulties. As mentioned in the ecology section of Chapter 02, one way that biologists have in the past attempted to define the extent of a particular ecosystem is by identifying the keystone species for that ecosystem. For instance there is the designation of the oak-hickory-pine forest of the Fall Line area of the Eastern United States, where the upland Piedmont meets the Atlantic coastal plain. These species have been used to identify the characteristics of the ecosystem in which they appear since they have been thought to determine coeval understory plant community probabilities, as well as ecosystem extent. However, it is now recognized that "ecosystems follow cyclic rather than linear paths of development, regularly punctuated by sudden, unpredictable and rapid episodes of change to a variety of other states" (Lister 1997: 141). This requires species to move between multiple operating states, suggesting that a species understood as a keystone species in one state may quite well be redundant in another. Plants, animals – and humans – are likely to perform different functions under differing circumstances, and therefore in different situations (Lister 1997: 141). The relative importance of different actors in a situation will be best discerned by using multiple scales to study the same phenomena. The particular role of individual components may only become apparent under particular conditions that trigger their individual structuring function. These conditions are often discernible only at different scales, which, for instance, demonstrate causal conditions that are beyond the capacity of one scale to apprehend. As a meshwork, an ecosystem's connectors and hubs (its keystone species and the species they connect to) may change depending on food supply, the presence of predators, disturbance by humans, drought and other perturbations. This means that it is not possible to get a permanent reading of the ecosystem at any scale. The ecosystem itself is scale-free. There is no characteristic node, no permanent keystone species; instead there is a continuous shifting of primacy with no single species that can be identified and claimed to be characteristic of all the species in the system. While a scale-free network is one that has several large hubs that define the network's topology, just which species serve as these hubs is a matter of the continual co-evolution of the species in the system and their ongoing responses to disturbance (Barabasi 2002: 70–71; Buchanan 2002: 138–155). As in a Fall Line forest, so in a Fall Line city such as Baltimore, Richmond, Columbus or Montgomery. When viewed from different scales different urban relationships emerge. It is therefore not sufficient to reduce information to that gleaned from a single scale. It is necessary to think about the situation from a variety of scales, to identify the relationships that those scales reveal, as well as other considerations that might aid in understanding the situation. But moving from scale to scale really only moves the problem of recording nonlinearity from scale to scale. We can notate probabilities at any one scale and regard these as "good enough," but then we have to recombine them with the probabilities we have recorded at other scales. The conclusions we draw with respect to the existing conditions with any situation are therefore going to be a simplification of complex phenomena, and in some cases the very phenomena we seek to understand will be lost. The realization that the data sets that comprise situations, be they a ball park, a river walk, a farm or a polluted wetland, are ultimately and inherently unrecoverable, shows that whatever method we use for discovering the dynamic components, processes, gradients and structures of open systems, the data we retrieve can only give us options.

How we deal with the fact that our knowledge of the world we wish to change is inchoate and contingent is a matter of the value systems we incorporate into our planning and design procedures, and just how we do that. Part of the project of this book is to use emergence theory to suggest ways of renegotiating values on an ongoing basis. People are aggregates of different types of multiplicities that coexist, interpenetrate and change places – machines, cogs, motors and elements that are set in motion at a given moment, forming an assemblage productive of statements: "I love you . . ." Let us invoke Merleau-Ponty once more. He places before us an intimacy with the world that is not primarily visual, but corporeal. The self is not simply in the world but of it. The human body is the medium of this intermingling, rather than the mind, Merleau-Ponty argues. Since we make the world by living it, we cannot map the world by means of visual codes. We are animals, and animals are in the world, as Bataille says, "like water in water." This is why Elizabeth Meyer rebukes Corner for his suggestion that aerial mapping processes may discover a site's essential qualities, as these qualities will always elude the "application of an external order" (Meyer 2008). Meyer suggests a mode of exploration that can engage with what consists within the structures and operations of the system in all its materiality (counting memories and histories as concrete things). The notion of the situation acknowledges the site as a moving target, distributed, entangled, de-centered, connected to other situations. Drenched with potential, it incorporates the plural assemblages of cultural and social histories that are not tangential but constitutive. If all open systems are sensitive to initial conditions and their ongoing historical circumstances they can be explained only in terms of these.

Understanding the situation means understanding one's own complicity in its construction, one's own stories, histories, predilections, dramas and fantasies. Mapping has the potential to engage with the interactive nature of multiplicities but needs, as Meyer says, to be supplemented by or at least include "forms of notation" that register the situatedness of the mapper and the shifting registers of the mapped. Meyer argues for walking as a way the designer might enter into an "embedded understanding" of a place. "The complex understanding of a place made possible through walking argues for the necessity of situated relational methods to accompany aerial mapping techniques" (Meyer 2008). W.G. Sebald, whose landscape writing has been an inspiration for cultural geographers, demonstrates this important source of understanding. In the late 1990s Sebald conducted a walking tour of East Anglia and wrote about it in *The Rings of Saturn.* This and his other books conjure what Wylie calls "a strange metaphysics of landscape." Sebald gets inside the landscapes he visits, is always in the middle, linking themes of exile and displacement, history, memory and forgetting, destruction and fragmentation. He moves seamlessly between concrete landscape elements such as estuaries, clouds, tree clumps, sandbanks and swiftly moving streams and the vast but personal themes of loss and deprivation that motivate his endless inquiries into the connections between landscape and individual becoming (Sebald 1998; Wylie 2007: 207). The maps artists make also offer a way to invigorate normative landscape mapping a là Corner, often insisting on an intense personalization of the material being mapped, such as a street corner or a garbage dump. Meyer, like Connolly, argues for a "middle scale of interaction" as this is the scale at which "social collisions" occur and at which community is expressed. She argues that the middle scale of experience "is actually the scale at which we develop and communicate both meaning and identity in the landscape" and that "large scale aerial views privilege those landscape relationships that lie outside of human perception, neglecting those that are experienced" (Meyer 2008).

Landscape architectural designers should be interested in what is actually functioning in an open system, and what the mechanisms of this functioning are. Once these have been identified the designer may effectively select the material with the greatest capacity to generate affective encounters with the world. As Connolly says, the self-organizational material regulates the act of discovery. The designer enters into his or her own affective relationship with the existing conditions and through this engagement discovers the specific generative mechanisms that comprise and envelop the situation. An ecosystem, as much as a city street or a coastal reserve, is always in the middle, even, or particularly, when it is far-from-equilibrium. From microbiology we know that the genetic algorithms driving mutation are also always, being open ended, in the middle. From the art of gardening we know that a garden is, as the gardener wearily complains, "never finished." The middle is not only important, in open systems it is inescapable. The question is how to use the representational notations available to landscape architects adequately to present the open systems of the urban realm, so that the work that is based on these representations can both participate in existing multiplicities and imaginatively achieve new ones. Connolly exhorts landscape designers to identify "the interacting sets of conditions (forces and relations) that produce functionings." In fact, "(f)unctionings, conditions, mechanisms, connectabilities and trajectories all require discerning, identification and abstraction in a manner useful for a designer." Once identified in this way – and this is the only *material* suggestion Connolly makes – "(r)esponding to openness involves plugging what can only be discovered on the ground into what can only be discovered in representations – plugging the affectivity of the world into the affectivity of representation . . . and the ground into the map – to produce a single affectual landscape urbanist material" (Connolly 2006: 210). This is a way of saying that designers should understand the processes described above very well for every landscape situation they are asked to "stir." Indeed, knowing in great detail the physical elements and processes to which Connolly's list directs us (functionings, conditions, etc.) as well as the thresholds, constraints, singularities and phase transitions to which the landscape is susceptible, would seem to be absolutely necessary in order to ensure that one's data and visualization (the ground and the map) are well founded in reality. Social processes are as difficult and complex as natural processes to discern, interpret, evaluate and reorganize. The best hope is to work from within, using conceptual frames whose artifice and contingency are acknowledged. What we isolate and frame, for instance, as erosion, is only part of a larger, moving assemblage that includes natural, social, cultural and many other "functionings, conditions, mechanisms, connectabilities and trajectories." This is the message of Shepherd's Birdscapes, outlined in Chapter 04, Projects and Practices, in which social and professional decision-making processes contributed for many years to the problem, and slowly, latterly, to the solution.

Assemblage

The notion of assemblage has been used in a number of disciplines to describe a collection of different types of objects and relations that act on, and with, each other to form a dynamic arrangement or organization of material conditions. In archeology, for instance, assemblage identifies a collection of diverse things unearthed by a dig, that "hang together" not only through their having been discovered on the same strata, but through their expression of some aspect of a life (or lives) that has brought the specific things together in the first place. Bowls, cups, bones and figurines express a certain condition of everyday life-ness in precolumbian Mexico. In geology, assemblage refers to "a group of fossils that, appearing together, characterize a particular stratum" (Wise 2005: 77). There is a contingency to the arrangement of fossils that might be discovered on a stratum, but it is not random, as only certain animals existed in that form in that location at that time. The fossils constitute a *group* and express a certain *character* (Wise 2005: 77). In art of the twentieth century the word assemblage is a term associated with collage, Dada and other avant-garde or pop art styles, designating works assembled out of diverse objects (like Duchamp's *Bicycle Wheel* and Jean Arp's *Trousse d'un Da,* an assemblage of driftwood, paint and found board). The later work of Frank Stella also has qualities of assemblage, using painted materials such as magnesium, fiberglass and aluminum on one canvas. Tinguely's drawing machines ("meta-matics") introduce the elements of time and productivity, as they actually make drawings (Hulten 1975: 80). They are mechanical ecosystems.

Landscape architects are most likely to have come across the word assemblage with reference to ecology, where it refers to a biological community of organisms living in a localized area of habitat. All animals and plants associate themselves with particular combinations of environmental conditions. In so doing they form part of a biological community that includes plant, animal, bird, reptile and insect species as well as themselves. A biotic community is an assemblage of organisms living together and interacting. A lizard assemblage (say) is a subunit of such a community. Biotic communities and their component assemblages are without rank and scale. A reptilian assemblage, for instance, could be as small as a dead log, or it could be the entire forest floor. It could even be the rainforest itself (Heatwole and Taylor 1987: 185). Important aspects of ecological assemblages are the numbers and kinds of species they contain, and how these structural characteristics change in space and time in relation to environmental conditions. Different lizard species can comprise an assemblage: at any one location it is possible to identify an assemblage that contains half a dozen species of lizard. The nature of the habitat is an important determinant of assemblage structure. The more diverse the vegetation, and the more complex the geometry of the habitat, the greater the number of spatial niches available, and the greater the species richness that is possible.

The idea of assemblage has been extended and enriched through the work of Deleuze and Guattari. Their work has particular appeal to landscape architects as it adds a temporal dimension to the physical, spatial and territorial aspects of the term. The French word they use is *agencement*, which does not refer to a static condition, but rather to a "putting together" or "laying out" or "fitting." An *agencement* is not so much an arrangement or organization as a *process* of arranging, organizing, fitting together. This is very much like the process of organizing landscape encounters. Moreover, for Deleuze and Guattari, assemblages create territories. They actually make the milieux in which they thrive. These milieux are much more than spaces, involving a sense of belonging to, and causing change within; they do something – stake a claim, express a relation or set of relations. Assemblages are as much about intensive qualities as they are about extensive qualities. That is to say, how something occurs is as important as what occurs. Not just: the lizard ate the insect, but a lizard-insect event occurred when the sun was at its highest and the wind was in the west. Or, in order to maximize heat gain the lizard perches on three legs on a twig in the morning and lies flat on a rock in the afternoon. A garden is a composition of nonmetrical spaces and properties defined by the speed of plant growth, the seasonal temperature gradient, the filtration rate of water through the soil and daily variations in light and shade, that produces metrical spaces and properties such as the area of lawn where the water pools, the distance between perennials in the understory, the massing of hedges and shrubs.

As well as being material, however, assemblages are discursive. They are systems of signs.

> Archeological assemblages are not just the things that are dug up and their qualities and relations, but the discursive assemblages through which the things, qualities and relations are expressed through nomenclature, jargon and the semiotics of the dig: the semiotic system that transforms a cup into a particular type of cup.
>
> (Wise 2005: 80)

The discursive element refers to the way in which the assemblage enters into discourse: how it is described, framed, allocated meaning. The evolution of an assemblage includes the discursive component in the ongoing rearrangement of connections. For instance, a landscape condition such as a small urban park, which assembles a wide range of species and their interactions (birds and bread; squirrels and spilled grain; residents in summer shade feeding birds and watching squirrels) may feature in a community discussion about housing development. Future states of the assemblage and its components could hinge on the outcome of this discussion, which may have to do with the evaluation of the terrain with respect to different criteria: tidiness, economic inputs and outputs (maintenance, use as a concession site), inhabitation by homeless people, and so on. Humans, in fact, just as nonhumans, may participate in many assemblages at the same time. They may make new ones, and transform existing ones, bending and twisting them into new conditions. Humans enter into an assemblage through a process of taking on the conditions that make it up. It's like going to stay with some relations for a few weeks. You take on their patterns of inhabitation, their speeds, their directions and connections, their patterns of everyday life.

Assemblages are always emergent. Their dynamic components can be characterized according to their material content, their expressive content, and the processes in which they become involved. A landscape assemblage might be composed in the following way (adapted from De Landa 2006):

Material components
soil, sunlight, trees, animals, humans, water, walls, paths . . .

Expressive components
forms, textures, colors, habits, trajectories of the material components, such as plant growth and shadow gradients . . .

Process components
food chains, adaptive traits, microclimatic requirements that maintain the components and their relationships, and thus the viability of the assemblage . . .

Each of these components is subject to disturbance. For instance, a specific terrain's soil profile can be disturbed by flooding and the resultant sedimentation or erosion; climate change can affect the range of species that are involved in an assemblage; an exotic species can enter an assemblage of native species and redistribute its structural characteristics. A new regulation can require the removal of trees, or force a new road through an urban park. Disturbance gives rise to the recombination or reformulation of an assemblage by eliminating, replacing or relocating relationships. Components can be unplugged from one assemblage and plugged into another without losing their identity, as when birds migrate, or seeds drift into the chinks between the bricks of a building, take root and grow. The relationship between an assemblage and its components is nonlinear. Assemblages are formed and affected by the populations and elements of lower-level assemblages, but may also act back upon these components, causing adaptive strategies to occur.

Landscape architects develop, transform and create assemblages. They do this by gathering, composing, redistributing, emphasizing and enhancing existing material, expressive and processual components. Included in this orchestration are human and nonhuman elements. Landscape architects may develop assemblages without prejudicing or privileging different components, with the objective of developing collectives of human and nonhuman objects and relations – with the objective, that is, of creating interactive, democratic ecologies of becoming. What is critical is that landscape architects assemble elements in such a way as to create a world in which it is possible for all components of the assemblage, human and nonhuman, to engender their own possibilities of existence – to become what they may become. Rocks, trees, insects, birds, animals and humans are regarded as having in common a fit relation to an assemblage that continually evolves through its encounters with disturbance. While landscape assemblages may include ecological assemblages, they are not limited to them. Some of the most articulate and expressive landscape assemblages use social objects, events, signs and utterances to contain or extend the physical ones. Parc Andre Citröen in the 15th Arrondisement of Paris swings simultaneously between closure and movement. Designed by Gilles Clement and Alain Provost in the early 1990s, Parc Andre Citröen collects a wide range of heterogeneous components to create an integrated composition of landscape architecture structures and processes that weave back and forth between old-fashioned and new-fangled constructions of nature and society. Like many late modern landscapes (Parc de la Villette, Paris; Yorkville Park, Toronto; Mahuhu ki te Rangi Reserve, Auckland), it lays out combinatorial, collagist techniques of disjunctive juxtaposition and geometrical layering and ordering in plan, and a sculptural massing of cubes, cylinders, cones and extruded lines in elevation. The result is a self-consciously art-inflected arrangement of mineral and vegetal rooms, edges, planes and walls that can readily be modeled in three dimensions. Surrounded on three sides by the city and bounded on the fourth by the River Seine, the park has a large rectangular central lawn contained by a long canal to the west and a series of six small gardens to the east. Each of five of these "Jardins sériels" is associated with a metal, a planet, a day of the week, some aspect of hydrology, and one of the five mechanisms of human sensory perception. The sixth is devoted to the sixth sense. Adjacent to this last garden is "Le Jardin en Mouvement" – The Garden in Movement – a wild garden where the rigid geometry of the main areas of the park gives way to rose bushes, cornflowers, poppies, bamboos, balsams, digitalis, thorns and weeds growing freely.

The sharp, figured delineation of static baroque landscape elements of Parc Citröen, such as the canal, the vast lawn, lines of trees clipped into platonic forms and the six dynamic garden ecologies at once restates the binary nature/culture divide that has haunted landscape architecture and western society at large, and comments on this division. Visitors weave in and out of the enclosed gardens, at times totally immersed in intimate sensorial realms of touch, smell, hearing, vision and taste, and at others acutely aware that these sensations are framed within a composition that presents them as so many carefully disordered boxes of light and shade, bird and bee (one even has hives in it). These emergent systems are set within a larger assemblage that produces a much wider range of affects. It draws on art history, botany, architecture, landscape design traditions; it pushes corridors and allées into its urban context of high rise apartments and offices, visually and spatially pulling these into its orbit: the city, the RER (the rapid rail system that services Paris and its suburbs), the River Seine, water moving in canals, splashing in fountains and pouring down sculptured walls. The six ecology gardens are simultaneously abundant and controlled. Cataloging, classifying, putting on display: a museology of open landscape systems. Parc Andre Citröen uses the material, expressive and processual components of a landscape assemblage spatially, formally and discursively. While at first the visitor sees division, between nature and culture, order and disorder, slowly this stratification breaks down and it becomes clear that nature is culture and culture is nature; order rises from disorder and chaos generates order. Emergence can be understood epistemologically, it reminds us, as well as ontologically.

Naturecultures

We start from the premise that landscape architects are interested in nature because they are concerned with the relations between humans and nonhumans. For many centuries this interest was guided by habits of thought and practices of making derived from garden discourse. Regulated and assisted by advancements in horticulture, husbandry, botany and many other practices which we now know as scientific, landscape architecture was also deeply influenced by aesthetic, religious and symbolic interpretations of the world (Debus 1978; Prest 1981). When natural philosophy developed into the natural sciences, landscape architectural discourse became even more influenced by what was going on in science, and its relationship with the natural sciences became both more pronounced and more intimate (Baridon 1998). As we now know, a turn to the sciences was not a turn away from myth, symbol and aesthetics. In recent years a revolution in the philosophy of science has shown that the triumphal histories of science which based its endeavors on a formulation of truth that was to be progressively revealed only by advances in science, has been superseded by a more nuanced reading of scientific practices (Jardine 2011).

In *The Politics of Nature* (2004) Bruno Latour develops a comprehensive attack on the division of the social, traditionally regarded as concerned with matters of value, from the realm of science, which has always concerned itself exclusively with facts. In the world of science there were objects (nonhumans like rocks and micro-organisms); in the world of society there were subjects (social beings). Planet earth was neatly divided into two types of being, two types of knowledge, and two kinds of reality. The problem of science was how to explain natural objects in such a way that social beings (of which humans were the only class) could make sense of them. Landscape architecture joined in the game: its problem was how to construct a world for humans from the realm of objects (nonhumans were not regarded as part of the collective for which landscape architects worked; they were regarded as a separate constituency). Part of the problem, well-known among landscape designers, was how to develop a visual language adequate to their complicated subject matter. Latour argues that things have changed. We realize that it is dangerous and misleading to think in terms of two assemblies, science (fact) and society (value). We must, says Latour, get rid of the notion that there will always be two blocs, nature as it is and the representations we make of it. He suggests that the distinction between nature and society should be blurred, and that we should think of the inhabitants of the earth as a collective of humans and nonhumans that are not considered epistemologically or ontologically distinct (Latour 2004: 41). This means abandoning the notion of nature and replacing it with the idea of naturecultures, an assemblage of humans and nonhumans. In doing this we still leave intact the two elements that matter most to us: "the multiplicity of nonhumans and the enigma of their association" (Latour 2004: 41). Accepting this idea as a working proposition, landscape architects can concern themselves with where this collective comes together, and how the collective becomes gathered into a whole. It implies a search for what makes up the common world. Instead of oppositional approaches to the gathering of humans and nonhumans, we can foreground the webs of cohabitation and encounter that landscapes comprise, permit and sustain. We can work on an account of human–environment relations that emphasizes the affective nature of human relations with the nonhuman and with other humans, in the context of constructed landscapes. We can consider, further, that landscapes encourage a subtle but passionate confrontation with what it is to be human in a constructed world. They do this by enabling specific kinds of encounter with other beings that share this world – such as bag ladies, weeds and urban coyotes.

While there is more than a single site for the collective encounter of humans and nonhumans in the constructed world, we will consider one that has concerned landscape architects in recent years: public space. The public realm may well, in fact, be the primary site of this encounter, if we extend the notion of the public to include nonhumans. An important consideration in the theoretical construction of public space as the primary site of human–nonhuman interaction is its operation as an assemblage, an open system, a multiplicity. Public open space is peculiarly sensitive to changing conditions, and changes along with these as it is continually constructed by its users. When we say that people construct space we can see by that very possibility that space is an active component of the social assemblage. Through the construction of space this assemblage can itself be constructed, or maintained, or challenged. And when we refer to open space we refer not only to the spaces of the city but all the spaces of our lives. In the contemporary urban realm the realization of new public space is not a matter simply of the establishment of a central plaza or core site. Public space now, as often as not, is distributed along multiple broken continua consisting of streets, parks and reserves, blurred terrains and riparian edges. Of course, this has always been the case, but these supplementary ribbons and patches were usually supported by a singular magnetic condition that encouraged a concentration of social and cultural capital at the intersection of significant urban flows. Now we find – in the towns, suburban precincts and urban fringes most people inhabit – that the spatial condition of civic freedom is a tangle of landscape structures whose role as public space exceeds their primary function as transit lines and water management systems. Public space is not only distributed and disconnected, but its potential as a collectively formulated civic terrain is obscured and ignored. The idea of an open civic realm was never high on the agendas of postwar suburban developers anyway. "To understand why public space is missing, Americans must remember that after 1945, most of the built environment was never planned or designed" (Hayden 2006). Federal policies devised to stimulate the real estate and construction sectors of the economy granted private developers lavish subsidies without concomitant incentives to create well-designed residential neighborhoods, transit and public space (Hayden 2006). With the current redevelopment of low-density residential tracts on the outskirts of cities – and in the inner fringes as well – we have the opportunity finally to build public space in the residential landscapes in which most Americans live – the suburbs, the "outtowns," the edge cities and the outer fringes.

The old cores and inner precincts of many industrial cities are now semi-deserted or occupied by the urban poor, low paid and blue collar workers (perhaps filled daily with white collar commuters and then bled of diversity every evening and weekend). The low-rise business districts, apartments and inner rings of genteel housing stock are continuing to empty out. In the once-great steel city of Birmingham, AL, for instance, a mapping study by the Regional Planning Commission of Greater Birmingham shows an almost continuous braided river of delinquent lots following the old incised valley in which the city was sited and along which the steel mills and worker housing subdivisions were laid out. Again, the opportunity to provide a new kind of public space is clear. The question is, for whom? What, or who, is "the collective" anyway? The very notion of community has long been problematized and now exists only as fabrication, or myth – an essentialist vestige of nineteenth-century social idealism. The concept of community is challenged, ambiguated. The identity or definition of (any) community remains open, like the idea of the site, as a scene of political struggle. The common notion of the community as a coherent and unified social formation has been put in question by the work, for instance of feminist social theorist Iris Marion Young and the French philosopher Jean-Luc Nancy (Young 1990; Nancy 1991b). Nancy, in particular, proposes that the idea of community is necessarily unstable and "inoperative," enabling us to think beyond formulaic prescriptions of community and to open on to an altogether different model of collectivity and belonging. The ideas of shared purpose and shared meaning are no more than Orwellian thought rules blasted out of loudspeakers atop cruising party political motorcars. The last twenty years of global interaction have shown clearly enough that the "we" who inhabits and manages the earth is no we at all. It seems pretty obvious now that the nonhuman portion of the earth is much more important on a global scale than the humans that have purported to manage it. Now, nonhumans have the greater part to play in the destiny of human life. Should we not agree with Latour that the collective, now, is an assemblage of humans and nonhumans, and that the idea of the public extends to animals, insects and plants?

The public realm offers an extraordinary venue for the unmediated assembly of beings or, as we should say, becomings. Actual urban terrain is no longer a single and privileged stage for political activity, but a location of multiple intersections, an assemblage of naturecultures. This condition is not coded according to rankings of fact (science) and value (society), nor is it coded aesthetically in the sense of a realm in which civic images are indistinguishable from general image-making (the hallmark of our consumer society). It is not a mix of social programming and private consumption. Instead, and this is the challenge for landscape architecture, public space becomes a designed realm in which the assemblage of humans and nonhumans is made visible and vocal. New conditions for the organization of public space are required, conditions far-from-equilibrium characterized by a kind of collective imaginary, a collocation of humans and nonhumans searching for whatever it is that makes up their common world. The public landscape becomes a terrain of encounter, a place where a collective of humans and nonhumans is *articulated* (to use Latour's word). To articulate the collective – to work inside the notion of naturecultures – the landscape architect must be tuned to the habitus of each party that composes the collective. An individual's primary habitus is the durable, transposable system of affiliations and correspondences acquired by that individual as a result of the conscious and unconscious practices of families. Subsequently this is transformed in to a secondary or tertiary habitus by the individual's passage through different social institutions (Bourdieu 1972: 56). The habitus is both structured and structuring, and it develops their possibilities of endeavor, and of freedom of endeavor. The designer of collective terrains successfully articulates the assembly of species that inhabit them when he or she mixes these parties while retaining and enhancing their freedoms. The degree to which the landscape is truly public is the degree to which potential trajectories and empowerments may be realized as a result – the degree to which the individual party's habitus is acknowledged and brought into sensitive, practical and mutually beneficial encounter with that of others. A landscape that abandons the division between nature and society includes all conditions necessary and sufficient for the ongoing self-organization of the assemblage. To do this it is necessary to explore the common worlds of the collective, not in the sense of a unified march into the future along a line of time, but by means of enabling the formation of intricate attachments and affordances between and among species and elements through affective contact and inter-affirmation. The landscape architect who works for and with the republic of human and nonhuman naturecultures envisages life as a contingent process of growth and change. He or she participates in this movement of perpetual differentiation through the invention of forms that bring tree, child, rock, sky, bee, cat, schoolteacher and butterfly into equivalence and association.

Public space, then, is the location of encounter and negotiation between constituencies that not only speak different languages but for whom life is a variable that often means nonliving. Whom, among humans, speaks the language of rocks? Who, now, is brave enough to declare that rocks do not speak? That rocks cannot enter the discussion; that minerality is a mute condition? Public space is no longer simply spatial. Nor is it simply terrain. It is, as well as being a physical, material place, a nonphysical site of assembly and interaction. These conditions merge when physical and virtual intensities intersect, as in the occupation of Wall Street, a gathering whose center was everywhere and circumference nowhere, across a landscape of transmissions and receptions that are material and spectral, intensive and metric, coagulant and dispersive.

In order to build this realm we need an account of generative or operational strategies that enable landscape architects to work in such a way that the communities they serve may be engaged and empowered in the process. But first we need a manifesto. Something like this? A truly sustainable public realm will be one in which difference is not only tolerated but is productive of difference; where the pressures of control and appropriation, consumption and politics are negotiated, enabled and carried out in a democratic manner by those who constitute the democracy; where safe, transgressive practices are not repressed or discouraged; where flexible and non-exclusionary cultural and social life is enabled. In conditions of natureculture, the human and nonhuman publics who construct the assemblage will be trusted and trustworthy. Species will be enabled to coexist and all species will have access to the resources that sustain them, just as earth-created nonhuman objects and processes will not be destroyed and natural systems not be compromised by development. The urban realm is not solely a human realm – it is a collective realm where the collective is an assemblage of interacting species, objects and processes that support biodiversity and the difference it implies. This space will always be vague, and ambivalent, and open to the sky under whose impenetrable and indiscernible operations the drama of life on earth can be both enacted and witnessed by all. An example of a design that investigates the notion of a public realm created for and by a republic of humans and nonhumans is discussed in Chapter 05, Encounters.

Field theory

Ultimately urban field theory derives from Einstein's revolution in physics, documented in his papers devoted to the special and general theories of relativity, published in 1905 and 1916 respectively. Einstein freed space and time from the metaphysical and absolute character that classical Newtonian physics had formulated. In his equations, which deal with transformations between stationary and moving systems, he reduced time to a dependent, or variable, coordinate. He made a number of theoretical innovations that are interesting and fruitful for landscape architecture. For instance, in Einsteinian physics, space and time are no longer absolute entities but rather a continuum – "space-time" – a field or set of fields with their own frames of reference and, as in Deleuze's notion of difference, no external or material substratum. Reality, according to Einstein, is a four-dimensional manifold, not a three-dimensional system evolving according to a separate and external one-dimensional time. Each inertial system now expresses its own particular time, determined as a mutual relation of events to the frame in which they are registered. Time flows at different rates depending on when and where it is measured. It is relative and contingent. Space-time becomes understood as a field that cannot be reduced to component dimensions or conditions. Why this is important in landscape studies is that any unification that might be observed within the field, any organizational coherence that appears in the manifold (and let's call it an urban manifold) may well occur in the absence of a supplementary or higher dimension or power. That is to say, urban systems, understood as fields, can achieve coherence within the terms of the system through the ability of the system to self-organize. We have already noted that natural systems do this by means of the matter-energy that moves through them. Movement and the time of movement – what Bergson calls duration, not clock time – these are the parameters of the field.

The notion of "programming the urban surface," the title of Wall's influential chapter in *Recovering Landscape*, edited by James Corner (1999c: 233), as both Connolly and Meyer note, presupposes a condition that awaits the hand of the maker in order to actualize its potential, and not just to actualize it but also to order, or unify it (Connolly 2006: 82; Meyer 2008). But the concept of a space–time manifold or field bears no relation to an urban condition organized by a horizontal spatiality. What Einstein's field prepares us for is the idea of multiplicity, an organization that has no need of unity in order to form a system. The major break here for landscape architecture is the emancipation by Einstein of the field concept from any association with a substratum as a bearer of forces and events. Forces and events are imminent in the field, they are part of it. Field theory made possible the scientific expression of the principles of immanence, dynamism and continuity, qualities of emergent systems with which landscape architects engage. Einstein's unified formulae not only linked space and time but also showed that energy and matter were one, or at least that energy could be created from matter and vice versa. This was a new concept: matter-energy. Landscape architects are orchestrators of matter-energy.

It will be recalled from Chapter 02 that the philosopher Henri Bergson, whose *Creative Evolution* was published between the dates of the special and general theories of relativity, shared with the science of his day a sense of the undividedness of the object-field. For him, as for Einstein, with whom he debated the nature of existence, the world is at once an aggregate of separate fragments *and* a materially indivisible whole. The substance of the world is not resolvable into pure or independent material forms. Rather, these latter shift and fluctuate in and out of formal arrangements (Bergson 1998: 11–15). Bergson re-introduced the importance of the notion of becoming, the ancient philosophical conception of the world that had fallen out of favor after Newton published his *Principia Mathematica* in 1687. In *Creative Evolution* (1911) Bergson repudiates the mechanistic view of time. Understanding becoming, he argues, requires living in it, and therefore in time. The mechanistic philosophy only holds "outside time": Bergson overturns this conception. Experience tells us that we are immersed in becoming, flux, field. We find that movement is not constituted of successive immobile states, as Newtonian science would have it, but that "the body is changing form at every moment; or rather, there is no form, since form is immobile and the reality is movement. What is real is the continual *change of* form" (Bergson 1998: 302). For Bergson, the world is a moving field of vectors of differing qualities and intensities and that form is "a mere cut made by thought in the universal becoming" (Bergson 1998: 318).

Wall describes the contemporary city in terms of this moving field. In the latter half of the twentieth century the idea of the core–periphery city was replaced by a polycentric and web-like sprawl. Here multiple centers are served by overlapping networks of transportation, electronic communication, production and consumption. The daily urban system is dynamic and temporal. There is a shift of emphasis from form to processes, and a corresponding move to a rhizomatic model of urban growth rather than a hierarchical tree-like growth. However, this description fails to take into account the qualities of "field" that are of most use to landscape architects. The kind of field that Einstein and Bergson are talking about is not Wall's "dynamic agricultural field" (1999: 233), though the thickened urban–agrarian condition is full of different becomings. Nor is it the famous thin "vegetal plane" of Rem Koolhaas' Atlanta for it is neither planar nor vegetal. The field of becomings is not a surface at all. If we look a little more closely at Bergson we find that the field is a configuration of qualitative, evolutionary and extensive forces occurring in time. These forces are to do with more than mobility, access, transit; more than postindustrial and interstitial sites; more than networks and infrastructures, plazas, squares, streets, railroad and stream corridors, and the polymorphous conditions that constitute the contemporary metropolis. These sites, which landscape architects wish to provision with multifunctional, flexible services, cannot become multiple unless and until landscape architects develop a love affair with nonmetric time and nonmetric space – the space–time field of Einstein and Bergson. The production of difference in this field can be accelerated, expanded, slowed, bent, deformed, twisted, layered, in short it can be subject to all the operations, instrumental and otherwise, that landscape architects wish to deploy as long as they realize that movement and duration are the primary components of their operational ontology. This is the message of the field. If, as is argued here, landscape architecture is an art of encounter, and that encounter in and with landscapes and all their affordances is our subject matter, then how time and duration operate within the field is very important. Bergson's non-chronological duration is the time of trees and animals, of wheat fields and ruderal urban landscapes together, not reduced to a single representation of becoming.

It was in the spirit of this remarkable vision of life and world that in the 1990s a number of writers and designers began to explore the potential of understanding urban conditions in terms of a field structure. Stan Allen and James Corner even named their design firm Field Operations. Allen wrote an influential article called "From Object to Field" in which he argued that a field condition is "any formal or spatial matrix capable of unifying diverse elements while respecting the identity of each" (Allen 1997: 24). Note that Allen confined field conditions to a spatial matrix. Field configurations, he said, are "loosely bounded aggregates characterized by porosity and local interconnectivity." The internal regulations of the parts are decisive; overall shape and extent are highly fluid. Field conditions, then, like open systems, are bottom-up phenomena: defined not by overarching geometrical schema but by intricate local connections. One of the outcomes of the new thinking was to redefine the relationship between figure and field. Figures (buildings, bridges, trees, seats on the one hand, and patterns, pulses, intersecting forces on the other) are not understood as demarcated objects but as effects emerging from the field itself – as moments of intensity, as peaks or valleys within a continuous field – as figure-*and*-field (Allen 1997). What is intended here is a close attention to conditions at the local scale, even while maintaining a relative indifference to the form of the whole. If we take the example of an urban forest, we see how this approach can focus attention on particulars. There are trees, of varying species, heights, density, canopy shape and extent; there are the full range of ecologies that interhabitate with them, including all the micro-organisms, the insects, birds and subcanopy plant species that have developed the structure of the field. There are the territorial human components: roads, curbs, drains, lawns, driveways, houses, flower beds, power lines and subterranean services including water and cables – all these contribute to the field. Then there are the intensive elements: predilections, and habits of walking, driving, gardening, burning, playing: kinds of movement that occur more in the spring and the fall than in the summer and the winter; flows, of groundwater, and those species whose privilege it is, as Levi Strauss once said, to live in the aerial medium, whose ecologies are aerial as much as they are terrestrial. There are the macrofaunal visitors, nomads and migrants: cats, dogs, coyotes and other wild animals. Vectors of pollination and gradients of temperature, pulses of breeze and moving shafts of sun. Children on bicycles, with air rifles, looking for balls, untangling kites, climbing small trees. With this list we have barely begun to describe the field that we call the urban forest. Any coherence we may attribute to this open system rises within the system. And its figures are fields too. Let's come back to

Einstein: forces and events are constitutive of the field. Bergson: what is real is the continual change, not the things that are changing. The unique spatial condition of the urban forest is a product of its field condition and this is a product of movement, of becoming, of duration and time. How we start to consider working with this forest field is by thinking of these things, not of a horizontal surface, much less a program. It has its own program.

Landscape architecture texts, such as *Landscape Urbanism, Recovering Landscape, Changes in Scenery* and *The Landscape Urbanism Reader*[1] are full of nomenclature and terminology related to field conditions, moving form, non-Euclidean geometry, temporality, self-organization, differential change and spatial organizations with active parts. Unlike texts on music, biology or mathematics, however, there are few terms to discuss the role of humans as active participants in the dynamic processes these landscape texts investigate, either transmitting or receiving, or simply engaging on different levels with these processes, and in different ways. For instance, the systems that are analyzed by Easterling in *Organization Space: Landscapes, Highways and Houses in America* (1999) are placed "outside" the skins of those who live and work in them. Within the author's fascinating account of "network interplay" and "relationships among multiple distributed sites" there is no corresponding account of how human bodies are aligned and realigned with respect to these networks, no sense of imbrication of body and site. This is not necessarily a failing of Easterling's; her attention is elsewhere. There is a need, however, for a theorization of the specifically human dimension of these multiple systems, for there is a lingering sense in Easterling, and in the books mentioned above, that if we are to frame sites in terms of fields, this could be done without reference to the fact that these fields are lived. When Easterling does incorporate the role of human subjectivity it is in terms of what she calls "housekeeping," the "comprehensive control" of architects, planners and managers over the distributed spatial systems under their care, rather than an active engagement with their possibilities for the imaginative construction of what might be called affective participation. What is needed is a new formulation of practices of human intervention in organizational protocols and systems, in order to spark critical realignments of subject and object in the felt experience of life in the landscapes of the city. Note however that, as an exception to the rule, Mostafavi's introduction to *Ecological Urbanism* seeks to found that project within a philosophy (derived from Guattari) that includes human subjectivity as one of its three registers, along with social action and environmental practices.

Part of the point of Deleuze and Guattari's field-like formulation of "the rhizome" is that with this notion they want to get away from analyses based on distinctions between individuals and their environments, whether these environments are physical or sociological (Deleuze and Guattari 1987). The arrangements in space and time to which Deleuze and Guattari point are not subordinated to taxonomic separations of bodies, places and times. Instead, the problem of making multiples is the problem of how a life is constituted, for as subjects we are never fully constituted but always passing in and out of the systems through which we compose our sense of ourselves. Rhizomatic fields are likewise never given to us but constructed through us, linked to us, and though they may always seem to precede us, nevertheless it is just their possibility that we express to one another through design. Bearing in mind that a field, if nothing else, is a network of relations, there is little to gainsay the proposition that landscape architecture is a relational art. If this is the case then it would perforce take as its theoretical horizon the realm of human–nonhuman interactions and their social, environmental and political contexts, rather than the assertion of an independent, symbolic or semantic space from which different groups were excluded, or included as collectives without a voice. In this case we would expect to find, among landscape urbanist projects for instance, the creation of free areas and time spans whose rhythms contrast with those structuring everyday life, encouraging interhuman and human–nonhuman commerce that differs from the habitats and zones that are imposed upon us.

The work of landscape architect David Hill provides an example of how the temporal dimension cannot be separated from the spatial field. His phenological research studies periodic plant and animal life cycle events and how these are influenced by seasonal variations in climate. In particular Hill is interested in how the structural, spatial, active and performative dimensions of landscapes can be formed through special attention to their phenological conditions. This requires careful observation of plant and animal communities, their relations, habits and operations. Hill considers that landscape designers too often imagine and present plants at one fixed moment in time. Instead he looks for design opportunities embedded in time and uses the ephemeral qualities of plants to register the changing patterns and structures of complex landscape systems. Hill's own research investigates the spatial and textural qualities of a specific plant palette through a series of photographs taken from the same vantage point at the same time each week through an entire year. In this way he can chronicle the exact changes that plants go through and the myriad ways these changes transform the landscapes in which they are growing through their relationships with light and shade, wind and water, and each other in terms of scale and spatiality.

Morphogenesis

Deleuze and Guattari introduced the notion that there are two fundamental structure-generating processes responsible for the development both of material form and social process (Deleuze and Guattari 1987). These are hierarchical structures and self-consistent aggregates. The former operate like tree diagrams and are therefore called arborescent; they include, among many other things, hierarchical structures such as social classes and castes and geological strata or layers of rock. The development of hierarchies occurs by way of sorting mechanisms and the consolidation of relations into more or less permanent architectectonic structures (De Landa 1999a: 60). Self-consistent aggregates, which De Landa calls meshworks, are completely different. They are dynamic networks, or "acentered systems," that Deleuze and Guattari illustrate with the metaphor of the rhizome, which spreads by producing stems and filaments in all directions (Deleuze and Guattari 1987: 15). Meshworks develop through the operation of feedback mechanisms where "the product that accumulates due to the acceleration of one reaction serves as the catalyst for yet another reaction, which in turn generates a product that catalyzes the first one" (De Landa 1999a: 63). An autocatalytic system can be self-sustaining for as long as its environment contains enough material for the reactions to proceed. Any system that evolves in this way by, as it were, drifting, can grow in directions that are unplanned. Like a rhizome, it adds new nodes when it encounters certain conditions, developing and complexifying, multiplying from point to point without beginning or end but always metabolizing from and in the middle, operating "by variation, expansion, conquest, capture, offshoots . . . becomings" (Deleuze and Guattari 1987: 21).

Morphogenesis is a term derived from developmental biology to describe the ability of organisms and natural systems to generate form from within. That is to say, the resources involved in the genesis of form are immanent to the organism or the system itself, and do not come from outside the system. De Landa gives as an example the way that the spherical form of a soap bubble emerges out of the interactions among its constituent molecules as they seek the point at which surface tension is minimized (De Landa 1999b: 120). However, the term has been adopted in architecture with rather a different emphasis. The Emergence and Design Group, who have established a Masters program at the Architectural Association in London, have investigated morphogenesis through two issues of the *Architectural Design* journal (Hensel et al. 2004 and 2006). "It is process," Weinstock writes, "that produces, elaborates and maintains the form or structure of biological organisms (and nonbiological things), and that process consists of a complex series of exchanges *between the organism and its environment*" (Weinstock 2004: 13 my italics). This definition separates the organism (or system) from the environment to which it connects. As noted in Field Theory, landscape architects are less likely to accept this distinction just because it distinguishes between figure and field. Of interest for landscape architects are the connectivities that enable form and behavior to emerge from process. Because in landscape architecture it is the system rather than the organism (for architecture read: the building) with which the designer works, the sense of a separate, or individuated, condition is much less strong.

All environmental processes work across parameters, or boundary constraints, that act as local principles for self-organization during morphogenesis. These conditions may be as simple as a legal boundary – a garden fence or park wall, for instance – or more complicated: a local authority by-law limiting public use of urban terrain in some way, as when skateboarding is prohibited. The parameters define the space of possible states that a landscape system may inhabit, the territory it incorporates, the resources it may call on. These prior formations are critical to the development of a new system, as it is these, in combination with a specific enactment on the part of the landscape architect, that create the initial conditions for a project. Because emergent properties and processes are sensitive to initial conditions, a small change to the original platform of a project can make a large difference further down the track. Landscape architectural interest in the notion of designing by setting up initial conditions, and then "managing" the process of realization, has increased in recent years. This development has gone hand in hand with the rise of open systems thinking in landscape architecture and with meeting the challenges of autocatalytic design (see Open Systems). As Waldheim notes, these projects are often described as "open works or infrastructural systems that are meant to distance questions of authorship in favor of an explicit open-endedness and indeterminacy in the face of future cultural contingencies or larger urban forces" (Waldheim 2007: 16). Waldheim implicitly harnesses morphogenetic – sometimes wrongly called performative – design techniques to the objectives of landscape urbanism, neatly described as "a modest, socially responsible and economically efficient urbanism" (Waldheim 2007: 16). But not all landscape design that uses nonlinear techniques of production is to be considered landscape urbanism. Designers who have investigated morphogenetic processes try to facilitate bottom-up form-finding processes that generate structure and organization. The difference, as Neil Leach puts it, lies in the emphasis on form-finding rather than form-making (Leach 2009: 34). This means that the outcome of the design process cannot be known prior to it coming into being. At first glance it might seem that "problematized authorship" is an abdication of intentionality, and that the final artifact is not a product of the landscape architect's imagination or careful analysis, so much as an outcome of a generative process that is blind, an issue Corner canvassed ten years prior to Leach in his essay on eidetic operations (Corner 1999d). It might even seem that the use of such an operating system is irresponsible, or at the least will make actualizing local virtualities a matter of chance. That this is not necessarily the case can be found in the work of many landscape architecture firms whose practices incorporate morphogenetic techniques. Vista is an interdisciplinary Dutch design and planning practice whose work on the remediation of the Volgermeer Polder has been cited favorably by landscape architecture commentators, for instance Poole (2004), who calls it a "creative schematization" rather than an explicit design, and Raxworthy (Raxworthy and Blood 2004). For this project Vista developed a strategy that guided the regeneration of a bog landscape by means of the creation of artificial ponds and the separation of toxic groundwater. Essentially a grading project, the landscape architects developed ponds of different depth that were lined to prevent contamination, and permitted water to enter the ponds over time in an uncontrolled fashion. Different plant species colonize the different aquatic environments that were established as initial conditions. Cattle and sheep interact with the developing water bodies. If a pond is isolated from fauna it can develop into a forest; if grazed by cattle it will turn into grassland. Vegetation will regenerate rather than being planted, its final configuration and composition being a matter of what seeds were already in the soil, and what seeds are carried there by wind and birds. The repetitive interaction of these agents over time will guide the life of the project.[2]

The proposition that initial conditions can be designed to direct the future movement of those conditions is illustrated by Vista designer Roel van Gerwen's figure of a stick in the sand. "To make a sand pile on the beach," he writes, "you can form a mound of sand with a bucket and shovel, then the mound will disappear with the wind over time. The alternative is to place a large stick in the ground where the wind will constantly form a pile, reshaping the pile every time the wind changes direction" (van Gerwen 2004). In van Gerwen's analogy placing the stick is less exhausting, gives a less predictable result, and is more dynamic. It is also autocatalytic. In process design, as Vista calls their morphogenetic work, the important thing is to use the right "sticks" in order to "unravel and manipulate" the landscape-forming processes that are already at work in both urban and rural situations. Van Gerwen makes the point that "(p)aradoxically, the more unpredictable the landscape proposal the more well-founded one's data and visualization has to be" (van Gerwen 2004: 233). The figure of the stick in the sand is particularly apposite because it demonstrates that a carefully chosen element is key to the actualization of local potentials through the engagement of specific forces. This is clear in Delwyn Shepherd's Birdscapes proposal (see Chapter 04), which uses a combination of carefully selected local tree species, local community intervention and indigenous bird-foraging and flight-line behaviors to develop an adaptive approach to coastal erosion on New Zealand's west coast. In a project that plays out over time, she restores coastal forest and creates designs for human habitation by working with sand dune formation processes and the prevailing wind, literally by placing sticks in the sand (Shepherd 2009).

These projects explore the productivity of chance as a form-finding device. But not just any form. Processes are set in motion and then, through autocatalysis, landscape conditions change and adapt as a result of the ongoing interaction of the processes with contextual (or sometimes internal) states. Because these states themselves are changing they cause further reactive events within the evolving system and between it and its contextual systems. Nothing commands the system to move in a particular direction and, owing to its sensitivity to environmental factors, it can move in many different ways, given the parameters of its possible forms. The interaction of each part with its immediate surroundings causes a complex chain of processes leading to a new state. A river system is intimately interconnected with the webs of topography, vegetation and surface conditions that comprise its watershed. This goes for urban as well as rural environments. In the urban realm, buildings, transport networks and pedestrian precincts (for instance) are understood not as singular and fixed, but as energetic and material systems that share their environment with many other processes. Within these processes, principles and dynamics of organization and interaction are at work, guiding and regulating emergent urban patterns. In order to work within these structuring processes, landscape architects need to analyze the micro- and macro-environmental conditions that moderate the systems of habitation and encounter that they find on the streets, in the parking lots, along the urban streams and through the fluctuating urban–suburban edge conditions that make up the urban field. This requires the development of particular analytical tools, methods and skills so that found conditions can be deployed in design approaches that integrate and regulate rather than relocate or substitute. Analysis is of central importance to morphogenetic design process. Self-organizing tendencies and interactive affordances need to be discovered, and the relationships between these processes and spatial orders revealed. Of special significance are the relationships between people and the dynamic conditions that they create and inhabit. Close observation of behavior is necessary. What are the specific assemblages that comprise these human–nonhuman interactions? How do we find this out? How do low-key social processes, for instance, cause major territorial disturbances? In landscape architecture we have to take account of the immersion of humans and nonhumans in the environments that support and maintain them, and find out how these environments are continually transformed on the basis of this immersive condition. The study of morphogenesis shows that discernible structural and organizational patterns underlie these interactions. When we understand these underlying blueprints we can start to seed urban processes with catalysts for the development of new patterns and forms.

Here is an example of how morphogenetic processes have been used to interpret urban form. In "The Nonlinear Development of Cities" (1999c), De Landa proposes that cities are mixes of "hierarchies of command and control" (strata) and self-organizing systems (meshworks). One of these structural elements typically predominates. Capital cities, such as Washington, DC, that are "state" cities of bureaucracy and governmental regulation, are hierarchically ordered, while metropolises like New York – that are commercial networks – are self-organizing. "A self-organizing structure [e.g. a beaver dam] typically emerges without central planning, as a consequence of a de-centralized process," whereas a command center is gathered around a state or royal seat of power. "It is not a matter of opposition of one to the other; nature is filled with these two types of structure. What does matter is to determine which structure predominates" (De Landa 1999b: 25). Cities, as mixtures of stratified and meshwork patterns of organization, tend to represent some combination of market (self-organizing) and state bureaucracy (command hierarchy). According to De Landa the state or royal towns of Cairo, Peking, Paris and Madrid may be contrasted with the commercial towns of London, Venice and Amsterdam (De Landa 1999b: 27). Commercial towns, often maritime, connected to the self-organizing seas and oceans, always exist in networks, which may lack a distinct center. State towns are centers that conquer other towns to enable their wealth to flow in. Like Peking, Canberra and Madrid, they are often land-locked and protected. While this example may seem a little stretched, it is useful as an indicator of how the notion of morphogenesis enables us to understand production and development without requiring a supplementary mover. All of the energetic resources for the evolution of particular urban typologies are found within the operating system itself. The identity of an urban square – its qualities and characteristics – and therefore what is appropriate and desirable in terms of proposed design interventions, will be a matter of its contingent historical process of individuation, rather than being reliant on a form-giver. While we have known this to be the case with respect to the organic development of medieval cities, the idea of morphogenesis fleshes out the word "organic" and describes the specific material principles of growth to which it refers.

For a long time the genesis of form and structure was based on the Aristotelian conception of matter as an inert receptacle for forms that are imposed from the outside. While ecosystems ecologists now see the ecologist as inside the system that is being observed, and landscape architects are similarly rethinking their ontological geographies, ecologies obviously do not await the designer in order to evolve structure, responsiveness and conditions for life. Designers do not create ecologies, though they do manipulate the processes, elements and conditions that enable ecologies to develop and evolve. As Corner has put it, landscape architects "stir" ecologies into different conditions (note, however, that stirring is a metaphor that requires the agency of a human hand). In ecological situations, design intervention, again, is often a matter of the reconfiguration of pre-existing conditions, as when water-borne toxins that pollute a degraded wetland are diverted or filtered. Can this also be the case for urban design? The best way to create landscapes that are dynamic, flourishing, productive, suitably scaled and self-sustaining, according to van Gerwen, is by utlilizing the processes that form landscape. "This counts for rural or natural circumstances but just as well for the urban environment, although the steering processes become more anthropological or democratic as the project becomes more urban" (van Gerwen 2004: 250). If cities are open systems it must be the case that urban processes, too, can be set in motion by means of a careful intervention and permitted to evolve in response to contextual change. It is a matter, then, of those "anthropological or democratic" steering processes. A matter of the landscape architect's ongoing regulatory activity, of his or her own performativity. Of the degree to which the hand stirs the process. Koolhaas and Mau's Tree City, discussed in Open Systems, illustrates the sensitivity of an urban project, rife with political challenges, to the stirring hand – or lack of it. If one of the advantages of an initial conditions approach to urban landscape architecture is that it sidesteps some of the difficult aspects of ongoing community participation in decision-making, while at the same time enabling participation in form-generation, it is clear that participation is critical. When, as Mitchell and Van Deusen argue in their overview of the role of public space in the Downsview Park entries, the "communities who will use – and continue to create – public space" are to be "drawn into the decision-making process at every step," then the designer must occupy a role that permits a "multiple, shared and combinatory" process. "It is not in design per se that publicness can be encouraged . . . but in how that design is administered and implemented" (Mitchell and Van Deusen 2001: 112). The designer's ongoing performance is as an orchestrator of the conditions for democracy.

Difference

The objects and processes with which landscape architects design change through time. A key concept when considering this is that of difference, developed by Deleuze in his *Difference and Repetition* of 1968. Difference relates the notion of emergence to the idea that everything that exists always becomes and never is. The changing dynamics of the natural, the social and the urban worlds can all be conceived as things differing continually from and in themselves. The human beings who encounter this unfolding difference in the world are also continually differing from themselves. We have, therefore, a becoming different of many diverse conditions, so that each encounter between these conditions is a novel encounter – unscripted, unpredicted and always open to further novelty. Deleuze does not mean by difference that some things are dynamic and ever-changing and some other thing – a substratum, underneath it all – persists, or remains identical to itself. The concept of difference does not entail a version of the one and the many, a multiplicity that bodies forth from an unchanging unity (be that God, self, nation or nature). Instead there is only difference. Everything is transforming itself – recreating itself – all the time. All things are engaged in an ongoing genesis of themselves. It is important to understand that morphogenetic processes produce the conditions of the here and now, and that it is these local processes that landscape architects deal with when designing open systems such as urban precincts, waterfronts, parks and other constructed landscapes.

The great beauty of this theorizing of change is that designers no longer have to ask what something is. Instead, and this is echoed by both Latour and Serres, the questions become "Where . . .?" or "How . . .?" or "When . . .?" or "How many . . .?" or "In which case . . .?" or "Under what conditions . . .?", or (mobilizing our own human enframement) "From what viewpoint . . .?" In other words it extends the exhortations of Allen, Corner and others who draw on Deleuze, to find out what things do rather than how they look. "How does this urban stream silt the flood plane? When?" "How many species of fauna inhabit this edge condition? Under what conditions does the species count change?" "From what viewpoint do these residents consider this tract of open undeveloped terrain to be waste land?" People, birds, soils, trees are not understood as having identities or essences. "Things are constituted by virtue of the differential relations they enter into, both internally and in relation to other things" (Patton 1994: 152). Change and difference, then, are not simply happening *out there*. Nature is in constant flux and humans are part of this flux. Human activity contributes as much to the development – the self-differentiation – of nature as nature does to human development. Thus, human beings are in a constant state of productive practice: the production of the concrete, physical open-ended world understood as a composition of forces in which other forces (including us) are always intervening, contributing and participating. The world, that is, what is given to us in experience, is a difference-driven process. Many phenomena, in geology, biology, economics and social theory – whatever – emerge from the interplay of differences. Matter itself processes its own imminent resources for the generation of form. Material systems are continually traversed by strong flows of matter and energy, and it is these flows that enable systems to self-organize. The field of our endeavors as landscape architects, then, is a spatio-temporal aggregation of people and things that are continually being brought into relation with each other. In his discussion of affect in the *Ethics* the seventeenth-century philosopher Baruch Spinoza suggests that humans are always seeking to combine their relations with others in order to empower and enrich their capacity to become themselves. We get better at this by developing our capacity to relate our feelings, conduct and intentions to different modes of being. The tasks of the landscape architect are both to develop milieux that enable this to occur in multiple ways, and to present different modes of becoming to all members of the republic of organic and mineral life. The landscape architect is engaged in a continuous experimental process of composing or constructing qualitatively different modes of existence and relating them to each other.

An especially critical aspect of the concept of difference, and the ongoing self-differentiation of things, is its overturning of the representational order of being. Plato initiated this order with his Theory of Forms that sought to understand the world as a copy of an ideal "heaven" in which the true forms of things dwelled. The worldly idea of beauty, for instance, could only be understood with reference to the ideal form of beauty. Courage in a human being was a copy of the ideal model of courage, and so on. This model–copy theory has characterized most thinking about how we know and value things since Plato's time. The notion that things self-differentiate does not require the model–copy order. Instead all things simply become different from themselves on the basis of the matter-energy flows that course through them. A particularly interesting aspect of Plato's theory is what he called the simulacrum. This is the case of the bad copy, or the copy of the copy of the model. The person who pretends to be courageous, for instance, to serve his or her own ends, rather than to serve the ideals of courage – rather than copying courage. Deleuze argues that the simulacrum destabilizes the reign of representation. The simulacrum exists in and of itself, without grounding in or reference to a model: its existence is "unmediated" (Deleuze 1994: 29). Each simulacrum is its own model. This analysis helps us reframe or reinterpret "artificial" landscape conditions such as a culverted or canalized river. The Los Angeles River, for instance, can be thought of as not a "not-river." That is to say, a natural river system is not a model that the LA River imperfectly copies; the LA River is not a river that is "in error" or "inferior" to a natural river system. Understood as a simulacrum the LA River exists in and of itself as a condition to which a range of attributes and functions can be ascribed in terms only of the performative parameters of the canalized river itself. (Do not ask what it is, but how it works . . .) Once we depart from the model–copy dialectic, we can begin not only to reinterpret, reanalyze and revalue the Los Angeles River as a functioning system in its own right, but we can also focus more clearly on its capacities. In this way, considering it as a prosthesis or artificial device, we can address inadequacies with respect to its own productivity as an organizational and distributive regime. We should attend more precisely to the becoming-different of the prosthesis itself. We should affirm, rather than negate, its individual condition.

The model–copy dialectic occasions a profound delimiting of the power of the prosthesis to inhabit and reorganize the forces with which it deals. If the LA River is regarded as a unity, as an entity that achieves and embodies a static condition it is very difficult to understand it as a process that continuously differs from itself in multiple ways. The denial of difference implies an *a priori* conceptualization, a telos, a meaning, and through these a logic of landscape architectural practice that crushes open practice under an image of the same, or similar, and betrays what it means to practice openly. Any kind of modeling of nature enters the orbit of the simulacrum. This is the problem, as yet seemingly not addressed, with biomimicry, which seeks to use nature as a model for design solutions. Biomimicry asks for any design problem, "What would nature do?" and then seeks a solution based on an animal or plant. This apartment building models a termite mound; this one is like a wasp's nest. In nature, however (wherever that is), a kingfisher is not like a mussel. The termite is not like the dolphin. Faced with a piece of unpalatable urban territory the model–copy landscape architect says "revegetate." Make it like nature. Part of the point of emergence theory is to get away from the re, from the model–copy enslavement. The notion that things self-differentiate does not require the model–copy order. Biomimicry requires us to accept a representation of nature projected through the models and codes of science. Should planet earth be neatly divided by science into two types of being, two types of knowledge, and two kinds of reality? As suggested in Naturecultures, the problem of science has been how to represent nature to social beings. Landscape architecture joined in the game: its problem was how to construct a world for humans from the realm of objects. Nonhumans were not regarded as part of the collective for which landscape architects worked. Are we not trying to find an alternative to this, the two-world problem that biomimicry supports? Considering the world as a natureculture assemblage enables us to communicate in terms of a single republic of humans and nonhumans, instead of a continual transmission back and forth across a dividing line. Andrea Brennan writes:

> Biomimicry re-enforces an overly-simplistic distinction between "nature" *over there* and us [humans, designers, architects, etc.] *over here*. If you read between the lines, the over-arching assumption is that "nature" is pure and logical and balanced and good and we are, well, somewhere in between *not* and *less so*, depending on your personal environmental politics. The implication is that as humans, we are flawed; therefore, we ought to look to "nature," the ultimate teacher, in an effort to understand and replicate the "right" [i.e. "natural"] way of doing things . . . you know, all of the things we have ignored and overlooked over the past few hundred years of our mis-guided modernity.
>
> (Brennan 2010)

Perhaps the most interesting parallel between biomimicry and emergence theory is to be found in the application of its principles to the development of new urban systems. It is important not to see the city as like an ecosystem. A city *is* an ecosystem, or a set of them (if we wish to see a city as anything other than a city). This way we can start to work with what cities already offer. We can respect them for what they are, and find in their perverse and ambivalent processes the conditions that, through careful and informed intervention, can become different from what they are. Even from this perspective, however, it is difficult to use biomimicry as a model, a measure or a mentor. The Biomimicry Institute says that the key to using the biomimicry taxonomy is forming a question. Instead of using high pressure and temperatures to manufacture tough, lightweight building materials, an engineer can "ask" a toucan how it manages impact with its strong and light beak. The problem is readily provoked when turned into a landscape architectural challenge. How would nature formulate some unproductive, ambiguous terrain (regarded as "wasteland") under an elevated freeway into a habitable urban landscape that connects two disparate parts of the city? How would nature remodel this homeless people's campsite into a terrain that does not patronize or marginalize them, much less remove them, at the same time as enabling opposing social forces to negotiate and collectivize it? How would nature marshal and activate new orders of communication here? How would nature deal with this territory, which is so denatured, so ambivalent, so *institutionalized*, that it cannot be discovered "in nature"? Thinking of design as already ethical, already inhabited by a moral purpose highlights a concern we should have that this conception of design – as a force for good – is in the service of political and moral forces that themselves require examination. These forces lead us away from design as a practice that responds to its unmediated encounter with the world, a practice that must create openly in order to cope with the violence and force of these encounters, their randomness and disorder. Design should be seen not as a bearer of moral purpose but as a self-engendering act of creation. This is design without image, without purpose. How should such design operate in the world?

Disturbance

Drawing on nonlinear mathematics and the work of complexity theorists, ecologists have advanced our understanding of how ecosystems develop their adaptive and responsive structures by being open to disturbance. But disturbance can be thought of as much more then a way that ecosystems evolve. Serres devoted an entire book to this subject. In *Le Parasite,* translated in 1982, he uses the figure of the parasite to show how disorder alters ordered systems in the direction of greater complexity, ensuring their continued existence.[3] In another work Serres invokes the first century Roman poet Lucretius (c.99–55 BCE), who wrote a long poem, *De Rerum Natura,* describing the orderly system of the world as a laminar flow of atoms raining in the void. The rain of atoms is disrupted by the introduction of the *clinamen*, of chaos or declination: the swerve by means of which nature is introduced into the universe (Serres 1982f: 99), and by which being becomes becoming (Prigogine and Stengers 1982: 145). Lucretius first describes the homeostatic interpretation of the universe that the clinamen disrupts. It is a world where:

> no visible object utterly passes away, since nature makes up again one thing from another, and does not permit anything to be born unless aided by another's death.
>
> (Lucretius 1928: Bk I, ll. 265–266)

Having described the "order of fate" as a consistent rain of atoms, Lucretius introduces the clinamen:

> As the atoms are falling straight down through the void owing to their weight, at undetermined times and places they swerve a little with just the smallest change of direction.
>
> (Lucretius 1928: Bk II, l. 216)

Lucretius then asserts the cosmological rationale for the swerve:

> If it were not so, all would go on falling like raindrops through the infinite void, there would be no collisions and no blows, and nature would have created nothing.
>
> (Lucretius 1928: Bk II, l. 221)

This gives us free will:

> but what keeps the mind itself from having necessity within it in all actions, and from being as it were mastered and forced to endure and to suffer, is the minute swerving of the first-beginnings, at no fixed place and at no fixed time.
>
> (Lucretius 1928: Bk II, l. 284)
>
> Again if every motion is the outcome of the last in an endless chain, and the atoms do not by swerving initiate a motion to break the law of destiny, and prevent an infinite causal sequence, how can living things have free will?
>
> (Lucretius 1928: Bk II, l. 251)

Lucretius makes it clear that the swerve is diagonal:

> Wherefore again and again I say the bodies must incline a little; and not more than the least possible or we shall seem to assume oblique movement, and thus be refuted by the facts.
>
> (Lucretius 1928: Bk II, ll. 244–247)

The clinamen describes a nature that can renew itself precisely because it is rich in disorder and surprise. A powerful surge occurs in the nature of things, and now vitality, movement and systemic transformation are a dance of order and chaos. Unaccounted for in Newtonian physics, the world of the clinamen is where trajectories are unstable, the irreversible unfolds – it is the open world in which, through fluctuations and bifurcations, things are born, grow and die. The world of landscape architecture we might say, where, as Lucretius puts it:

> stars and meteors fall to earth
> And the sun also from the height of heaven
> Throws its heat out and sows the field with light.
>
> (Lucretius 1999: Bk II, ll. 210–213)

Lucretius was an Epicurean, and his poem is a narrative retelling of the philosophy of his master Epicurus, who founded a school of philosophy in Athens teaching from a garden until his death in 270 BCE (readers will recall that in Chapter 02 Freeman regarded the philosophy of Epicurus "less profound" than that of earlier thinkers). Lucretius uses his beautiful and compelling poem to explain the philosophical system that Epicurus had developed and to spread it around the known world. Serres brought Lucretius to the attention of modern thinkers, linking his work specifically with the rise of complexity theory and the development of emergence. The clinamen, reformulated by Serres as the parasite, is to be found in all landscape systems, rural and urban. Typically, it first presents itself in a negative guise, viewed perhaps as a malfunction, an error or noise within a given system. For example, the presence of artifice within the pictorial system of a natural landscape, or a weed in a cultivated garden. Its appearance elicits a strategy of exclusion. Epistemologically, the system appears as primary, and the parasite as an unhappy addition that it would be best to expel. Such an approach, however, misses the fact that the parasite is an integral part of the system. By experiencing a perturbation and subsequently integrating it, the system passes from a less to a more complex condition. Thus the parasite ultimately constitutes the condition of possibility of the natural system. Serres' formulation of the parasite offers a way of theorizing the transaction between nature and city without privileging the latter as fallen, pictorial or architectonic, and without treating the former as a moral force. Both Lucretius' clinamen and Serres' parasite advise us to consider nature, not in terms of its laws and regularities, but rather in terms of perturbations and turbulences, in order to bring out its multiple forms, uneven structures and fluctuating organizations. As Prigogine once said, we used to think of nature as ordered and regular but now we see instabilities and fluctuations everywhere.

Serres emphasizes the occurrence of unexpected novelty, the calling card of emergence. "The parasite," he asserts, "invents something new . . . It expresses a logic that was considered irrational until now, it expresses a new epistemology, another theory of equilibrium" (Serres 1982a: xxvii). Through the operations of the parasite, then, new things are made. This making is physical, real, but always different, new. The clinamen directs us back to the material world and its composition, putting relations between things at the heart of doing, acting and becoming. It shows how landscape systems can be driven to reveal new orders of agency. Serres' work enables us to see the city as a system of complex spaces linked by passages from realm to realm, from one local singularity to another. It makes possible a regionalism – a construction of place – that is not built on the notion of identity but on difference, for the "regional epistemologies" that the travelling subject navigates, while specific and local, are themselves travelling. Local processes, formations and operations undergo "regional evolutions, partial accelerations, temporary regressions, alterances, equilibriums, finite transformations" (Serres 1982a: xxvii). The continuing transformation of regions is itself plural. Serres attributes these transformations to interference. He argues that the declinations opened up by interference make new passages through systems, following a language of paths, routes, movements and planes. This is the language of landscape architecture. Does not landscape design open up paths, cut routes, enable transitions, develop planes of interaction across which beings engage each other in intensive, novel becomings? Is this not achieved by means of a magisterial interference in the order of things?

Extending the idea of disturbance to the modern city, urban theorist Henri Lefebvre argues that the complexification of space and the objects that occupy space cannot occur without a collateral complexification of time and the activities that occur over time (Lefebvre 2003: 167). "This space," says Lefebvre of the urban realm in general, "is occupied by interrelated networks," and is characterized by a tendency towards homogeneity (of logics, intentions, codes, values and systems) that is simultaneously disturbed by differences; "subsystems, partial codes, messages and signifiers that do not become part of the unitary procedure that the space stipulates, prescribes and inscribes in various ways" (Lefebvre 2003: 167–168). It is a space whose relationships, according to Lefebvre, are "defined by interference." For Lefebvre, the city is entirely relational – and parasitism is, as Serres says, "the heart of relation," joining order and disorder in a never-ending dance. The parasite, the noise in the system, through its presence and absence, produces the system (Serres 1982c: 52). The contemporary city, then, is a space of transformation. Its metabolic processes move at varying rates, and the parasite is the cause of this fluctuation, it is the catalyst of metamorphosis. The vast "interrelated network" of the city incorporates the processes of the social, economic, cultural, physical, mineral, organic. It includes all forms of matter, image, movement, light. Sometimes urban designers and administrators try to expel the parasite, and it is through just this attempt at exclusion that an infinite number of possibilities appears as the irruption of a disorganization into the system. The modern city is therefore impossible to organize – despite insistent attempts on the part of planners, designers and political authorities to do just this. Serres and Lefebvre demonstrate how resonant an "ecological urbanism" could be if it could adopt the transgressive concept of disturbance as the condition of possibility of the city, as its space of transformation. If we could design passages, routes and pathways to permit the turbulent stream of natural process to enter the city and disorder and disrupt its operations of control, and the "logics of surplus value" that Lefebvre says deny natural process (Lefebvre 2003: 35).

For Lefebvre, the urban square, the plaza of modernism, is the zero degree of multiplicity, where everything is part of an order that hunts down disorder with overwhelming repressive force. According to geographer Edward Relph, who documents the decline of mainstreet as an organizing spatial program, and the subsequent rise of the plaza in the 1930s, the plaza, or urban square, which is situated at the heart of the modernist city, is the inverse of nature

(Relph 1987: 85). Neutralized by democratization, reduced to Lefebvre's "blind field" where the urban simply becomes the industrial and a re-presentation of fictive nature dominates, modernist urban public open space rejects particularity and ravages anything associated with natural systems.

But the new polycentric and contradictory urban field has become truly multiple. However, we can only intuit its movement and its demands on human life. The organizational figure most suited to the contemporary city is not the plaza but the dynamic meshwork, a truly differential space–time web whose constituent elements can only be perceived from place to place and moment to moment, never in totality. This urban realm is not mappable, it has no fixed coordinates or causes that can be easily traced. In its movement from modern to postmodern the city created a milieu in which the public realm is uncertain and dynamic, a realm for which transgression, understood as disturbance and interruption, is a way of life. It is a force field of social and political, aesthetic and cultural, individual and collective intensities that construct spatial orders and code, distribute and organize the flows of human desire throughout these orders. The meshwork symbolizes the decentered, multiple, indeterminate, open-ended and capricious twenty-first century city. The movement of bodies in space, and of relations between them, a movement that occurs in different time frames in different sets, which jumps spontaneously from level to level at different intensities, this is the habitus of the contemporary subject. This subject is in the grip of larger forces, the laws of fate, which are always disturbed by chance, by the clinamen.

It is the tipping point that provokes transgression, bifurcation, the phase transition, the swerve from order that introduces unexpected consequences. This is the basis of adaptability, of resilience. What does not kill me will make me stronger. Bataille understood transgression as a gift. The notion of the gift emblematizes a transgressive, sometimes violent act, a transgression that is necessary in order to introduce into the reified world of things the illuminations of sacrality. This is why Lefebvre and Serres separate the city from nature. So it can come back as a gift. They are explicit about this. Serres says the city is founded on the exclusion of nature. Thus when the garden has appeared in the city it is as a sign of an absence. Likewise, for nature to appear in the city it must become artifice. Olmsted underlines the artificial character of the park when he justifies Central Park as art. But nature is always already in the city. Natural process is the "infrastructure" of the city. Human beings are natural beings and if, as Lefebvre says, "there is a production of the city . . . it is a production of human beings by human beings" (Lefebvre 1996: 101). The city depends on that which it excludes. This dependence takes the form of a continual renegotiation of its relationship with nature. Nature therefore is the parasite of the city. It is the noise in the system. But because the city itself is naturalized, this nature cannot be seen. The city appears denatured even though it is permeated by representations of nature. Therefore it must be revealed. Urban landscape architecture is thus a matter of revelation. Landscape architecture, however, has given away the key to the revelation of nature under the guise of human accommodation. Is this not the same reason that, despite the existence of the parasite being the condition of existence of the city, we try to expel the parasite? Lefebvre describes what happens when we do this. "The city," he says, "has a hole which is sacred and damned, inhabited by the forces of death and life, times dark with effort and ordeals, the *world*" (Lefebvre 1996: 88).

04

PROJECTS AND PRACTICES

This chapter explores some built and unbuilt projects that negotiate the concept of emergence. A number of professional firms demonstrate a serious involvement in the protocols of emergent systems; some of them have developed a keen awareness of the potential alignments between emergence, affect and ecological performance. The best are dissolving the boundaries between function and feeling, society and ecology, representation and expression. Most of the work is well known, and frequently discussed in recent texts and reviews. This discussion rarely broaches the degree to which such work exemplifies emergence, however, and this may partially be because to do so requires a deep immersion in the processes of design articulation in addition to real familiarity with the physical and social ontologies that these processes let into the world (Elizabeth Meyer is one critic who is sensitive to the world-making potential of landscape architecture, and recognizes these new ontologies when she sees them). It has not been possible for this author, within the constraints of this book, to visit a wide range of built work and interview the design teams and the stakeholders, a process of analysis necessary to build a comprehensive overview of the current state of play. To undertake a review of several built works would require another book. Therefore this chapter describes projects with the more modest goal of demonstrating how emergence may feature in landscape design as, on the one hand, an organizing force, and, on the other, an interpretative device.

When assessing the exemplification of emergence in contemporary work two sets of criteria suggest themselves as evaluative guides. The first, laid out in Chapter 02, comes from what scientists and philosophers of science have considered the defining characteristics of emergence. These professionals do not think that any one situation needs to exhibit all the qualities listed here, although in most cases if there is one quality present some of the others are too, but the conditions of irreducibility and downward causation are regarded as critical.

- Irreducibility
- Supervenience
- Downward causation
- Unpredictability
- Novelty

It will have been noticed that these properties are connected and can, in fact, be considered as aspects of one primary condition that emergence theory defines. The requirement of irreducibility simply states that emergent properties are not reducible to the lower-level phenomena from which they emerge. If properties are irreducible, it does not follow logically that they are supervenient, or autonomous from the basic phenomena that give rise to them, though they are likely to be. The use of a parking lot by guerilla gardeners to grow pumpkins in the middle of a city cannot be reduced to the constituent urban food discourses, casual labor, materialities and traffic ordnances that gave rise to it (if it could the gardening would not be *guerilla*). The pumpkin patch in this case is autonomous from its base, in that its success or failure is not a condition of the establishment of the parking lot in the first place. Neither of these qualities necessitates taking the further step – the move from weak to strong emergence – of affirming that higher-level entities affect their lower-level constituents by means of the process scientists call downward causation, and which we have been referring to as autocatalysis, or feedback. Guerilla gardening's spread, through marginalized urban spaces is, however, just this. The effect of whole–part influence is crucial to the role of emergence in landscape architecture for many reasons. For instance in community participation it is a way of minimizing subjective decision-making about social–environmental design interactions on the one hand, and of reducing program rationales to the influence of social conventions on the other. Now, if developments that occur in a system are not the result of interactions at a more basic level, it follows that it is impossible to predict these new features on the basis of the more fundamental phenomena. That is to say, the outcomes of these basic interactions are indeterminate. Landscape architects such as Walter Hood often leave many decisions to the people who have a stake in the terrain being developed. Hood will not know the outcomes of their decision-making processes until they have been made. Sometimes what the stakeholders develop is emergent, supervenient, autonomous from the sets of decisions that Hood Design has established in their studio (Hood 2012). Aspects of an overall design process that have reached this level of autonomy cannot be explained without recourse to a set of descriptors or concepts that the landscape architect has not envisaged. Scientists would say that a new theoretical apparatus must be utilized to account for the spontaneous outcome of community input. The design can be regarded as incorporating novel features, that is, emergent phenomena, that were not able to be determined at the outset. All of this adds up to a process of individuation, or actualization, through design that is contingent (which individuals will be involved, and when, and how?) and historical (what histories, of involvement in the site and its community, will be instrumental in the process of change?), and cannot be thought of simply as fulfilling generalizable or pre-formed goals, whether these be formulated on the basis of design principles, aesthetic rules or even civic virtues. What is more, the process of design cannot be regarded as closed or finite, and nor can the productions of this process – designed landscapes. The world itself emerges as full of divergent processes that yield novel and unexpected properties, structures and entities. No description of a landscape, in this scenario, can be regarded as a full, final or total one.

The second set of criteria is that assembled in Chapter 03, Key Concepts, and it is an extension of the first into a more open and designerly language. Like all translations it is only one among many possible restatements of the themes and propositions of scientistic constructions of emergence. Indeed this set of key concepts is not so much a direct interpretation of ideas prevalent in current scientific literature, as a reconstruction of them using different theoretical resources to extend them into concepts that may be of use to landscape architects and designers who wish to bring the insights of emergentists to bear on their work. This set therefore is neither exhaustive nor prescriptive.

- Open systems
- Situation
- Meridian
- Assemblage
- Naturecultures
- Field theory
- Morphogenesis
- Difference
- Disturbance

It should be said at the outset that there is a provisionality, even an arbitrariness, to the establishment of both sets of criteria. The first set is an amalgam of issues that are still being debated by scientists, both as to the substantive and the performative capacities of the concepts. The discussion of these concepts in Chapter 02 noted that while some physicists and chemists do not agree with the concept of emergence, particularly as expressed in the idea of supervenience, others do. This division does not necessarily follow disciplinary lines. Adherents of soft emergence, which repudiates supervenience and downward causation, are not all reductionists, and not all hard emergentists are ecologists and social scientists. The role of these definitional terms in this chapter is to test landscape architectural discourse around emergence against a – not empirical – but material, instrumental and practical tribunal of effect. To what extent are ideas of emergence being articulated in current landscape architectural practice?

The first list is generally agreed by scientists of different stripes to consist of the primary definitives for emergence, whether they agree with them or not. These are the terms being debated in scientific journals and books. The second list, elevated here to key concepts of emergence, has been assembled for this text. It is not a personal list, in the sense that it includes topics and ideas only to the degree that they fascinate the author. Instead it is an assemblage of ideas that have been developed in, or are instrumental to, discussions across a range of disciplines for which emergence has become a significant heuristic or intellectual framework. In a very obvious way the scientific ideas have been filtered through ideas developed by Deleuze and Guattari, Serres, Badiou and Bataille. The concepts have been created for, adapted to or deployed by architects, ecologists, historians, sociologists, urbanists, economists and landscape architects, among many others. Their presentation here does take on a specific task, and that is to bring out the affective potential of landscape architectural practice. This brings us to what is perhaps the most significant point of this chapter, hinted at earlier when it was stated that no single description is full, total or final. It goes to the heart of the inquiry. Humans are in the system. More than this, the system can be known to humans only through operations of perception and cognition that themselves seem to be beyond knowing because they are part of knowing. Landscapes are enacted by our immersion in them. They are not something that exists apart from our feelings for them. This is not an issue purely of perception and cognition, although these faculties are instrumental in our experience of becoming human in a world that is itself emergent. It is as pointless to "review" landscapes for their affectual and emergent capacities as it is to argue that one emotion, anger say, is more or less emergent and affectual than another, say love.

It would be specious to interpret constructed landscapes in terms either of the criteria that are currently debated in scientific circles or the key concepts developed specifically for this book. The traditions of landscape making show an integration of technical and aesthetic performativities. But it would be a mistake to think of science as simply the technical side and the arts as the aesthetic side. It is not the case that science is devoid of aesthetic formulations, assumptions and experiences; instead, as many historians of science have shown, it is built on them. Conversely, art practices have always required technique, facility, instrumental ways of knowing. Indeed artists such as Paul Cullen have developed their practices through investigations of the artificial separation of epistemologies of science, technology, aesthetics and art. To take one step further, the phenomenon of emergence does not recognize the distinction between the arts and the sciences, nor do the phenomena whose behavior it describes. Emergence shows us that the design activities of landscape architects, while necessarily sensitive to measurable performativities, must understand these very outcomes as having effects that engage the world and its inhabitants in very different ways.

A birdscape in New Zealand

A rocky headland juts out into the Tasman Sea, two thousand miles of turbulent waters that suck ice-cold currents out of the Southern Ocean and push them on to the west coast of New Zealand. Here, in the Roaring Forties, the combination of wind and water whips up mountains of ocean that find their way to the black sands of Muriwai where they crash and grind at the coastline as they have for millennia. Under the headland there is a rock shelf from which fishers pitch their lines for kahawai, kingfish and snapper. Every season at least one of these adventurers is knocked from the rocks by a Big One and drowned in the roiling sea. It's safer to take the green-lipped mussels from the inshore rocks, but even then you can be pulled way out and never get back. Muriwai is a coastal community like so many others around the globe. It has its residents who have chosen to live there and commute to the local metropolis, and it has the weekend and holiday visitors – surfers and swimmers – whose cars and vans crowd the parking lot above the beach. Giant pohutukawa trees cling to the rocks and the dunes, their red flowers ablaze over Christmas. They provide the shade for holidaymakers who picnic in the back dunes. From here they can look inland to the high hills, also covered with pohutukawa. The shiny black tui live among these trees, feeding off the flax flowers that grow beneath them. Their call is tuneful, melodic. The enormous kereru, or wood pigeon, gorged on puriri flowers, flaps drunkenly through the picnickers whose heads turn at the sound of its wings beating the air.

Out at sea, the large persistent swells are full of energy that cannot simply disappear. Wave watcher Gavin Pretor-Pinney notes that "waves on the surface of the sea lose remarkably little of their energy to the surroundings once they are up and running. This means they can travel enormous distances" (Pretor-Pinney, 2010: 33). The swells come in groups of larger waves, separated by gaps in which the waves are smaller, and sometimes barely even there. What is so strange, Pretor-Pinney writes, "is that each individual wave crest travels faster than the group it is in. The crest appears from the calmer water at the back of the group, travels through it and disappears again in the calmer water at the front." As they approach landfall, entering shallower water, the wave base makes contact with the seabed and they slow down, bunch up, steepen dramatically and, at Muriwai, form what oceanographers call "plunging breakers," the kind that surfers love. These are the waves that are slowly but surely dragging the community of Muriwai into the sea.

Coastal erosion, caused by this combination of Antarctic swells and westerly winds, has been a "problem" at Muriwai since European settlers first cleared the land in the 1870s. Since then, a range of solutions has been tried, with no success. When, in the 1970s, the local council culverted the stream that runs out beside the headland it made the erosion worse. Ocean waves crept up the water way and grabbed great hunks of land. When the council constructed tightly engineered walls along the beach below the parking lot in order to save this infrastructure, it made the erosion worse again. The walls were undermined and pitched into the sea.

Prior to the development of the Muriwai settlement the erosion, of course, was not a problem. Muriwai is no different from other seaside communities. "More than 80% of the world's shorelines are eroding at rates varying from centimeters to meters per year. Erosion only becomes a problem when a human-made structure is threatened" (Pilkey and Hume quoted in Shepherd 2009: 24). The reason current methods of coastal erosion control have had no success at Muriwai is simple: complexity. The accelerating erosion is a direct result of the continuous interaction of open-ended, unpredictable, prevailing weather systems with structural engineers' use of hard forms to change these dynamics.

Landscape architect Delwyn Shepherd has lived at Muriwai for sixteen years. From her deck she has watched the wind and water shift the sand dunes around for that long. She has participated in the community meetings where local residents constantly discuss the problem, throw their limited financial resources into ever-futile engineering solutions, and wonder how they will ever stop the erosion of their hard-won infrastructure, like the car-park, the public rest rooms, the surf club perched above the beach, the bridge across the culverted stream that lets vehicles down to the beach. Drawing on complex systems research and a thorough analysis of the conditions specific to Muriwai Beach, Shepherd has developed a design technique that offers a reinvigoration of this shifting terrain. Significantly, her method does not aim to stop the erosion but rather to manage it in such a way that it contributes positively to the ongoing development of the Muriwai community. The key to her approach is that it is "bottom up." Like the Tasman Sea weather system, it emerges from the mixing of physical processes and evolves over time in an unpredictable, open-ended way.

There are three components in Shepherd's system: pohutukawa trees, birds (tui and kereru), and the local people. Pohutukawa (*Metrosideros excelsa*) is a climax tree species forming indigenous coastal forest. These trees produce vast quantities of wind-dispersed seeds that have a preference for open sites, particularly close to seawater. They are colonizers of bare land, their adventitious aerial roots penetrating cracks in rocks and establishing multiple connections in coastal sand. The leaves are coated with cuticle to resist desiccation, giving them resilience against salt-laden winds and long dry periods. The wood comprises thick-walled fibers, narrow elongated cells with limited storage and conducting ability, resulting in a dense, hard timber with the ability to coppice from both trunk and branches after damage. If a pohutukawa growing in sand becomes partially buried, it can continue to grow by forming new roots along the buried trunks. The resulting tree looks like a cluster of separate trees (Simpson 2005: 83). When conditions are favorable their bright red flowers produce vast amounts of pollen and nectar which are transferred from tree to tree by birds, especially the New Zealand tui (*Prosthemadera novaeseelandiae*). "The spectacular flowering that is a feature of pohutukawa . . . is an unusual strategy in New Zealand, reflecting the link between bush-blossoms and birds that is the hallmark of the myrtaceae" (Simpson 2005: 90). Shepherd draws on a critical characteristic of the robust and genial pohutukawa – its ability to form "vegetation islands" which coalesce into contiguous coastal forest. Her proposal uses the co-evolutionary bird–tree dynamic in a simple and slightly mad way. In the first phase of the project stakeholders and landowners are consulted and the terrain between the sand dune system and the hills behind them is planted with vegetation islands of pohutukawa trees – by the landowners and the community. Specific areas of accelerated erosion on the beach are identified, especially those that threaten infrastructure or places of community value, and trees are planted here too.

Shepherd has identified what she call "focus trees" in the hills above the beach. Interestingly, these enormous climax specimens are often found growing in or around Maori pa sites, the terraced, defensive fortresses of pre-European tribes. Shepherd projects tui flight-trajectories from the focus trees to the selected beach zones to connect the eroded areas with mature pohutukawa growing on the hills. As it happens, tui are very directional in their flying habits, and they can cover distances of up to 35 kilometers per day (Shepherd 2009). As the pohutukawa planted in the vegetation islands grow, tui will visit them when they are flowering, transfer the pollen from tree to tree and eventually the coastal forest will return, colonizing the sand dune system and forestalling the erosion of the beach. This process will take between ten and twenty years.

Shepherd calls her proposal Birdscapes. The aspect of it that I have outlined does not comprise the whole project. As the planted and adventitious networks of trees grow they will form what Shepherd calls "a nonlinear maze" throughout the terrain between the hills and the sea. This maze of pohutukawa creates different spatial arrangements within the landscape. A second project phase has been developed, in which permeable spatial zones formed within the interstices of the pohutukawa islands are subject to what Shepherd refers to as "a disturbance of the system." These areas are disrupted by a range of different procedures, or operations. The forms of the disturbance that she has investigated are wide-ranging, including structural, extractive, hydrological, horticultural and topographical operations. By means of close observation of how people connect with, and use, the sand dunes, the pohutukawa trees and the beach, Shepherd has identified community behaviors and preferences. Picnickers, holiday-makers and everyday users have been observed and beach–dune sightlines and view shafts, for instance, analyzed to find the various ways in which landscape affordances are incorporated into their lives. Boundary and edge conditions have been carefully studied, types of function and use investigated. The pohutukawa trees are prominent in the complex ecological needs of humans as well as nonhumans, and Shepherd has turned this human–environment interaction information to the development of spatial landscape structures that draw on the adaptive and emergent conditions that she established with the Birdscapes project. The idea is to add to the ways humans use the beach, contributing new ways for them to connect to the landscapes they inhabit. Shepherd's hope is that the resultant patterns will be effective in altering peoples' perceptions about erosion in their social and economic systems. The design operations she has developed create resilient ecological networks that work with the erosion rather than against it, and in the process form self-managing landscape structures that contribute to the daily lives of the people of Muriwai. The structures link upland forest to the coastal sand dune system, provide access to cultural heritage sites (the Maori pa) and contribute to the ongoing development of the Muriwai Regional Park.

Fluvial landforms in Birmingham, Alabama

Described, prior to settlement after the Civil War, as "a vast garden as far as the eye could see," irrigated and fertilized by "one beautiful clear running stream," the valley in which Birmingham, Alabama was built is now a post-industrial city and the stream a polluted urban water-course (White quoted in Craft 2011: 3). One hundred and fifty years of industrial development have taken their toll. The quick growth of Birmingham, named after the English coal city, which also developed very fast, earned it the nickname "the Magic City," and its enormous steel production caused it be known as the Pittsburgh of the South. Its population went from 3,000 in 1880 to 178,000 in 1921. Much of the city's industrial success is due to the "clear running stream." Spring-fed, Village Creek, as it is called, flows forty-four miles from Roebuck through the Jones and Opossum Valleys where Birmingham is located to the Black Warrior River. It rises in and flows through the limestone that underlies the 100 square miles of Jefferson County that it drains. Founded in 1871 at the intersection of two railroads in a cotton field at the center of Jones Valley, Birmingham's strategic location placed it close to the raw materials required for steel production: iron ore, coal, limestone – and water. Village Creek provided not only drinking water for the nascent city, but also huge quantities of the water that was needed at each stage of iron production. The Tennessee Coal and Iron Company estimated that by 1910 they were using 400 million gallons of water a day to cool their blast furnaces in Birmingham.

Today Village Creek is a dirty river. Surrounded by development, its riparian corridor no longer exists and run-off from the industrial and other sites along its reaches, which include a gravel quarry and rail yards, is not filtered before it enters the creek. Birmingham landscape architect Clay Craft drily notes that "polluted run-off combined with source point dumping from industrial sites and sewage outlets has left Village Creek extremely impaired" (2011: 25). Craft, however, has a plan. He has identified a particular reach of the river for an ingenious redevelopment scheme.

Because they were especially low-lying and prone to periodic flooding, many areas adjacent to Village Creek were not developed until the 1920s, when the pre-crash boom put pressure on developers to build mill villages in the water meadows. These residential communities were subject to frequent inundation from the beginning. Those flooded most often and most heavily included Ensley, the area Craft has subsequently selected for his project. Attempts to prevent the flooding of the residential areas were many and varied, including straightening the bed and installing culverts and gabion walls. These measures have removed most of the sinuosity of the original creek. "Engineering structures," Craft writes, "combined with increased development, have led to 80 + years of flood problems for Village Creek floodplain residents" (Craft 2011: 27). After repeated flood events, including two within four years of each other that crested fourteen feet (four feet above flood stage), Congress finally passed the Water Resources Development Act 1986, which authorized $29.6 million for the Village Creek hazard area. Birmingham City contributed a further $7.4 million, and 642 properties were purchased and their houses removed from the floodplain. In 1995 another major flood triggered a FEMA Hazard Mitigation Grant Program that gave $5 million to remove further properties. In 1996 two more floods hit and another FEMA HMGP grant enabled 135 homes to be removed. In 1999 a further acquisition took place, bringing the number of structures removed over twenty years from the Ensley area of the Village Creek floodplain to 735, at a cost of $37.5 million.

The Village Creek Human and Environmental Justice Society (VCS) has planned a new facility, Moro Park-Ensley, as a commemorative landscape to mark the site of a neighborhood where many thousands of people grew up. The park site is a seventy-five acre open space between Ensley township and Village Creek. The infrastructure of the subdivision still exists – only the houses were removed. Curb and channel streets, footpaths and light poles are there and the lots are still clearly visible. Pecan trees dominate the vegetative cover. Since 1977 this site has flooded more than twenty-five times. The VCS, focusing today on creek restoration projects and water quality issues, continue the efforts to lessen the regular inundation of the floodplain. These involve gabion walls to stabilize creek banks, and the removal of a bridge whose twelve support columns impede water flow and contribute to flooding. The Society's initiatives are estimated to cost $8.3 million, and are directed toward the establishment of the park as a permanent feature, with a skating rink, mini-golf, an outdoor classroom, a historic museum, rose gardens and a picnic area. It is within this context that Clay Craft wishes to develop a landscape project that, like Delwyn Shepherd's in New Zealand, works with the natural processes that cause so much strife, rather than against them.

Craft uses a disturbance–colonization–succession model of landscape design. When natural or human disturbance occurs, he says, nature will respond. When this disturbance is taken into account in design, conditions can be established to respond to it using the natural processes already inherent in the situation. Since the Village Creek floodplain is a dynamic (i.e. ever-changing) condition, a master plan is inappropriate. Craft wishes to use "calculated design interventions" as the initial conditions of a sequence of processes that will respond to those conditions and through this create a landscape that both commemorates the neighborhood that used to be here, and provides – let's not call it a park – a new terrain that itself continually evolves. This new terrain will enable a range of social and environmental functions not envisaged in the VCS plans. It's safe to say it will not include rose gardens, but Craft argues that what it will provide will be more appropriate and more interesting.

His approach has three key components. First, flood relief channels will be dug into the footprint of the existing residential roads (see Figure 4.1). These channels will serve to filter storm-water into the creek during normal rain events, and act as flood retention areas during heavy storms, thus dissipating the waters that breach the banks of the creek in these events. This strategy includes removing 50 percent of the roading to allow the framework of the once residential roads to organize a new pedestrian system for the site. The street signs will be retained as remnants of the former neighborhood. Craft envisages vegetation colonizing the roadways as deposition occurs through under-use. Eventually the roads will become paths with adventitious vegetation on either side (see Figure 4.2), and pedestrian trails permitted to filter through (see Figure 4.3). Second, Craft will install low "sediment deposition walls" in the floodway to collect sediment, litter and woody debris. Over time, and after many flood events "fluvial landforms" will begin to emerge on and around these structures, thus revealing the geomorphological process of flooding. Adventitious biota will colonize the sediment and grow, slowly increasing the drag-effect the walls have on the floodwaters, causing more sedimentation as the waters slow down. These deposition walls, which also serve as seating, will be constructed from asphalt slabs that are recovered from unnecessary roads. As mentioned, some of the roads will be partially removed leaving only a small walking path for pedestrian movement across the terrain. Other roads will be completely removed to make way for the flood relief channels to be dug in their beds. The removed asphalt will not only be used for the deposition walls, but also spread in crushed form to make pedestrian paths and used in the remediation of creek banks. Craft's third intervention is the setting of vertical "fluvial rods" or poles along the stream banks (see Figure 4.4). These, he says, will encourage on one side of the creek the natural deposition process, and on the other reveal the process of erosion. When placed in lines that are tightly formed the rods induce deposition. When scattered loosely along the banks, they reveal erosion processes. On the far side the rods will reflect the natural effect that tree trunks have on flood banks by encouraging deposition of sediments entrained in either stream flow or run-off flow. The banks will build up, vegetation will colonize the new soil deposited on them, and together the rods and the vegetation will begin to stabilize the stream banks. They will provide an ecological and economically sustainable alternative to gabion walls, Craft says. On the near side of the bank the rods, organized in a looser manner, will be painted white to the grade at which they are set. As erosion occurs the unpainted subsurface portion of the rods will be revealed, indicating the amount of erosion that is occurring.

Clay Craft considers his project as "designing with initial conditions." In this approach, decay and degradation are not only welcomed and accounted for, but they contribute to – and regulate – the form-making process. Landscape degradation leads to new series of events and these lead to new opportunities. Designing with initial conditions requires the designer to stay with the project well after these conditions have been established. The designer makes periodic evaluations and adjustments based on what is happening. Like a gardener, or an arborist,

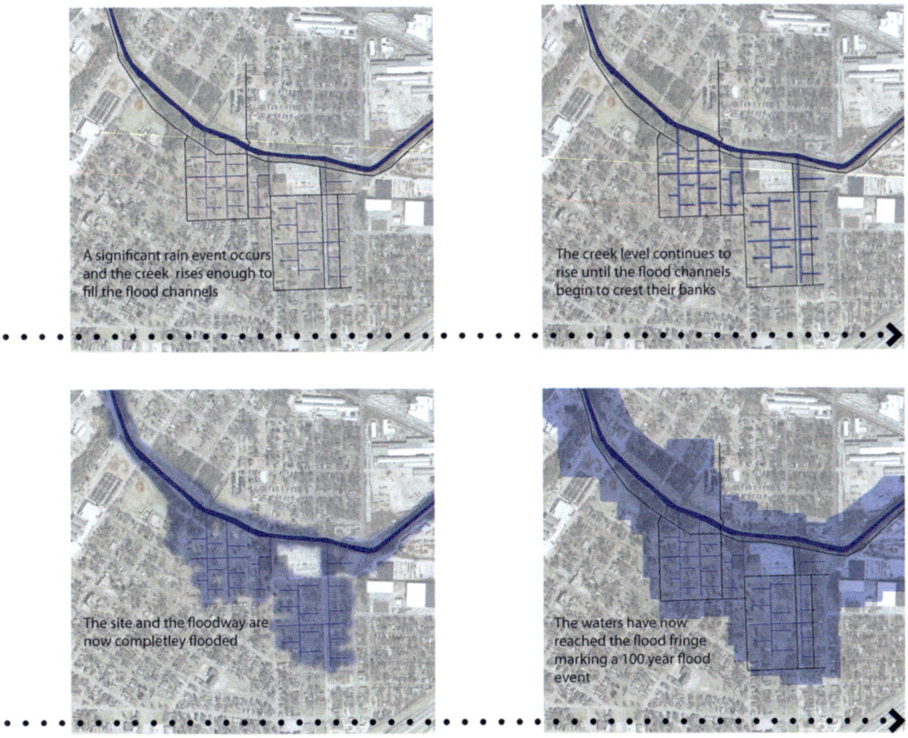

FIGURE 4.1 Because the subdivision was built in the flood plain, the successional roadway "planters" swell with water periodically, marking the advance and withdrawal of the flood waters. Courtesy of Clay Craft.

FIGURE 4.2 Clay Craft's Village Creek strategy enables the emergence of aquatic channels that trace new biotic communities along the plan of an old stream-side subdivision. Courtesy of Clay Craft.

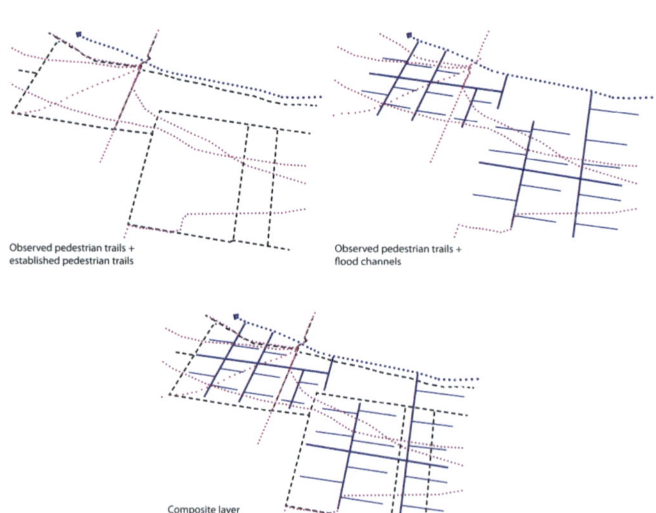

FIGURE 4.3 Pedestrians have already etched pathways across the vacant subdivision. Craft's proposal enables these to intersect with the orthogonal channels, and new paths to form along the channels. Courtesy of Clay Craft.

FIGURE 4.4 On one side of the stream the fluvial poles encourage sedimentation where the water flow slows as it bends around a curve, and on the opposite bank they both slow erosion and mark its progress. Courtesy of Clay Craft.

the landscape architect can adjust conditions in response to natural processes. And just like the arborist who must know the biological functions of a tree in order to know how her actions will affect its final form, the landscape architect needs to understand the ecological functions of a specific terrain in order to understand how the landscape will respond to the conditions he establishes.

Urban assemblages in Birmingham, Alabama

Another project, not far from Craft's, looks for solutions in a related but different way. There are over 18,000 tax delinquent properties in Greater Birmingham (see Figure 4.5). Most of these are vacant and abandoned residential lots, some are brownfield sites (for example Ensley Works and US Pipe); others are grayfields such as Bellview Mall. There are multiple lien holders on many. The unoccupied residential sites are often overgrown, run down and uninhabitable, and contribute to negative perceptions of affected neighborhoods. All the properties add to the growing problem of "urban blight" (see Figure 4.6). Birmingham City needs to find solutions to this problem. It is an ongoing issue, exacerbated by the continuing evacuation of neighborhoods perceived by residents to be increasingly devalued, unmanageable and unlivable. In concert with a range of urban authorities the author has developed a proposal that attempts to answer the question "What to do?" It suggests a comprehensive analysis of an especially affected area of Birmingham, followed by an adaptive design strategy that leads to regeneration by means of a long-range redevelopment plan. The emphasis of the proposal is on sustainability, flexibility and self-regeneration achieved through new environmental structures and relationships developed over time. It includes community-based programs for building

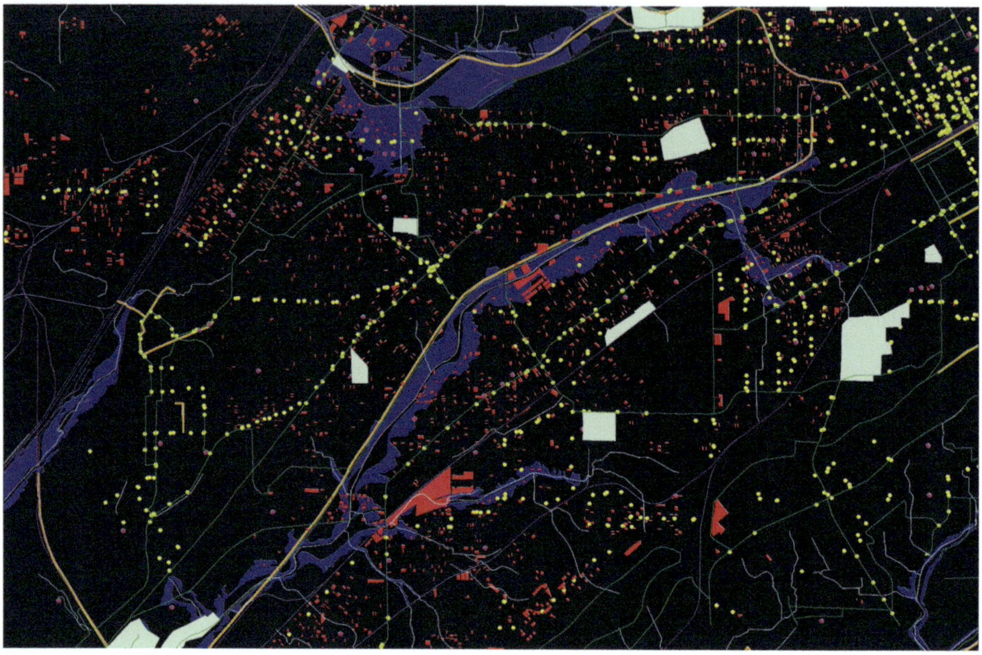

FIGURE 4.5 Valley Creek runs through West Birmingham almost parallel to Village Creek. In Birmingham's boom years worker housing was constructed in the floodplains of both waterways. Now many whole streets are filled with vacant houses and abandoned lots. On the map these "tax delinquent properties" (whose owners no longer pay their mortgages) are marked in red. Courtesy of Kelly Homan.

local amenities within an overall urban landscape framework. The goal is an interconnected series of useful and imaginative design interventions that contribute to the social and physical rehabilitation of neighborhoods affected by urban blight.

The proposed project area is in the Upper Valley Creek district of West Birmingham. It comprises a six-mile stretch of urban fabric that runs from the Germania Park-Jones Valley neighborhoods to downtown Birmingham, mostly along the flood plain of Valley Creek. This area of Birmingham, which has a large number of tax delinquent properties, is connected by the creek's continuous riparian zone, a thick tissue of blended conditions that hosts a range of development opportunities. The network will extend from the Upper Valley Creek neighborhoods to downtown Birmingham, terminating in a specially designed pocket park. An important feature of the project is the repurposing of materials from demolished buildings in the creation of new landscapes that can be accessed and utilized by the communities who are linked by this new green urban infrastructure. For example, one of the first priorities is the development of the pocket park in downtown Birmingham. It is envisaged that this will make use of materials harvested from selected properties to create a well-designed, convenient urban space that both provides public open space and draws attention to the project. The properties from which the materials are removed will be turned into productive healthy ecologies that link residential neighborhoods to redesigned landscapes along Valley Creek.

The project will roll out over a period of twenty or thirty years. This requires a managed, staged process (a lesson learned from Downsview Park) that does not so much work towards

FIGURE 4.6 Some houses in these areas are completely vacant, many are informally occupied, others are falling down. Birmingham City is systematically demolishing most of them. The City is also looking at ways to integrate and repurpose the new kind of urban landscape that is being brought into existence here. Courtesy of Kelly Homan.

a goal that has already been established but, being self-organizing, develops in such a way as to permit unforeseeable events to change its direction, or help it move to new levels of organization. Since the initiative is based on the re-appropriation of tax delinquent properties as sites of ecological and social recovery, these parcels need to be returned to City ownership in an expeditious and open way. Instead of tearing vacant buildings down and taking them to the dump it would be necessary to reclaim materials from them before they are disposed of. Once a building has been deconstructed and the materials harvested, they could be repurposed perhaps on adjacent sites for a small ecological park or garden near a community hub, such as a retail corner, or as a link to a newly designed reserve on the Creek.

In 2011 the author asked Birmingham landscape architect Kelly Homan to become involved. She was charged with developing a set of plans for a site at a strategic location within the proposed armature of creek, abandoned sites and declining communities. Homan focused on a neighborhood of abandoned duplexes between Valley Creek and a small neighborhood of historic worker cottages. Her design initiates the development of an urban assemblage, an interactive community "of objects, bodies, expressions, qualities and territories that come together for varying periods of time, ideally to create new ways of functioning" (Deleuze and Guattari 1987: 311). The subdivision of four dozen vacant and abandoned duplexes, two stories high, built of brick, mostly intact, is served by two primary roads (Monroe and Madison) which break them into four linear groupings, and a third, secondary road which runs up the middle of the two internal rows. Faced with the necessity of demolition, Homan plans to have this occur in a way that involves local communities in the act of harvesting. But the key component of her approach is her refusal to lay out a determinate outcome for the site. Instead she identifies critical elements of the Monroe/Madison assemblage and uses a strategy based on multiple event-planning to suggest how the futures of the assemblage may unfold given different contextual eventualities. Homan investigates four speculative scenarios that might eventuate from four divergent sets of initial conditions. Each of the scenarios involves different components of the assemblage in different ways, and each requires a different originating operation to kick-start the process of actualization. Before turning to the operations and their evolution into new kinds of community-involved event-spaces, it is important to look more closely at how Homan uses the notion of assemblage. For her, an assemblage is not simply a web of objects and processes. She includes words and their meanings, the ways in which things fit together and move apart, how structures are arranged across territories, the myriad influences of ephemeral and contingent causal agencies, divergences in value-systems, intangible shifts of position in physical and non-physical meshworks – a "complex constellation" of dynamic material and immaterial conditions. Homan is aware of the dangers of nominating keystone species to stand in for these categories (see discussion in Chapter 02, Narratives) and so develops her scenarios in a way that prioritizes relationships rather than individuals or species (see Figure 4.7). Homan's Urban Assemblages project seeks to empower

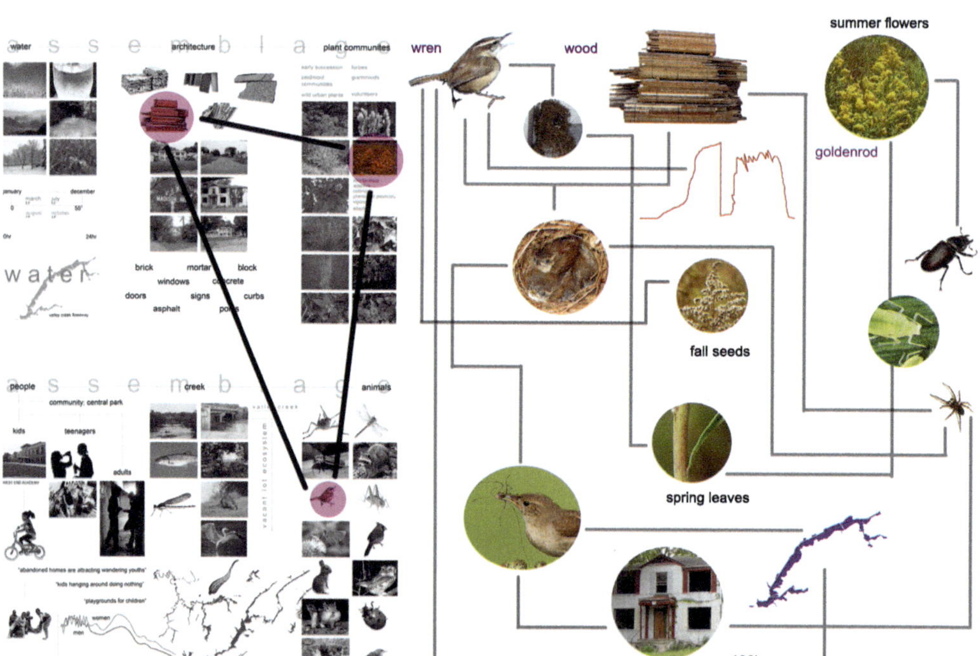

FIGURE 4.7 Kelly Homan has developed techniques for identifying and relating the various elements of an assemblage (in this case, wren, liquidambar and lumber) so that their specific material, expressive and processual components can enter into the design of novel conditions in appropriate and unique ways. Courtesy of Kelly Homan.

local communities to become involved in the creation of assemblages of people, wildlife, plants and building remnants to develop an engaging public green space. She explores creative deconstruction techniques, repurposing of existing materials, and establishing worker-training programs focusing on the trades of deconstruction and material salvaging. The project further proposes fostering an ongoing relationship between local art organizations and local communities as a strategy to re-establish neighborhood enthusiasm for the potential of abandoned sites. To do this Homan has focused on site-specific dance and community theater. She uses the operations of orientation, adjacency, displacement and dispersal to regulate the reconfiguration of the foundations, walls and other existing assemblage components into performance spaces (see Figure 4.8). Each operation is a type of deletion or erasure and removes portions or entire buildings. Many of the exterior walls, as well as sections of the buildings' interiors remain and are reconfigured in ways that enable such components to be used as play spaces for children, nesting locations for wildlife and debris catchment piles for flooding. Conceiving of the neighborhood as an assemblage has enabled Homan to rethink the role of community organizations. Birmingham's West End communities are working communities with a higher than average unemployment rate (citydata.com). The neighborhoods in the West End also have organizational structures, neighborhood associations and neighborhood council members which all have an interest in helping these communities organize around the reuse of abandoned sites. With respect to the materials and wall construction, it was discovered that many of the valuable materials within the duplexes have already been removed. Elements such as counter tops, copper wiring and pipes, hardware and so on are no longer in the buildings. So the bulk of the materials that could provide value both for reconstruction, and in potential resale are mostly brick, pressure-treated wood, concrete block, steel I-beams and shingle roofing. It was also determined through the design explorations that the site probably contains more materials than would be needed for construction. These materials could be stored, sold or donated to local non-profit organizations such as Habitat for Humanity, which currently needs roofing materials.

Homan notes that a key characteristic of an assemblage is its ability to adapt and change with the environment. In this way, the assemblage is unpredictable. Ongoing explorations of the Valley Creek site entail scenario investigations as to how the assemblages will shift, change and reorganize over many years. Homan has also raised the issue of the role the landscape architect has in the formation of the assemblage. The landscape architect is part of the assemblage, she says, but to what extent? Is it a job for life?

A plaza in Montevideo

The Plaza de Deportes Numero 7 is in a barrio called Prado Nueva Savona in the Paso Molino district of Montevideo, Uruguay (see Figure 4.9). It is flanked on one side by a park and on two others by residential areas. On the fourth side of the Plaza a busy road separates it from an active row of locally owned and operated shops and an outdoor market which sells knock-off items to all and sundry. Soaring above the local road an arterial viaduct carries buses, cars and trucks from one side of Montevideo to the other. The viaduct creates a median zone between the two carriageways of the road below. In this otherwise inactive strip mobile food franchises have established, serving the requirements of people who shop in the market and stores nearby (see Figure 4.10). The Plaza de Deportes is, as the name implies, a sports park. Soccer fields, basketball courts and a swimming pool are haphazardly arranged within a chaotic

FIGURE 4.8 Homan's ideograms show how an urban assemblage can combine many different elements, social, theatrical, imaginary, as well as natural processes and diverse species. Here the walls of abandoned houses are reconfigured as stages for community-run dance and theater events. Courtesy of Kelly Homan.

FIGURE 4.8 *continued*

framework of eucalypts and pine trees. Youths and their admirers congregate in the park to play formal and informal games and smaller children drift about while their mothers stock up with groceries from the shops across the road. Montevideo's city council has developed a program of redesigning local parks and outdoor open spaces, and expressions of interest in Plaza de Deportes have been called for. Landscape architect Marjorie Woodbury has submitted a proposal that explores how humans and nonhumans can together generate a democratic civic space that is open to appropriation in unmediated, but negotiated, ways. She thinks of de Deportes as a place where animals, plants, minerals, humans and other beings can interact or "associate" as they see fit, forming various collectives by coming into contact with each other in an arena where they have a chance to express their voice, by which she means "do what they do." She describes her work as "exploring how we can live together" even with "the things that scare us" (Woodbury 2012: 110). It is critical for her project that the emphasis is not on what things are, but what they do, how they act. For this reason she uses Latour's word "actants" to refer to the members of the collective, as it is a nonanthropogenic word for actor. Democracy, Woodbury says, is not about an ideal harmony of all things – it's a messy process through which we learn, bit by bit, how to live together. "As the collective expands, there will always be more actants on the scene, disturbing the peace . . . this is a good thing, and this process should happen in public space" (Woodbury 2012: 111).

Her design therefore removes all amenities that serve exclusively one set of members of the collective – the sports fields and swimming pool, for instance – to enable a reassembling of nonhierarchical conditions that support a much wider range of species (see Figure 4.11). Many of the trees remain, since their provisions are important for nonhumans and humans alike. Using the inherently value-free organizational potential of grids, Woodbury develops a ground plane order that maximizes access and the interpenetration of actant groups (see

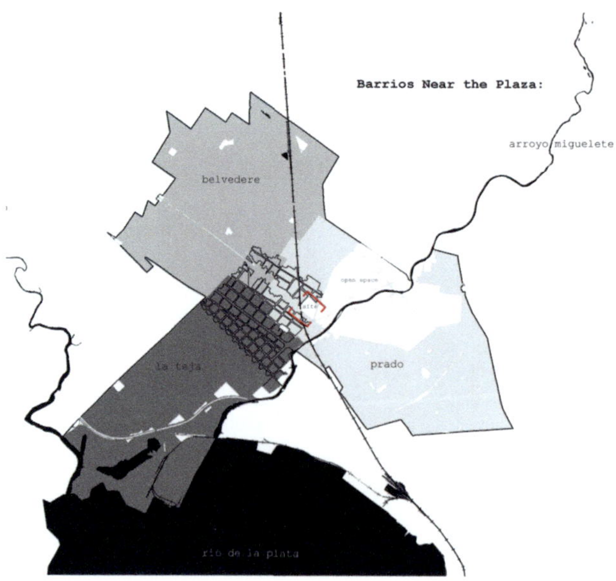

FIGURE 4.9 The Plaza de Deportes Numero 7 is one of many sports plazas in Montevideo. Located in the old part of the city it is crossed by freeway, railroad, waterways, teenagers, animals, insects and birds. This is the heart of the barrio Prado Nueva Savona. Courtesy of Marjorie Woodbury.

FIGURE 4.10 Food and clothing franchises have colonized the parking strip beneath the elevated freeway. Courtesy of Marjorie Woodbury.

Figure 4.12). Grids both permit and emphasize difference even as they provide shape, define spaces and refer to traditions of urban planning – in this case, Spanish and Jeffersonian typologies of order. It's the same point Stan Allen made in "From Object to Field" (1997). The Plaza de Deportes grid, inspired by the political landscape of the famous Plaza Independencia, where Uruguay's modern social history began, arranges the surface of the terrain in a legible fashion, but enables the thick, complex republic of soils, plants, micro-organisms, birds, insects and other beings their own kinds of expression. Thomas Jefferson's purpose, Woodbury observes, was that each square of American land divided by his grid would be equal, but the consequence was the obvious fact that every square of land is *not* equal.

> Each square of land is unique and some squares offer more benefits than others. A grid allows a comparison of each piece, framing collectives of actants, which can then be

> compared, even as the actants move between the squares and influence each other and change each other. Each square is constantly changing, and constantly becoming different from every other square.
>
> (Woodbury 2012: 111)

Woodbury has discovered that, by setting up a simple set of rules, an infinite variation is possible, while still conforming to a legible conformity. The Spanish grid achieved this by requiring buildings on the edges of lots, connecting with each other, creating unified facades. "Within that formula," Woodbury writes, "personal expression is limitless. More importantly, it's legible. The personal expressions don't fade away into the field: you can see them." To achieve a democratic layout the street system of the neighborhood was laid over the terrain of the park. Two pathway systems were introduced. Larger, twelve foot wide pathways bisect the Plaza and immediate context, creating a new organization of space and also connecting destinations. A smaller system of six foot wide gravel pathways helps movement around the Plaza elements and offers easy access into the central areas of the new spaces. These two

FIGURE 4.11 Marjorie Woodbury's "actants" include cotorra parakeets, elderly people, sunflowers, bees and eucalyptyus trees, among many others. Courtesy of Marjorie Woodbury.

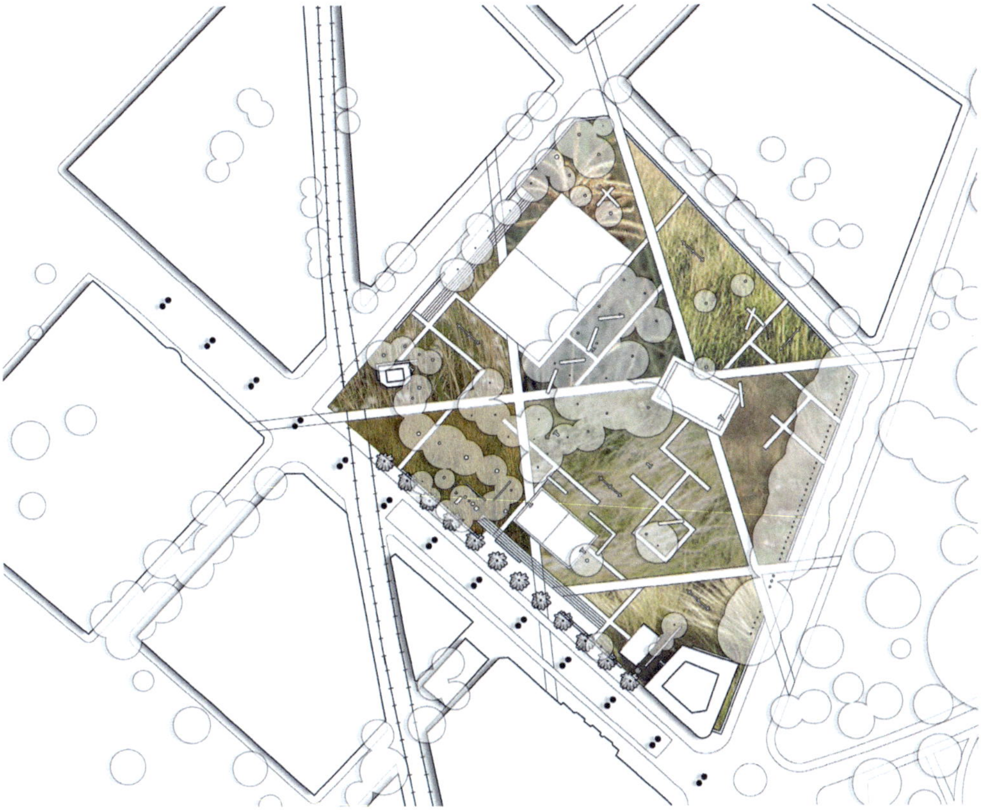

FIGURE 4.12 Careful study of actant behavior and requirements (including those of the children for whom the sports plaza was developed) has enabled Woodbury to create a "republic of diverse becomings" that connects to its surrounding conditions and enables humans and nonhumans to interact in a truly democratic way. Courtesy of Marjorie Woodbury.

systems are an abstraction of the city grid. (While remnants of broken orthogonal and symmetrical grids remain stitched together throughout the city of Montevideo, large sweeping avenues and railroad tracks bisect these grids, further disturbing and fragmenting them.)

To further develop what she calls "a common world," Woodbury scattered human-made elements associated with particular functions throughout the Plaza. She explains:

> When humans make things, they are used by nonhumans (unless we specifically prevent them from using them), and when nonhumans make things, humans also use them. Obviously, humans use things other humans make, and nonhumans use things other nonhumans make, such as a bird making a nest in a tree out of branches, which is later decomposed by insects. There is no fundamental difference here: they are all merely associations between actants. Yet, when nonhumans use manmade elements, we usually attribute that to abandonment.
>
> (Woodbury 2012: 101)

A swing set in sunny, grassy portions of the Plaza might not be used often by children due to the heat. However, it might be used by morning glories, a vine species that also colonizes the railroad and would have direct access to the study area. A basketball hoop in the forest may not be suitable for practicing shots. However, due to the strong support of the metal ring, it may be perfectly suited for a colonial parakeet nest. In a different kind of association, lighting fixtures, which imitate the viaduct lighting, attached to seatwalls, create an intense condition where insects, toads, bats, children, adults and stray cats can all come together.

Everything Woodbury has done has been with the purpose of connecting nonhuman and human systems through the establishment of corridors, transition zones, gradients and multifunctional elements. In order more explicitly to connect the Plaza to the shops and market on the other side of the road, Woodbury introduced a 200 foot wide set of steps to the sidewalk. She also lifted the franchise zone under the viaduct to prevent cars from parking there, which they do now, and afford safer, less encumbered pedestrian access to the Plaza, as well as locations for street vendors. The steps are of particular interest as a filtration device. Not only are they able to be colonized by adventive plants, but the hollow nature of their materiality enables small animals to move through and beneath them, from street to Plaza and back again. Woodbury says she is not interested so much in "creating habitat" but sites for interaction. The tall palms that she has introduced along the sidewalk between the Plaza and the road filter exhaust particulates from the overhead traffic, extend the garden-like qualities of Montevideo's urban plaza traditions to Verde and, most importantly, provide nesting sites for the Monk parakeets that form a significant actant group. By uncovering different actant groups and providing neutral sites of interaction, Woodbury is establishing the initial conditions for an ongoing re-inhabitation of the plaza by species that can then participate in its continuing development through that interaction. The lighting that she has designed not only enables a much-needed extension of occupation, but attracts frogs, bats and spiders that incite further action.

> Outdoor spaces are already inhabited by plants, animals and minerals. As landscape architects our designs affect them. Therefore it is our responsibility, when we design public space, to make it truly public for all the beings that our actions affect. Moreover, by bringing them together and giving them a voice, the infinite variation of future possibilities is instantly exploded.
>
> (Woodbury 2012: 101)

Plaza de Deportes No. 7 no longer feels like a sports plaza. While there are specific things to do, such as swing, skateboard, bike, play basketball, handball and soccer, it is no longer a place children and teenagers go simply to do those things. The real opportunities now afforded to all actants are the possibilities to share events, interact and associate with others: with senior citizens sharing yerba mate (a local beverage), for whom are provided ample seating and accessibility; with people who live across the city and are merely changing bus lines and now can wander curiously through the Plaza; with mice that you might catch sight of scurrying below the path; with bees and other insects pollinating the flowering plants; with neighborhood residents passing through the plaza; with retail employees eating lunch in the shade; and with parakeets foraging in the grasses instead of hiding far above (see Figure 4.13).

FIGURE 4.13 Woodbury is interested in a nonjudgmental approach to the interaction of species. In this way, she believes, Latour's vision of an unmitigated assembly of beings can be enacted in public open space. Courtesy of Marjorie Woodbury.

Open systems design in Baltimore, Maryland

Coastlines are critical to the ongoing health and vitality of human settlements. Since most people live in cities now, and many major cities worldwide are located in sensitive coastal environments, these metropolitan ecosystems are under increasing threat from urbanization: pollution, subdivision construction, transit corridors, industrial and commercial development. Conversely, urban systems are threatened by natural forces such as hurricanes and flooding. As a result of these interactive processes, the edge condition between land and sea has become a critical factor in urban design. The big issue is how to create waterfront conditions that promote biodiversity and resilience at the same time as providing appropriate environments for the millions of people who live and work in these conditions.

Landscape architect Qian Deng has developed an open systems approach to the design of urban coastlines. Readers will recall that open systems are created and informed by the matter-energy that continually flows through them. A crucial feature is their ability to deal with disturbance. Open systems do not just recover well from disturbance (hurricane, pollution, commercial development) but actually integrate it and evolve to more complex levels of operation. They are adaptive, resilient and creative, using turbulent conditions to produce novel features. An important component is the feedback mechanisms that enable new conditions to influence the material and organizational structures of the system, thus entrenching resilience.

Deng attempts to connect a series of complex coastal landscapes in Baltimore, MD by designing an open system of landscape structures and processes through a chain of wetlands, estuaries and river deltas, all of which are either developed or semi-developed. The designs for these landscapes are tested against a range of possible disturbances (flooding, economic decline and inappropriate urban development) to ascertain whether they will exhibit the degree of resilience necessary to reorganize into novel terrains that increase the potential for human and nonhuman inhabitation. The success of the designed landscapes therefore lies in their ability to exhibit both environmental and social adaptability. Deng wants to show that resilience to urban and natural disturbance can be designed into coastal landscapes for the betterment of all species and the habitats that support them.

Located less than one mile south of Baltimore's famous Inner Harbor, the Middle Branch is a shallow, expansive stretch of water that serves as habitat for a wide range of aquatic and estuarine species (see Figure 4.14). Masonville Cove, along the estuary's southern shore, is one of the best waterfowl staging areas in the state of Maryland. Currently the harbor suffers from water pollution, trash dumping and contaminants that leach from former industrial developments. However, the Middle Branch is on the brink of major change. The majority of industrial sites that ring the harbor are either vacant or being relocated, and Baltimore's 1978 Middle Branch Park plan has been dusted off as the city ponders the revitalization of an area still characterized by an awkward relationship between industrial and recreational functions. The ecology of the area is a central factor in its redevelopment. The Patapsco watershed now drains urban, rural and residential areas, and habitats in many parks and wetlands are affected by its waters. Its condition is important for the migratory bird and fish species that use the Middle Branch as a stopover and feeding area, but past development has meant that submerged aquatic vegetation, which provides habitat for numbers of species, along with removal of forest cover, has reduced habitat for migratory species as well as native birds. While

FIGURE 4.14 Baltimore's Inner Harbor has for many years been slowly degrading. It is a complicated matrix of urban and natural systems that has extraordinary potential for reformulation as a vast network of emergent landscapes. (See www.beslter.org/ for the amazing work being undertaken by the long-term research project, the Baltimore Ecosystem Study.) Courtesy of Qian Deng.

over 240 bird species have been counted in the marshes of Middle Branch, untreated sewage overflows and storm water run-off containing soap, motor oil, road salts and animal wastes continue to contaminate and cloud the water (Deng 2012). The harbor's human habitats are also fragmented. Westport, on the western shore, is separated from the water by industrial development, and cut in half by the I-295, while hundreds of public housing units in the community of Cherry Hill at the southern shore have either been demolished or are awaiting demolition. Major railroads and highway systems traverse three sides of the estuary. These, combined with local and arterial roads, physically and psychologically divide communities from the harbor and from each other. Nevertheless, subsistence fishing and crabbing continues from existing piers in Middle Branch Park and from the Hanover and Potee Street bridges. Members of the Baltimore Rowing Club scull and canoe the waters of the harbor (it is too polluted for swimming). There are many sports parks and the Gwynns Falls greenway extends fourteen miles from Middle Branch to the City Line in West Baltimore and two miles to the Inner Harbor.

The first intervention Deng made was a perforated membrane that reconnects the Westport community with the coastline by soaring above the rail transit system and enabling pedestrian and wildlife passage (see Figure 4.15). The next step was to address the Carroll Camden Industrial District. Over recent years, the real estate market has adjusted to the contamination of parcels in this area by lowering selling prices, but this has not been enough to stimulate redevelopment. Outdated parcel sizes, inadequate road widths for modern truck access, aging infrastructure, incompatible land uses and unrealistic assumptions about the land's possibilities have all exacerbated the district's decline (Howland 2002). Deng devised a remediation plan that over time regenerates the industrial landscapes by unblocking the Glynns Falls watershed of which it is a part. Deng says the brownfields will gradually transform into clean landscapes with a higher capacity, providing economic and social value and creating opportunities for future investment. Her program chart shows the phytoremediation process that slowly increases the ability of this once performative wetland zone to again act as a coastal sponge (see Figure 4.16). Deng's open system approach increases resilience and provides appropriate development opportunities within and alongside the recovering tidal terraces. She actually sets the phytoremediation terraces up as an initial condition for future development. The first stage is cleansing water and soils. This process establishes an easily accessed waterfront park planted with marshland species chosen for their ability to target contaminants, tolerate salt and withstand inundation by stormwater. Within the conditions becoming established, Deng situates a new housing prototype that cannot only cope with frequent flooding, but evolve to new levels of responsiveness and resilience as a result of it. Her plans show design propositions for human living systems that include low-rise dwellings with functional courtyards set among productive coastal farms (see Figure 4.17). All of these endeavor to create a flexible and adaptable living model. For example, it is proposed that the apartments are clustered around sunken courtyards. Deng writes:

> Ecosystems have multiple possible operating states, and may shift suddenly from any one of them, even at the small scale of a residential courtyard. Following a sudden flood event, the micro-ecosystem in a sunken community courtyard can reorganize to "renew" itself to a similar or perhaps different state.
>
> (Deng 2012: 19)

FIGURE 4.15 Qian Deng proposes a continuous webbed surface to act as a living membrane that reconnects a residential area with the coastal edge. The membrane provides access, habitat, shelter and gathering spaces, as well as great views of the harbor. Courtesy of Qian Deng.

Depending on the degree of toxicity, volume and the length of time water remains in the courtyards, different scenarios will develop (see Figures 4.18 and 4.19). Deng's plans explore the economic potential of open ecological systems that thrive on disturbance. She develops a set of criteria for the introduction of new industries. These criteria require future economic use of these sensitive wetland zones to be ecologically and socially "friendly" by being nonpolluting, by enriching habitat rather than destroying it, by supporting local market demands and by providing employment for local people. Her designs show the integration of algae ponds, urban orchards, aquaculture and aquaponic systems and a biorefinery plant with vegetated barrier islands and estuarial wetlands. All this along a coastline that merges ecological and social temporal cycles with the activities of everyday life and the unique water events of Baltimore City.

The High Line in New York City

It is only when a slave dies that he or she becomes free. Is it not fit that Michelangelo's sculpture of the *Dying Slave*, with its consoling and at the same time painful message, should be the lasting impression that one takes from the Promenade Plantée, the elevated concourse of gardens on the old Vincennes railway in Paris? Created for the tomb of Pope Julius II in 1513–16, Michelangelo's famous meditation on mortality seems to promise that we are all freed when we go through the turbulence of death. And here it is, glimpsed from the Promenade, as a massive frieze of multiple slaves in the throes of death all around an art deco building (see Figure 4.20). The promenade, thirty feet in the air, brings the pedestrian to the height of the slaves, all of them, so appropriately for Paris, exerting their multiple rights to be free. Already we are in the air, surrounded by the gardens of paradise.

What is the High Line's *Dying Slave*? If you walk down New York City's aerial ecology in search of the answer to this question, a number of options present themselves. Perhaps the official Friends of the High Line signs, with their logos and brands (Google; Toyota; Bloomberg) fit the bill. Take the Spring Cutback sign, which imparts the information that it is time for the winter foliage, left in place "to provide structure, beauty and habitat" through

FIGURE 4.16 Deng has developed a phased process of remediation and generation that integrates invasive plant management, habitat enhancement and wetland construction with social programs, historical preservation and, most significantly, industrial development. Courtesy of Qian Deng.

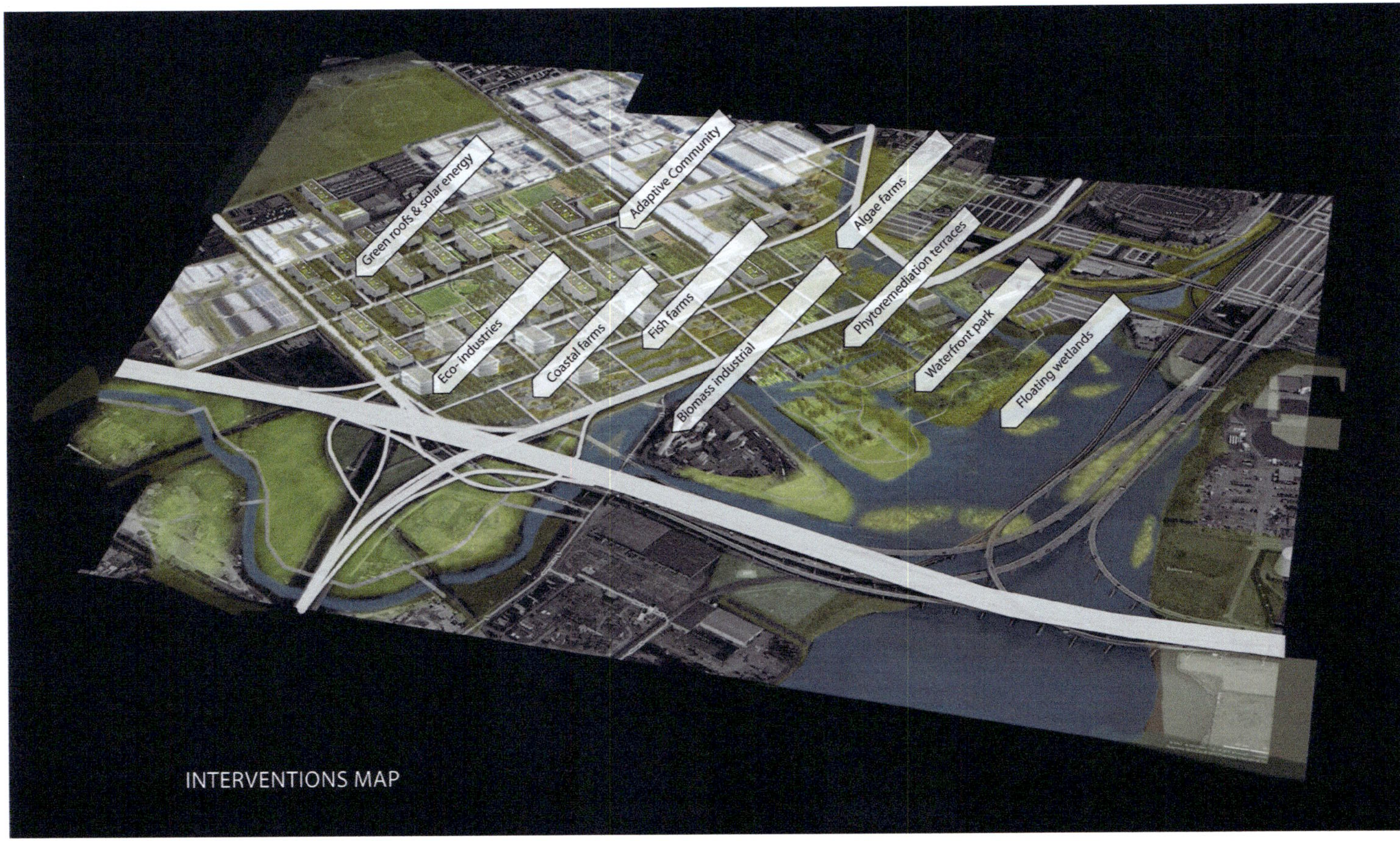

FIGURE 4.17 She has planned a low-rise housing development that operates as an open system, socially and ecologically, permitting flooding to occur and involving residents in the selection of possible futures once the floodwaters subside. Courtesy of Qian Deng.

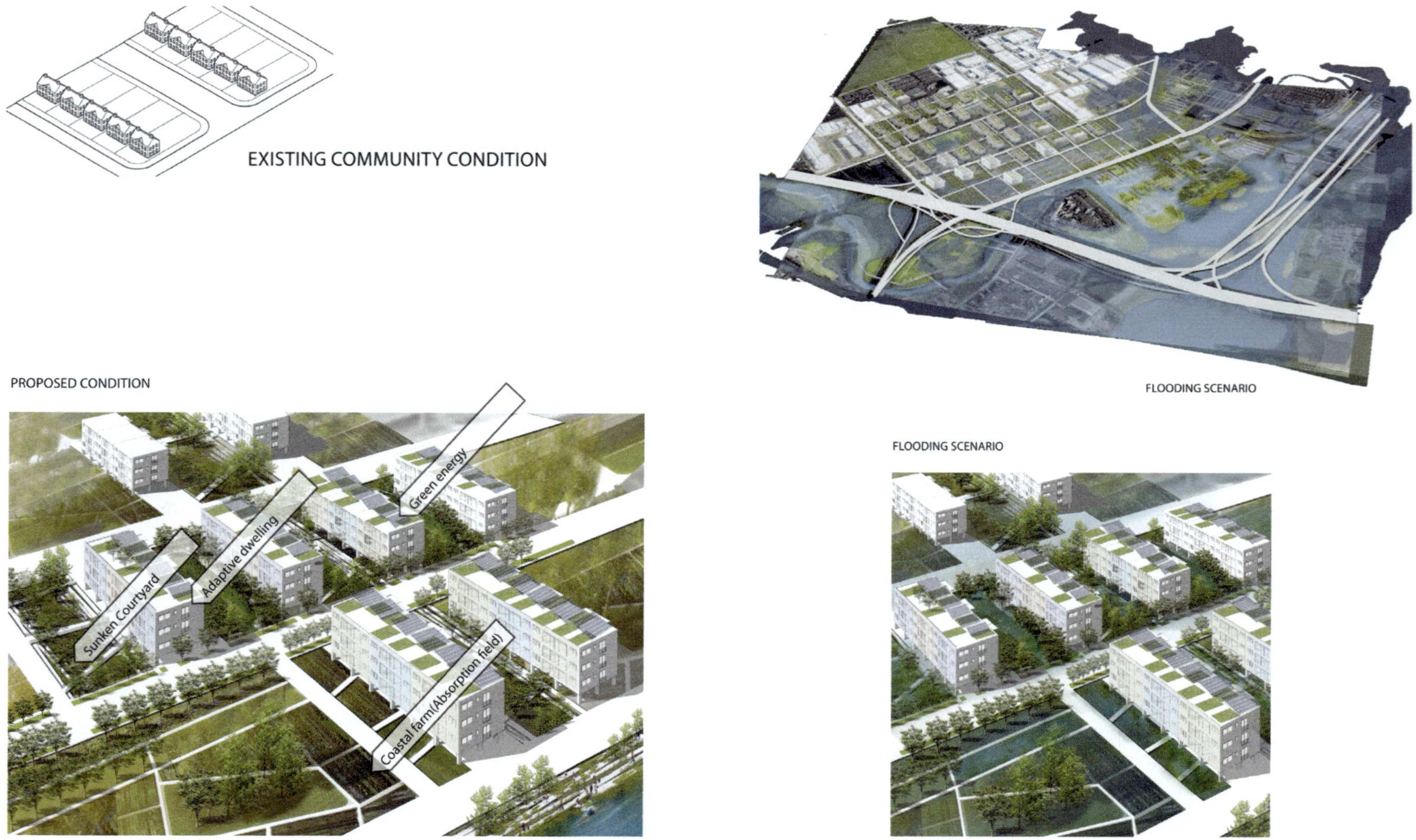

FIGURE 4.18 Deng sees the periodic flooding of the harbor's low landscapes as a positive element in the progress of the city towards an integrated and resilient built condition. Courtesy of Qian Deng.

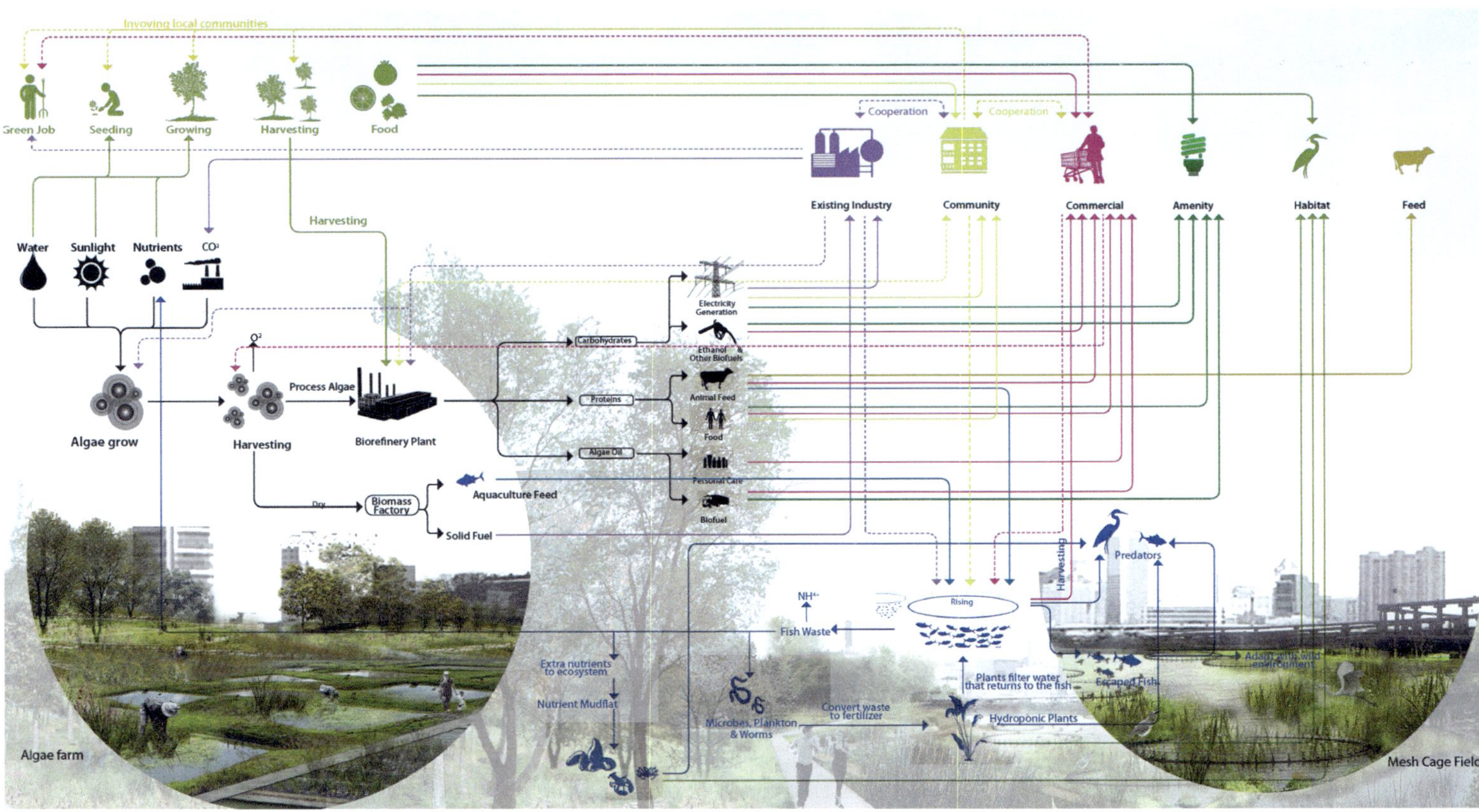

FIGURE 4.19 The future, she says, lies in intelligent and appropriate use of natural resources as economic and social incubators. Her design explores the use of different types of green industry in the reordering of the Camden Industrial District as a form of ecological urbanism. Courtesy of Qian Deng.

FIGURE 4.20 The Promenade Plantée offers spectacular views of Michelangelo's *Dying Slave* (1513–16) as the frieze of a modernist building in Paris. Photos by author.

the winter months, to now be removed. "While the High Line is meant to look like a wild landscape," the sign reads, "it requires an extraordinary amount of work to maintain the plant life." Like the dying slave, these plants also go through a transformation brought on by a bifurcation, a phase transition, a movement from one kind of existence to another. Life is being maintained, with a huge effort from Google and Toyota, and the seasonal awakening into newness is, appropriately for New York, a consoling and painful blend of biotic and commercial energies. This is not a self-organizing ecology, then, unless Google is included in the assemblage. Should it be, or not?

Like Central Park (and the Promenade Plantée) the High Line is both part, and not part, of the city. Separate, artificial, picturesque, a feat, primarily of visual imagination, its promise of freedom is gloriously ambivalent. "Some people think of the park as being an escape from the city," Robert Hammond writes,

> but the High Line works because it never takes you away from New York. You are not in a botanical garden. You can hear horns honking. You can see traffic and taxis. It's knitted into the city. And you're not alone. You're walking up there with other New Yorkers.
>
> (David and Hammond 2011: 128)

Perhaps it is this sense of the dissolution of borders that is the High Line's *Dying Slave*. It is an assemblage that draws together diverse geographies, ecologies in which humans (and their bank accounts) necessarily participate, along with the birds and insects that distribute seeds up and down the track. Part of and aloof from the commercialism that drives the city, the High Line eerily recalls Olmsted Snr's injunction, in a letter to New York City administrators, that Central Park should be free of commercial encumbrances. Olmsted and Vaux had often to defend the setting aside of developable land for "unproductive" park. Their argument against social and cultural incursion on the naturalistic landscape of the Park, an argument that inextricably links landscape architecture with artistic production, is prophetic. As Central Park became engulfed in urbanism, city development surrounding it on all sides, there was more and more pressure to develop areas of it (into cemeteries, railways, places of worship, etc.). Olmsted and Vaux argued that the "only solid grounds of resistance to dangers of this class" is to understand the park as a total work of art that is "framed upon a single, noble motive, to which the design of all its parts . . . shall be confluent and helpful" (Olmsted 1928: 248). In such a motive the reason for the park's existence is to be found: to relieve the citizens of New York from the "irritating effect of urban conditions."

Olmsted regards art + nature as the combination which "makes the Park the Park." It is through the application of "artistic principles" to the design of naturalistic planting that the work of art is achieved. Naturalistic planting is crucial because it is most removed from urban conditions. There are two kinds of naturalistic planting, Olmsted says: "open pastoral landscapes," and "picturesque sylvan scenery." The "artistic principles" are deployed when these typologies are "localized or adjusted" to the "various elements of the natural topography of the Park," for which there are three rules. First, screen the park from deleterious views – in this case urban views – by planting wherever appropriate a "wood-side." Second, give over as much as possible of the remaining ground to "tranquil, open, pastoral scenes." Third, assign the rest of the site to "scenery contrasting in depth of obscurity and picturesque character of detail with the softness and simplicity of the open landscapes." It was these three rules that gave the park its "essential elements" by providing a retreat from "urban oppressions through the eye." Everything else was secondary – all the appurtenances of the urban systems which had to be included in the park: roads, wires, bridges, towers, drives, walks, tables and chairs, refectories, statues, orchestral events, entertainments, lighting and so on (Olmsted 1928: 247–251). Here is the establishment of nature as antidote to the city through the practice of landscape architecture as art. Here, also, is the commitment to naturalistic plantings as constituting a different order from the city: artificial like the city, but in a different way. It is an interesting juggle of perspectives. While, on the one hand, the High Line is "not an escape from the city" like Olmsted and Vaux's Central Park, being "knitted in," on the other there was always concern, during its long gestation, about ensuring its differentiation from "life below." The Friends of the High Line went to great lengths to toggle between exclusivity and collectivity. They argued against people being able to get a cup of coffee on the High Line "because there are a million other places in New York to drink coffee." They did not want doors from adjacent apartment buildings opening straight on to it. "In New York there is no tradition of private properties enjoying private access to public parks and we didn't want to start one" (David and Hammond 2011: 97).

Olmsted's careful alignment of landscape architecture and art is present also at the High Line where the association with arts practices is open and ongoing. There is a deeper connection to Olmsted, however. When NYC parks commissioner Adrian Benepe was confronted with typical Parks Department treatments of paving on the High Line – hexagonal cement tiles – he objected, saying, "The High Line is going to be much more than a park. It is going to be a work of art" (Benepe quoted in David and Hammond 2011: 85). In what does its artistry consist? Perhaps it lies as much as anything in the planting by Piet Oudolph, an evocation of the adventive plant colonies that entranced David and Hammond, whose vision got the whole thing built. On first seeing the emergent ecologies on the railway, undisturbed for decades, they were staggered. It was the beautiful photography of this ruderal ecosystem by Joel Steinberg that often galvanized contributors into donating funds to the project. Now, in a moment of irony so typical of our post-postmodern world, the park is art because it has been returned to what David calls its "original landscape." Art + nature makes the park.

But neither the successional native planting schemes, nor the determination to preserve and promote a different environment from that to be found in the streets of New York City, nor the ambivalence about what is original and what is not, nor the involvement of organizations, economic flows, advertising and commerce, nor the goodwill of the art world, constitute the High Line's *Dying Slave*. While all these are truly emergentist, they do not by themselves add up to that extraordinary submission to the transfiguration of life into death that Michelangelo's piece brings to the Promenade Plantée. Is it simply too fanciful to suggest here what the High Line's model of transfiguration could be?

Let us push back further than the renaissance to the classical era. Had not Ovid (43 BCE–17 CE), the Roman apostle of emergence, investigated both the process of metamorphosis in his cycle of poems of that name, and the art of love in his *Ars amatoria*? The whole of the *Metamorphoses* implies that the very nature of reality is to be grasped fully only during the process of metamorphosis, before the final form is actualized, before, as De Landa puts it, the difference disappears (De Landa 1998). Indeed Ovid denies the possibility of closure at all, a theme taken up in the *Ars* where eroticism is portrayed as a fact of human nature, not a choice, and therefore always with us. Things for Ovid can always change into other things: as with Heraclitus, reality is fluid. The *Metamorphoses* addresses the flux of nature through an examination of the emergence of new forms from old, the continual creation of new structures from the complex webs of matter and energy that make up the world. The whole presocratic contemplation of the problematic of generation and decay, of morphogenesis, of the growth of the child, the decomposition of organic material and the metamorphosis of butterflies and moths is present in Ovid. It is in the very process of change, between becomings, that Ovid locates the moment of transformation.[1]

His rendition of the myth of Daphne and Apollo is an example. The myth tells how the beautiful Daphne, who wishes to remain "unnaturally" celibate, is changed into a tree in order to be saved from a violation of this celibacy at the hands of the god Apollo (see Figure 4.21). Through her crisis a new species of plant, the bay laurel, is brought into being. But there is nothing about Daphne's behavior or condition that implies the species *Laurus nobilis*, or even tree. What she might become is unpredictable, and yet it seems already present in her name.[2]

Ovid prises open a fleeting moment. The focus for his treatment of the myth is the becoming-different of a powerful, implacable force, a force of nature. With the actualization of the virtual, difference is encountered in the moment of transformation.[3] Which trajectory will this movement of matter take? This kind of invention is always genuine. Just as the palm

FIGURE 4.21 Bernini's *Daphne and Apollo* (1622–25) is a remarkable dramatization of the moment of becoming different. Both god and mortal are changed forever. Courtesy of Galleria Borghese, Rome.

of a person's hand can be read as an imprint of the virtual in the life of the individual, which qualifies but does not determine, Daphne is already imprinted with her transitions from mode to mode. The palm reveals the morphogenetic diagram of the person. The paintings of Daphne and Apollo are also diagrams, not of godliness or mortality, but of the viscous movement of nature through all possible instantiations. Like the folding of nature through the pleats of matter. Ovid's tale of Daphne and Apollo combines god and mortal, male and female, natural and cultural processes in an event that places the transformative, open-ended, stochastic process of change within a regime of fundamental encounter. A whole series of relations is caught up in an unlimited becoming, a perpetual movement of contraries, in the movement from virtuality to actuality, and all of it energized by the eroticism that pushes the god towards the girl. Is not the Highline Daphne captured by the god of commerce?

Teardrop Park in New York City

Located in Tribeca, adjacent to the Financial District on the south western tip of Manhattan, surrounded by four massive brick and glass apartment towers, Teardrop Park is the opposite of the High Line. It has taken its place in a string of new parks along the Hudson River connected by the green cycleway that circumnavigates the island, the catalyst for which was the Battery Park development of the 1980s. Speaking of the *primum mobile* for Teardrop Park, Tom Carey, former CEO of the Battery Park City Authority, who grew up in the Hudson River Valley said, "My vision was that those four buildings would be like hills and you would get the experience of walking through a glen in the Catskills and feel closer to nature. I wanted to provide children with an opportunity to come to a place that leads them to another place."[4] Built on sand fill pumped from the river in the 1960s, Teardrop Park is an urban dream of rolling hills, stone outcroppings and willowy marsh. The High Line is an urban dream of nature too, but these two parks operate in different ways. While the High Line floats above the compromised ground plane of the city, Teardrop Park is buried in it – a low spot, internal, enclosed, a *hortus conclusus*, it is an organic condition within the mineral life of the city. Even though it incorporates quarried stone, it bears this element as quite obviously a framed and situated nature, not a revelation of primordial geologic foundation. The rocks play the part of decoration: stones mounted in a beautiful setting. No, not a discovered *locus amoenus*, but a garden in the machine, a reversal of Leo Marx's famous figure of nineteenth-century American landscape imagination.

Unfortunately, a simple reversal of Marx's metaphor is neither adequate nor appropriate to twenty-first century urbanism. The true metaphor for this century is: the garden IS the machine, an integration of ecology and urbanism that is structural and functional, and which forges a new aesthetic. This is not the garden AND the machine, the cyborg figure of late-twentieth-century critical theory found, for example, in Donna Haraway's early work. Complete and utter integration does not yet happen with even the best schemes like Teardrop Park. It is not a failure of the design, however, but of landscape architecture discourse to communicate the format of twenty-first century urbanism to twenty-first century urbanists. Both Teardrop Park and the High Line are still working through the cyborgian, hybridic phase of landscape urbanism. They bring together elements that are always already framed as city and not-city, and these elements remain in their original, separate purity. If strictly artificial means are used to enable a natural "look" these means are mostly invisible, as with the geofibers used in the soil mix at Teardrop to mimic plant roots and stop surface erosion. This does not have to be the case. Parc D'Atlantique in Paris is a late-twentieth-century example of

FIGURE 4.22 Teardrop Park brings together in a very tender way themes of lightness and gravity, presence and absence, dislocation and place. Photos by author.

how the support systems and extreme artificiality of infrastructure may be incorporated into the aesthetic and become part of the show. Even though Parc D'Atlantique does not purport to perform ecologically as well as aesthetically, its suggestive ambivalence is rooted deeply within a determination not to separate base and superstructure (one and many, form and content). This park, on the roof of the Gard du Nord, would always be to a certain extent "artificial" but the postmodern tropes it engages play with this as a thematic element. While the retrospective insertions of ecological constructions into pre-existing urban fabric is more difficult than having fun with ideas, for this reason Teardrop Park runs the risk of serving neither purpose very well.

This does not mean to say that Teardrop Park is not an example of emergence in landscape architecture. Recall the desire of Battery Park City's Carey to "provide children with a place that leads them to another place." Put this together with what a principle of MVVA, Matthew Urbanski, said of the design: "We needed a strategy that allowed for some sense of exploration or unfolding in this very small space. Otherwise people would just come in, see everything, and leave."[5] And natural playground consultant Robin Moore: "It didn't take much to convince Michael [Van Valkenberg] and everybody else that the whole landscape should be a playground."[6] Emergence theory foregrounds the idea, prevalent in contemporary materialist philosophy, actor-network theory, systems ecology and other disciplines, that everything that exists only becomes and never is. As was argued in the section on Difference in Key Concepts, the changing dynamics of the natural, the social and the urban worlds can all be conceived as things differing continually from and in themselves. The human beings who encounter this unfolding difference are also continually differing from themselves. We have, therefore, a becoming-different of many diverse conditions, so that each encounter between these conditions is a novel encounter – unscripted, unpredicted and always open to further novelty. Teardrop Park fulfills, at least partially, this promise. It engages us visually and kinesthetically, combining reference and embodiment, and in winter and spring especially it is a landscape of movement and materiality, of touch and time and the penetration of sight into tunnels of darkness and light. A shimmering array of colors and textures, deep, long depth-effects, abrupt changes of direction and sudden shortened perspectives combine to accumulate such encounters (see Figure 4.22). The contrast on the one hand between stacked rock stratified with almost orthogonal fissures, smooth slopes of lawn and bare trees gilded with light, and on the other with great slabs of architecture is physical in its intensity. Neither is the architecture in any way inert. Spatially the buildings are highly active, forming enclaves of shadow and bright recessive planes that capture the sky in vertical pools of blue. The apartments are used as a contrasting geometry to the park's winding topographies. Their regular grouping and massed volumes cause morning light to fall dramatically, forming jagged edges of dark and light on the ice wall that divides the park, and across the bowl of lawn that spreads between the beech grove and the marsh. The extreme contrast between deep shadow and bright surfaces plunges people in and out of focus, and their activities seem to take on a heightened significance, though they are only reading or playing with their kids. Two teenagers sit by a sand pit against a wall of rock: she in darkness, he in light. They are leaning together quietly sharing words. Slowly he places his hand on her knee. The hand disappears into darkness. She gently moves the hand back into the light, returning it to his own knee. A child at the bottom of the shiny aluminum slide looks up at his mother who turns and breaks a small twig of witch hazel off a branch. Its bud is a spot of fire. She casually draws a circle in the sand at the child's feet. The boy stands the twig in the center of the circle where it casts a long shadow to the rim. Mother and child move off. The teens continue their ardent discussion. This landscape is "leading them to another place."

05

ENCOUNTERS

Serpent of pleasure: emergence and difference in the medieval garden of love

Our exploration of emergence now broadens significantly. The medieval garden of love plays a historically critical – but undervalued – role in the imaginative repertoire of landscape architecture. The garden of love calls attention to the condition of emergence, or becoming, as a specific and distinctive characteristic of landscape discourse. It appears in medieval art as a visual and verbal *topos* that bears associations because of its appearance in many diverse works as a memorable or highly charged cluster of meanings. As with rhetorical *topoi* in general, these meanings are not delimited but can broaden and deepen as the figure is used and re-used by different writers and artists. To medieval artists, garden *topoi* such as the *locus amoenus*,[1] the biblical *hortus conclusus* and the earthly paradise were available from a genealogy of garden depictions that include Theocritus, Homer, Ovid, Virgil and the Bible, as well as nonclassical mythic and oral traditions. The idea of becoming, as opposed to being, forms part of this genealogy. Through an analysis of the erotic and transgressive nature of the garden, it is shown how landscape architecture fits into this web of meanings.

As argued in Chapter 02, Parmenides' formulation of the One has underwritten philosophy and science since the time of the presocratic philosophers. If natural philosophy has privileged being, however, the idea of becoming has always destabilized this emphasis, and many thinkers, from Heraclitus to Einstein, have considered becoming primary.[2] Landscape architecture has only recently begun to take advantage of this counter-tradition of flux whose tenets and formulae have the qualities required for a design practice that works with both structures and processes. If landscapes are products of human negotiation with the world, then landscape architectural discourse must encompass the fundamental operations of human becoming in the world. These operations occur by means of a deep immersion in the structures and processes in which the human subject engages, and which make of landscape architecture a means of encounter, a possibility for world-construction and an art of inhabitation. Ideas of emergence and becoming are critical to this project, for human experience is a continual manifestation of these ideas. Some landscapes – and the pleasure garden is one of these – suspend the order of nature and permit the entry of a transgressive element through which a link is made with something beyond the everyday. These landscapes are fields of emergence and difference. Landscape design can place humans within these fields in such a way as to enrich and expand the possibilities of existence. By means of an analysis of medieval paintings and poems, the intention here is to demonstrate how the garden of love explores the condition of emergence in the cultural imagination of western Europe.

Emergence refers to conditions, processes and behaviors that change as a result of their continual encounter with other states and processes. Such change often, but not necessarily, includes or implies an increase in something: capacity, responsiveness, ability, sensitivity, interactivity, even sophistication. Deleuze uses the term "becoming" to describe the continual production of difference immanent within events themselves (Deleuze and Guattari 1987). The human subject, constantly in interaction with landscapes, is itself an event. A continually changing assemblage of forces, it arises from chance confluences of languages, behaviors, social values, laws, expectations and so on. Emergence, then, may be considered as a continuous unfolding that cannot be subtracted from the field of action or thought which it unfolds, as it actually constitutes the field and provides the flows that organize and regulate all activities and events within it. Similarly, in this discussion difference is understood as the uniqueness that makes such things as landscapes specific and local, as well as the processes that constitute this uniqueness. All aspects of reality demonstrate the operations of difference which, ultimately, are all there is. There is no background unity behind difference against which it should be conceived. The specific and unique development of each individual, be it person, city, wetland, bird or garden – its emergence – is an unfolding of its individuation "determined by actual, specific differences, multitudinous influences and chance interactions" (Stagoll 2005).

These concepts require us to take seriously the idea of the garden of love as a material depiction of the world by those who experienced it in the Middle Ages. A geographical trope that exists only in works of art and craft, the garden of love is a powerful and enduring cultural metaphor that begins to appear in a small range of visual arts practices around the ninth century and reaches the height of its popularity in the quattrocento. By this time it was featured in manuscript illustration, painting, sculpture, in bas-relief on marriage chests, and in small personal items such as combs and mirrors. From the thirteenth century it also appears in literature; in French poems and the *chansons de geste* of the troubadours; in the works of the "three crowns," of early Renaissance Italian literature, Dante (1265–1321), Petrarch (1304–1374), Boccaccio (1313–1375); in Chaucer's long poems (written between 1369 and 1387); and other literary forms. The garden of love is not, however, a sudden invention of the Middle Ages, for it draws on a *topos* that has a long tradition in literature and art from ancient Greece through Persia, the Holy Lands and the countries of the southern Mediterranean in the time of the Roman Empire (Pearsall and Salter 1973; Watson 1979; J. Harvey 1990; Landsberg 1995; Howes 1997).

What is a garden of love?

Although the figure features much earlier in poems and *chansons*, the earliest known illustration of a garden of love is a small late-fourteenth-century painting (Watson 1979). Painted by Maestro Ladilao di Durazzo and entitled *The Court of Love*, it shows several young men and women gathered around a marble fountain (see Figure 5.1). A woman plays a viol and a young man strums a lute. The background is arrayed in golden light against which a single tree rises from a meadow scattered with tiny, bright flowers amid clumps of grass. Three lovers dance

FIGURE 5.1 Maestro Ladilao di Durazzo (fourteenth century). *The Court of Love*. Courtesy of Museum of Chartreuse, Douai.

before a fountain, accompanied by a prancing dog and a dwarf bearing a falcon on his arm. Most of the motifs in which we are interested are here. The garden is enclosed, a world apart. Music fills the air, celebrating and encouraging the art of love, and animals, an important part of love's visual language, provide symbolic cues. In this painting the braided ancestry of the garden of love – classical, oriental, Christian – is apparent temporally in the reference to eternal spring and spatially in the traditional topographic theme of the *locus amoenus,* a pleasant, secret place where no harm can enter.[3] By the time the painting was executed, the garden of love was already established as a literary motif imbued with the spirit of courtly love, a theme whose development also grew out of classical and Christian investigations, albeit as inquiries into the nature of love itself. Discussing the tradition of the *jardin d'amour,* French critic Marie-Thérèse Gousset refers to its physical locus as "an idyllic place where perfect harmony reigns between the creatures of nature," a figure that resurfaces in medieval *chansons* "tinged with a lyricism which goes hand in hand with the development of the courtly spirit" (Gousset 2002: 82, trans. author). But it is more than this: the *chanson* poet Gautier de Chatillôn (c.1135–1182?) begins his *Verna redit temperies* with a note of perfect accord between nature, spring and love:

The season of spring returns
with her thick dusting of flowers,

the earth in her new aspect
smiles at our ways
for whom love is everything,
food for our hunger.
(Gousset 2002: 86, trans. author)

The garden of love occurs throughout Provencal love poetry "as the secluded, ordered, beautiful setting for the seizure by, or the loss of, love" (Giamatti 1966: 86). "Love," as Chatillôn sings, "is everything." Its relationship to gardens is a critical condition both of the affective potential of landscape architecture and that special sensitivity to its entanglement with things that landscape invites from human subjectivity. Within the confines of canonical landscape imagery, great variety of expression is possible (even though in the scenography of medieval poetry and painting there is remarkably little local realism). The flowering plants – of many European species, such as roses, lilies and daisies on the one hand or species of Eastern origin such as ginger, cinnamon and tumeric on the other – lend themselves to a wide range of semantic deployment. Conifers, hawthorns, mulberries, olives and palm trees provide regional as well as aesthetic frames, while the meadows and lawns are sometimes cropped, sometimes so long they wave in the gentle breezes that seem always to drift through the gardens. The enclosing element is often a high brick or stone wall, sometimes a wicker fence or a wooden trellis. In virtually all depictions of the garden of love there is a fountain, most often located in its center, usually octagonal, always flowing. For the classically educated medieval viewer/reader running water may have recalled the natural springs of Virgil and Theocritus, the fecundity of nature as well as Christian precepts and values, such as generosity, baptism and the superabundance of God's love. The springs and fountains of medieval imagined gardens are fed, however remotely, by the waters of paradise that rise, the German monk Honorius of Autun notes in the twelfth century, "just as Christ, fountain of all good things, springs from a chaste virgin. The four rivers which flow thence signify the four evangelists" (Turner 1910). According to classical myth, the fountain traditionally belongs to the god of love. Cupid sits beside it carving the arrows that pierce one and all, turning all those who gather about the fountain dancing, singing, playing, into prisoners of love. In later paintings these engrossed young men and women are usually richly dressed in brocaded fabrics, long trailing sleeves and parti-colored hose, the men more numerous than the women, obviously well-born, having the leisure to indulge in love and the education to admire its complexities (see Figure 5.2). Di Durazzo's painting emphasizes the separateness of the garden of love from its surrounding territories; in so doing it intensifies its rarefaction – artificiality, even – and consecrates its various components (music, dance, horticultural abundance) as marks of difference from the processes of the outside world. It is an opulent, rarified realm by means of which, upon entering, the individual is transfigured because transgression is permitted. In other depictions people are half naked, they fondle and kiss; time, law and death lie outside the wall, while the garden enjoys a fertile, eternal spring vouchsafed by the operations of difference.

The ambiguity of love

The medieval period is generally considered a time (roughly from the 800s to the 1400s) when the peoples of Europe and Britain understood the world and their place within it as primarily spiritual (Gurevich 1985; Eco 1986; J. Harvey 1990). A central feature of medieval

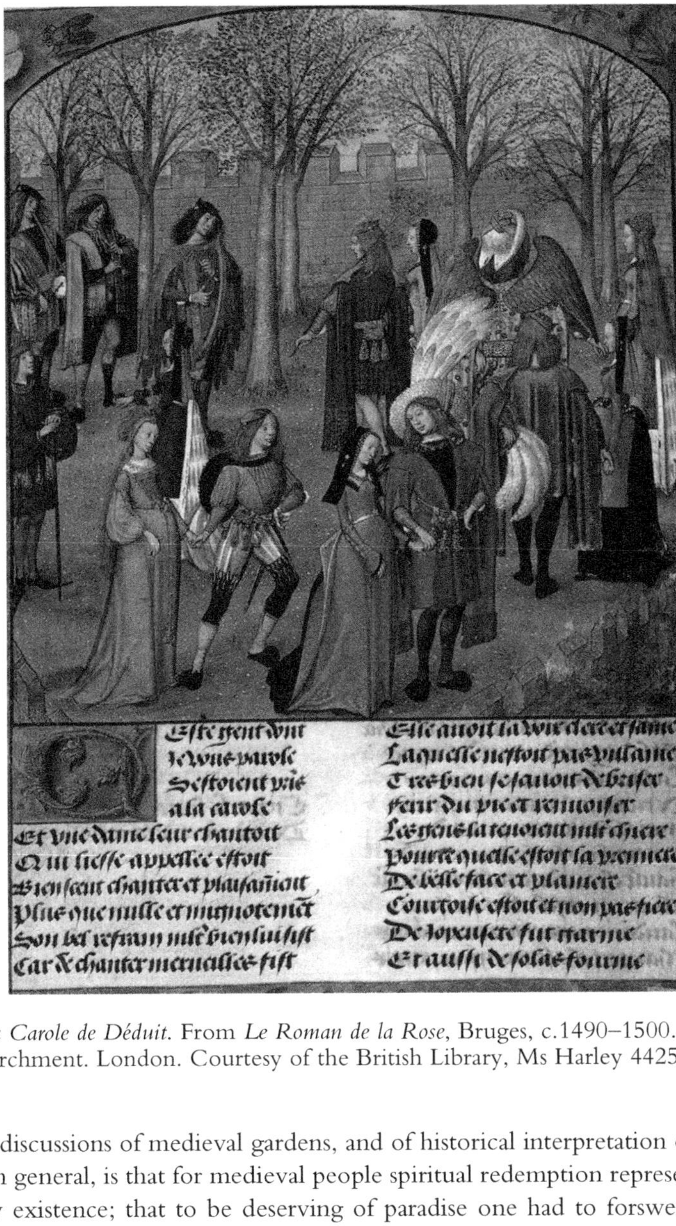

FIGURE 5.2 *La Carole de Déduit.* From *Le Roman de la Rose*, Bruges, c.1490–1500. Painting on parchment. London. Courtesy of the British Library, Ms Harley 4425, fol. 14v.

discourse, of discussions of medieval gardens, and of historical interpretation of medieval life and culture in general, is that for medieval people spiritual redemption represented liberation from worldly existence; that to be deserving of paradise one had to forswear all pleasures: "In medieval thinking redemption meant above all liberation from sensual, earthly existence" (Aben and de Witt 1999). The eschatology of the early Middle Ages regarded the pleasures of landscape not merely as an irrelevance but a positive distraction from the battle for salvation. The Dominican and Franciscan orders conspicuously gave up worldliness and luxury, and renounced all possessions in order to come closer to God, a devotion emblematized by St. Francis' marriage to "Lady Poverty." There was, therefore, a dramatic tension between the fascination and beauty of music, nature and human love associated with earthly gardens and the contemplation of the inner soul as a pathway to eternal peace. "The sleek looks of beauty are fleeting and transitory, more ephemeral than the blossom in spring," writes Boethius in *The Consolation of Philosophy*,[4] while St. Anselm simply states that "the delight of the senses is rarely good, mostly bad."[5] How, then, does the *jardin d'amour*, with its seeming glorification of earthly love, feature in this culture of renunciation?

It is worth asking another question. To what extent was the promissory condition of saintliness actually aspired to, or believed in, by the people who in the Middle Ages made gardens and enjoyed them, or by the artists, writers and sculptors who depicted the gardens in use and explored their imaginative possibilities? Garden historian John Harvey provides a huge range of evidence for the proposition that many of the extensive gardens constructed throughout western Europe between 1000 and 1500 were pleasure gardens and that these were much loved and appreciated for their sensual delights (J. Harvey 1990). Albertus Magnus (1206–1280) wrote a treatise on the planting of pleasure gardens in which he describes "places of no great utility or fruitfulness but designed for pleasure" (quoted in J. Harvey 1990: 6). The pleasure garden was an established feature of noble life from the early Middle Ages and there is ample proof that ornamental gardening flourished in England and Europe from the late eleventh century. From both literature and art it is clear that the medievals took a special delight in the appearance of plants and their perfumes, in the sight and sound of running water, and that they loved gardens and trees. In his book on medieval aesthetics, Umberto Eco argues that the view that the Middle Ages rejected the sensual world shows a basic misunderstanding of the period (Eco 1986). They were as much fascinated by the physical world of things as we are, he says.

While it is a critical convention (evidenced by the writings of ecclesiastics) that people were exhorted to turn from the fascination of the world in order to attain eternal life, the situation is not as simple as this easy distinction between the sacred and the profane. It is the very point of the garden of love to show how the separation of the world, life and love, into sacred and profane (and all the implications of this separation) is a misprision; that the rift between sensation and understanding (between Eve and Adam) is an illusion; that the message of the *jardin d'amour* is that paradise is to be found on earth. The misprision can be traced to an ambivalence, always present in medieval thought, that stems from Plato, whose dichotomy between sacred love and profane love profoundly affects all subsequent metaphysics, including Christian theology. Plato regards bodily desire as secondary. The body, he reasons in the *Phaedrus,* is merely an instrument (Plato 1952b), whereas spiritual love stems from the powers of the soul, a theme taken up by Augustine in *The City of God*, where he links the calamities of Rome to the pleasures of the flesh (Augustine 1931: Bk III, 1, 104–105). The consequences of this duality are manifest throughout medieval life, but the art that depicts this life in the garden of love is often misunderstood, by critics and historians. It is just this antinomial character of love in human affairs that the idea of the medieval garden of love explores, through Chaucer's poems, in Dante and Boccaccio, and in the famous garden text the *Roman de la Rose* (1279), as well as in innumerable paintings and illuminated manuscripts.

The world of difference

Diffused throughout society by means of the pervasive and powerful monastic culture, the worldview of educated people in the Middle Ages was shaped partly by the Bible and partly by the learning inherited from the Greek and Roman world (Grant 1996). Plato's division between sacred and profane was mixed with the philosophical theology of the Christian church (De Burgh 1953). God had created the universe and everything in it for man. The cosmos was still understood in terms of the Greek formulation. As discussed earlier, there were two zones, the celestial realm – beyond the sphere of the moon – which was eternal and perfect, and the sublunar, human world, which was subject to change and corruption. As far as theologians of the time were concerned, Aristotle had shown that the sublunar realm of human inhabitation, the world of opposites, differentiation and emergence, is the principal object of human knowledge (Aristotle 1952b). Chaotic, paradoxical, finite, the worldly domain of knowledge is the realm of *conjunctionis oppositorum* – the conjunction of opposites. Humans are caught in a web of transformation through which they must pick their path. The outcome of the expulsion from the Garden of Eden is that both the world and the existence it permits are contrary, antinomial and multiple. "The world is not our homeland," says the narrator of the Old English poem *The Seafarer*, "but our place of exile" (Gordon 1960: 40–41, ll. 55–66). To negotiate it – not to correct it – is the task of humankind. The antinomies cannot be harmonized, they are not resolvable into unity. Plato himself was clear on this point. For ordinary humans there is only multiplicity (Plato 1952d). The continual production of difference – that is, of uniqueness and particularity – is implicit in the singularity of things.

Christian theology determined that this realm of multiplicity and difference must be renounced. Jesus had told his disciples that those who, in his name, forsake worldly things will inherit everlasting life (Matthew 19: 29). But for the Greeks, the division between the perfect spiritual world and the imperfect physical world inhabited by humans was understood in terms of connection, not renunciation. The means of this connection is explained by Plato who, in the *Timaeus*, asserts that it is God's love for his creation that inspires "the divinity within [man]," depending on the degree to which the individual human is earnest in his own love of knowledge and true wisdom (Plato 1952d). "What is love?" asks Socrates in the *Symposium*. Diotima answers:

> Love is . . . neither mortal nor immortal, but in a mean between the two. . . . He is a great spirit, and like all spirits he is intermediate between the divine and the mortal. . . . He interprets . . . between gods and men. . . . For God mingles not with man; but through Love all the intercourse and converse of God with man . . . is carried on.
> (Plato 1952c: 163)

Therefore, although Plato divided love just as he divided the world, he reconnected the profane with the sacred by means of love, through the practices of "the prophet and the priest, their sacrifices and mysteries and charms."

Emanation and becoming

The philosophies of Plato were extended into the late Classical and early medieval eras by the neoplatonists, for whom the mediating figure of love became the principle of emanationism. Plotinus (204 BCE–70 CE) cast the whole of God's creation – the generation of the world by the One – as an emanation of love, summarized in the scholastic maxim that "good diffuses itself." He argued that the One, the most perfect of beings, would not grudge others a share in its perfection. Its superabundant power therefore overflows without dissipation or diminution (Plotinus 1884: V 2, 1). This emanation of love not only suffuses the world of creation, bringing the diversity of creatures into being, but carries on doing it forever. The emanationist interpretation of nature was transmitted to the influential cathedral schools of learning of the

FIGURE 5.3 *Jardin D'Amour*, Florence, c.1465–80. Engraving on copper. Courtesy of Staatliche Museen zu Berlin, Preussischer Kulturbesitz, Kupferstichekabinett, inv. 169–24

early medieval era by means of translations from the Greek, and was absorbed into a naturalistic understanding of the cosmos and God's creation as simultaneously alpha and omega, beginning and end. The result was that in the imaginary garden of love – that of the *Roman de la Rose* as much as of the painted chambers of the Palazzo Davanzati, Paolo Schiavo's panel painting *Realms of Love*, or the Florentine *Jardin d'Amour* – the sublunar realm of multiplicity and disorder (bodies, trees, flowers, carnality, desire) is itself an actualization of divine love, of a continual benediction of emergence conceived as *erotica* and *amoenita* (see Figure 5.3). The trope of the garden of love, it would seem, requires that the realm of the garden is one from which all finality has been excluded. The endless springtime, the overflowing fountain, the *millefleur* meadow with its starry carpet of flowers: within the bounded space of the *hortus conclusus* phenomena are transmuted into a vision of heavenly infinitude. The logic of neoplatonism demands that the process of generation continue. The essence of God's love is an immeasurable and inexhaustible energy.

A celebration of unconditional becoming can be found in the work of the Roman poet Ovid, dicussed in Chapter 04, who investigated the process of metamorphosis in his cycle of poems of that name, and the art of love in his *Ars Amatoria*. The impossibility of closure is a theme of the *Ars*, where erotism is portrayed as a fact of human nature, not a choice, and therefore always with us. The gardens that Chaucer (c.1343–1400) describes, echoing Ovid, in *The Parliament of Fowls* and *The Book of the Duchess*, are also places of "transition and transformation," where his characters discover the ever-changing and ambiguous field of love (Howes 1997). In these poems the garden serves as a "discordant threshold" that the lover must cross in order to encounter "the multiplicity, richness and variety" of the various traditions on which Chaucer draws in his explorations of the central paradox of the love experience – that love is joy *and* sorrow (Howes 1997: 55). In Chaucer's gardens, Howes writes, "the narrator encounters a multiplicity of voices and varying points of view" (54). Likewise awash in multiplicity, Amant, the lover in the *Roman de la Rose*, eventually discovers his sacred goal in the continual becoming of the world. It is exactly because they are immersed in life, this, the most famous text on the garden of love, seems to want us to understand, that the unfolding of beings achieves intelligibility through the active participation of the imagination. Humans are active intellectually in the realization of the world. "Creation," says St. Thomas Aquinas, "is in the creature and is the creature" (Aquinas 1963: Vol. 8, Bk1a, art. 45, qu. 3). This is critical, for difference is immanent, particularly in the operations of love. Images that depict gardens of love are not simple genre scenes. Within the fabric of generic elements of fountain, tree, bird and wall, the lovers are always fully fleshed out individuals with recognizable features and very human marks of difference. They are real people, and although each lover draws on God's blueprint for love, each also develops a complicated personal and interpersonal network of emotions and behaviors that characterize the singularity of *that* love. In the *Roman de la Rose* it is love that transforms and is transforming, always emergent, always gifting heaven to humans.

Both carnal and holy

In the *Timaeus* and the *Parmenides*, Plato does not base the highest objective of humanity on a renunciation, or transcendence, or resolution of opposites. The idea that transcendence is a source of "balance" because it reveals the coincidence of opposites – resolves them into an overall unity – cannot be traced to either book. In fact, Plato makes a clear distinction between a conjunction and a coincidence of opposites. A theory that would let the dialectic be absorbed into unity, "no matter whether defined as 'mature' or 'transcendent,' would be, in Plato's terms, an admission of defeat" (Wind 1967: 106). Misreadings of this *contrario* appear in many commentaries on garden images of the medieval era. In the anonymous *Livre des échecs amoureux moralisées* (1370–1430), an allegory based partly on the *Roman de la Rose*, an image shows Nature standing at the gate to a garden with the key (see Figure 5.4). The approaching lover sees within the garden the goddesses associated with Discord: Juno, Venus and Minerva (the subjects of the Judgment of Paris). According to art historian Michael Camille the painting shows the lover looking past Venus and Juno, who beckon him from the garden with their eyes, to Minerva and the claims of reason against carnal delight (Camille 1998: 81). From the Platonic separation of love into carnal and holy, Camille draws the moral of the subsequent desacralization of the former. This reading is also a feature of C.S. Lewis' interpretation of courtly love. For Lewis, the garden of love is only an imitation of the sacred garden, paradise; and not only a copy, but a misleading copy, a simulacrum. The courtly garden, the world and all that is in it, are but painted things, appearances on the outside of the wall whose inside no one has seen. The garden in this case is not a realm apart, where difference and transgression provide a window on to the sacred, but an allegory of human turpitude. Since Juno, Venus and Minerva, however, are all depicted within the enclosed garden of *Le livre des échecs amoureux moralisées*, we can understand this as suggesting that, rather than being mutually exclusive, carnality and perfectibility are, in the trope of the garden of love, reciprocally

FIGURE 5.4 Robinet Testard. Illustration from *Le livre des échecs armoureux moralisées*, Anon. (1496–1498). Bilbliothèque Nationale, Paris. Ms Fr.143, fol. 198v.

necessary. "The real paradox," says troubadour scholar Cherchi about the ambiguous nature of courtly love, "is that eros and moral perfection, far from being contradictory, live together in a subtle process of amalgamation" (Cherchi 1994: xiii). This process does not indicate a delicate balance between sensuality and the holy so much as what Cherchi calls an "integration of eros into the ideal of courtliness," a transformation "of nature into culture to such an extent that the origin of love . . . remains ambiguous" (Cherchi 1994: xiv).

In the *chansons de geste* of southern France (from the second half of the eleventh century), nature is used to reflect both psychological moods and social assumptions. The open, erotic landscapes of Theocritus and Virgil (the *locus amoenus*) are "spiritualized" into the terrestrial paradise of Genesis and the Song of Songs. A potent blend of the spiritual and the erotic emerges in Latin and vernacular poetry that explores themes of love in paradisal settings full of birds, fruits, flowers – symbols of mutability and transitoriness, of emergence and multiplicity. It is this fusion of sensual enjoyment and natural processes on which an anonymous troubadour lingers in his song *Les Adieux de l'Amante,* where the influence of the Song of Songs is clear:

In an orchard, under the hawthorne tree, the lady has kept her lover in her
arms until the lookout hails.
Lord! It is dawn already! It comes so fast!
Sweet, beautiful friend, caress me again, in this garden where birds sing until
the watchman sounds his horn.
Lord! It is dawn already! How quickly it comes!
My beautiful lover, courteous and sweet, is like the breeze that drifts through
the trees; of his breath I have drunk a long sweet draught.
(Gousset 2002: 87, trans. author)

These imaginary heterotopias provide a locus not for the renunciation of emergence and difference, but for a complete acceptance of the challenge they present: difference exists not to be overcome but to be understood. The tension between spiritual and sensual passion is imaginatively examined by artists, not condemned. The theme of the garden of love in the Tuscan art of the early Renaissance could not have appeared, as it does, in a wide range of works of art commissioned for marriages, if it depicted a life or culture of profanity. Nor would it have become a recurring theme in Italian vernacular poetry, especially the amatory verse of Dante, Petrarch and Boccaccio, in which allegorical gardens flourish, if the realms of love described in these works were to be despised as sublunar webs of deceit. Dante's description of a garden pond connects earthly emergence and spiritual love in a surprising way:

As in a fishpond still and clear,
The fishes draw near to anything that falls from without
In such a way as to make them think it something to eat,
So I saw more than a thousand splendours draw towards us,
And in each was heard: Lo!
Here is one that shall increase our loves.
(Dante 1814: Canto V, ll. 97–102, trans. Eliot)

Roman de la Rose

In Dante's *Divine Comedy*, set in 1300, God's presence is reflected in Dante's love for Beatrice. The image, above, of the fish pond reflects the teeming multiplicity of the sublunar world, and frames Dante's rapture as the goal of human beings caught in this multiplicity. The human drama of love connects the heavenly and earthly realms. The *Roman de la Rose*, composed between 1230 and 1300, also explores this connection, tracing the passage of a lover through an imaginary garden towards the object of his love, symbolized as a rosebud. Some readings of the *Roman de la Rose* interpret its story as concluding that the garden of love is dangerous and irrational, and informed only by the debased currency of courtly love (e.g. Gunn 1952; Seward 1960; Fleming 1969). They suggest that the *Roman* is saying that the divine can be approached only through the practice of Christian love, not carnal love that leads humans away from the One. But the movement of the *Roman* can be read as supporting the idea of the garden of love as a locus for the embrace of multiplicity energized by human love. The first part of the story establishes that it is within the relentless cycle of becoming and fading away that the lover achieves his goal, and that this goal is an equation of the rosebud with the sacred. The second and final part provides the corollary of this equation; that the sacred is to be found in the continual becoming of the world. A typical reading of the *Roman de la Rose* is that the garden of Déduit (its keeper) is the garden of Venus, where all who seek the delights and vanities of the world amuse and enjoy themselves in folly and sin. Amant, the lover and dreamer who enters this charged landscape, mistakes the rose garden for the Garden of Eden and is led into error.

> For wel wende I ful sikerly
> Have been in parays erthly.
> So fair it was that, trusteth wel,
> It seemed a place espiritual.
>
> (Chaucer 1987, "The Romance of the Rose": ll. 576–570)

And so, the *Roman de la Rose*, in this interpretation, "is the chronicle of a hero who is neither wise nor admirable, a young man who, overcome with carnal infatuation . . . rejects Reason and embraces false courtesy, hypocrisy, and wicked counsel to achieve the sordid 'heroism' of a seduction" (Fleming 1969: 50). Accordingly, to follow the door-keeper of the garden, Oiseuse, is to abandon Reason. In denying Reason, Amant (man) denatures himself, ceases to be human. Man should attend to the divine rather than the flesh. However, it is possible to find in the story of Amant's journey through the garden of the rose a different *significatio*: that it is within the world that Amant can find the redemption he so ardently seeks (see Figure 5.5). This is a new kind of garden that has developed according to the interaction of a number of metaphysical trajectories that intersect in the thirteenth century. St. Francis of Assisi (1181–1226) had praised nature. For him God's love was suffused throughout creation, including the natural world, where birds and animals, the sun, the moon and all four elements themselves were products of the loving energy of God. A more refined sensitivity to nature was developing.[6] Pagan mythology also features significantly in the garden of Déduit and not just as a workaday reference but as a living force. Additionally, the Christian tradition, powerful in the court as much as in the church, provides the context and representational order for the apotheosis of the lover in the garden: salvation was possible only in and through Christianity. When Amant enters this garden through the wicker gate opened for him by Oiseuse he finds himself in a dangerous terrain. In Fleming's interpretation, it is a postlapsarian paradise full of "earthly delights" that turn men away from truth and righteousness. Fleming argues that it is because Eden is so dangerous that God did not repopulate paradise after the fall (Fleming 1969: 60), and this is an undeniable feature of Judaeo-Christian theology. The dangerous qualities of the garden are exemplified in the biblical Song of Songs, an anthology of erotic metaphors:

> I am come into my garden, my sister, my spouse: I have gathered my myrrh with my spice; I have eaten my honeycomb with my honey; I have drunk my wine with my milk: eat, O friends; drink, yea, drink abundantly, O beloved . . .
>
> My beloved is gone down into his garden, to the beds of spices, to feed in the gardens, and to gather lilies.
>
> I am my beloved's and my beloved is mine: he feedeth among the lilies.
>
> (Song of Solomon 5:1, 6:2–3)

FIGURE 5.5 *Le Verger de Déduit*. From the *Roman de la Rose* by Guillaume de Lorris and Jean de Meun (c.1460). Painting on parchment. H.:39,4;1.:29,2; 190 fos, London. Courtesy of The British Library, Ms Harley 4425, fol.19v.

With its sexual ambivalence, the identification of garden and woman, the senses of secrecy, transgression and sequestration, we find in this imaginative correlation of the garden with love the forerunner of the *paradys d'amours* of the French court poets. There would seem, as well, to be little doubt both of the connections and the ambivalent contrast between the two biblical gardens. We find this carried through to the *Roman* where Guillaume de Lorris makes of the garden of Déduit a place so wickedly rapturous that Amant feels that it must transcend even the beauty and promise of heaven:

> Yet I doubt I much if heaven can give
> A place where I so soon would live
> (de Lorris and de Meun 1983: ll. 653–654)

The idea of the garden as a landscape of love, as a site for the exploration of human temptation, and therefore of the human condition of "passion versus reason," inaugurated in the story of the Garden of Eden, and eroticized in the Song of Solomon is, in the rose garden of the *Roman,* brought to its fullest expression. But the many wonderful and beautiful things contained in the transgressive garden of the rose are transitory and mutable. Flowers wither, day turns to night, birds and insects live their brief lives, and all things are accompanied by labor and pain intermingled with sorrow, just as in Eden after the fall. It is within just this relentless cycle of increase and decline that de Lorris situates the allegorical locus of the tale. Amant dives, as it were, into the world of visible things, *visibilia,* where what is seen is supposedly separate from what really is. While the Ovidian tradition accepts this world and its transformations, Christianity is customarily regarded as excoriating it. The garden of Déduit is dangerous precisely because it is the site of real redemption through love. Ambivalence and double-coding are present in all the features of the garden, to be sure, this multiplicity is not in itself evil. Flesh and spirit, beauty and danger apply to different things in different respects, in a continual dialectical movement that is not characterized by a progressive or teleological quality, but a continual inter-animation. It is the unstable relationship between the four elements enumerated by Plato in the *Timaeus* – the perpetual and shifting struggle between them – that makes the sublunary world a world of generation and regeneration.

Landscape architecture has not always heeded the message of the garden of love as portrayed in the *Roman de la Rose* and the other medieval productions explored in this section. This message is that Nature, the world of the garden, is to be honored not by transcending the condition of becoming, but by an immersion in the *conjunctionis oppositorum*, a commitment to the singularity of each individual thing, moment, perception or conception; that is, to the movement of difference in the world. The rose garden is the site of the fulfillment of human becoming, by means of an engagement – through God's love – in the very things of the world that are animated by that love. The garden of the spirit is thus achieved in this life, in this world, through the natural functioning of very human faculties. When Amant looks into the fountain in Déduit's garden, the mirror image of love is disturbed by surface ripples that suggest its multiple, refracted nature. What the lover finds at the center of the garden of desire, then, is not an impossible beloved object, but the ambivalent subject himself, in whom the object of desire is present; not identical with him – internal, but different. The ambivalent and transgressive garden of love is a reification of the conjunction of opposites. It is defined by authority *and* equivocation: carnality *and* moral perfection; desire *and* renunciation; acceptance *and* rejection, the opposites moving together in a subtle dance. This is a terrain, then, where the operations of emergence can be played out through the active role of cognition, vision and passion in the acceptance of the paradoxical nature of love itself. The lover, for whom the garden lives, is a kind of holy sinner, a joyful consumer of the delights of human desire who accepts and rejoices in the antinomial condition of becoming human in the shadow of God (see Figure 5.6).

This reading of the medieval garden of love places garden-making, the first and founding practice of landscape architecture, at the forefront of these rediscoveries. In the *jardin d'amour* the fluid immanence of love spreads throughout the garden as the world, uniting humans and nature in an ambivalent relationship that it is part of human destiny both to suffer and enjoy. The garden is a space connected to a poetic understanding of the world and the place and destiny of humans within it. Accordingly, in gardens humans discover that they are a part of nature, and at the same time separate from it, and that they are therefore always in *conjunctionis oppositorum.*

FIGURE 5.6 From Boccaccio's *Decameron*, c.1475. Painting on parchment. Courtesy of Bibliothèque Nationale de France. Ms. Fr.129, fol.1.

Sacred groves: sacrifice and the order of nature in the landscapes of ancient Greece

This section examines the sacred groves of ancient Greece and Turkey (see Figure 5.7). It argues that these treed precincts were designed for the specific purpose of making a link between the sacred realm of the gods and the profane world of human lives. The sacred grove is a sanctuary dedicated to a god. Within its carefully delineated spatial arrangement the violent and transgressive ritual of animal sacrifice takes place. The section uses French writer Georges Bataille's reading of the gift to explain how, by means of a radical declension from nature, the sacred grove signals man's control of the world. Essentially, the open system of the agrarian landscape is disturbed by an act of sacrifice that rearranges its productive elements to create a link to a sacred condition. The sacrifice, or gift to the gods, is a transgression that occasions a cascade of downward causation and reorders the structures of exchange, hierarchy and rationality that generated it in the first place. In Homer's Greece the wilderness is turned into productive land by labor. Through sacrifice, the products of this labor are turned into a gift. This gift exemplifies what Bataille calls the "endless and profitless consumption" of the divine order (1988: 59). It re-establishes the link with untrammeled nature that has been lost through the work that animal and plant husbandry requires. Sacred groves therefore mediate between two worlds by focusing and aligning the significance of special places through care

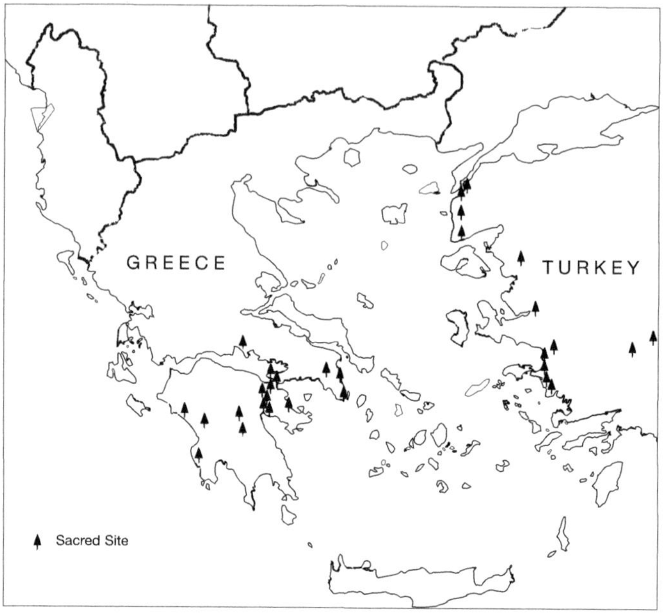

FIGURE 5.7 The map shows sites visited by the author 2004–2005, in the Peloponnese and along the east coast of what is now Turkey, but was once a string of Greek city-states such as Ephesus, Miletus, Didyma and Priene. Courtesy of author.

and propitiation. Trees are planted as an operation that differentiates the site from the natural order. Sacrifice reconnects with that order. By permitting the passage of the sacred, the grove functions as a portal between the sacred and the profane.

The landscapes of ancient Greece

The rural landscape in which the ancient Greeks lived their daily lives was primarily a cultivated terrain forged by three types of agrarian production: pastoral (grazed territories), agricultural (cropped fields) and horticultural (orchards and productive gardens). The basic geographical unit of Greek life was the *deme*, a territory which focused on the *polis* but included both the town and the productive countryside that supported it with food. In ancient Greece the countryside and the town were not separate worlds. The political center itself was permeated by the country at every level. Robin Osbourne observes that "[t]he countryside absorbs the time and energies of the majority of [the town's] inhabitants, directs its politics, and drives its calendar of activities" (Osbourne 1987: 21). Importantly, the periodicity of the religious year is also driven by the rhythms of the cultural landscape. In the countryside, vital farming activities occurred at particular times. The annual pattern of sowing, cultivation and harvest provided a framework for the understanding of long durations and uncertain periodicities. The life of the farm afforded a model of human change, a framework for understanding the transformational qualities of human life. It contextualized these transformations by contrasting the daily labor of farming families with the freedom and caprice of the gods. The network of Greek deities – the Olympian pantheon – provided a logic which had a profound connection to agrarian production: the farms with their terraces gathered from the wild, which order and domesticate the living landscape by means of agricultural diagrams of stone; the plantations of pines, cypresses and olives that surround shrines and temples and provide sanctuary for devotees of the gods; the fields of grain which stock the ships; the productive urban gardens whence politicians and nobles obtain herbs and summer vegetables for the table; the terrain that pastoralists cross and re-cross, moving back and forth along the ancient routes of transhumance in daily, and seasonal, ebbs and flows. It is through the various activities associated with agriculture, horticulture and pastoralism and, importantly, the festivals that celebrate the georgic cycle of the year, that the connection is made from the profane world of laboring humanity to the sacred world of the gods who are eternal (but can transform at will).[7] The timeframe with which this section is concerned begins, roughly, with the Homeric texts (the *Odyssey*, the *Iliad* and the *Hymns* attributed to Homer, 800 BCE) and ends with the Peloponnesian War (431 BCE) and the rise of philosophy in the classical period. The three hundred years from 800 to 500 BCE covers the movement from the geometric to the archaic period, an era which saw the development of the *polis*, the flourishing of mythic narrative and the advent of democracy. The reason I focus on the geometric/archaic period is that it is during this time the practice of ritual sacrifice flourishes in the cultures and societies of ancient Greece. Sacred groves are often the religiously charged settings for this practice.

In the geometric/archaic era unfarmed wilderness prevailed outside the domestic terrain of the cultivated landscape. Wilderness was where hunting occurred, but not much else. It had been the ancient home of the people, but was now the contemporary home of wild beasts. It was also the place of certain gods, such as Apollo, Artemis, Dionysus and Pan. Unstable, therefore, and uncontrolled. For those who lived in towns – and even the farming people – the wilderness of Dionysus, like that of Pan, son of Hermes, lay at the limits of the world (Detienne 1979). Pan, worshipped as a master-of-the-animals figure, controlled wild beasts and protected domestic creatures. The cult of Pan, associated with Arkadia, focused not on temples but on cave sites well off the beaten track.[8] The wilderness Pan inhabited was a place of marginal experiences and transgression in the form of drinking and dancing in the presence of the gods, a savage and violent possession. Artemis the huntress was also associated with wilderness, a goddess who "often dwelled in the mountains" (Cole 2000: 475). She had shrines on the margins of Attica, way out where the wild lands began on the edge of the territorial terrain. At her sanctuary at Brauronia the focus was on girls entering puberty, passing through their wild years, preparing for womanhood. Apollo, likewise, was distant, dwelling "far away" in a wild landscape. Apollo's main sanctuaries were at Delphi and Bassae. The temple of Apollo Epikouros at Bassae is particularly remote – Pausanius describes nearby Phygalia as "mostly precipitous and high in the air"(Pausanius 1971: Bk VIII, 39).[9] Though it stands only a few kilometers from this ancient Arkadian city, Bassae is located on an isolated natural plateau near the top of Mt. Kotilion, at a height of 1,130 meters. Constructed between 420 and 400 BCE, it was sited close to an even more ancient temple of Artemis Kotilion, erected at the end of the seventh century BCE. A bronze tablet found at the site mentions the deities worshipped on Mt. Kotilion as (among others) Apollo, Pan and Artemis. Remote sanctuaries of Apollo are also found on Mt. Parnassus and Mt. Lykaion (Pausanius 1971). In geometric and archaic Greece, then, while towns and agrarian landscapes comprised a symbolic and social continuum, a boundary was drawn between the wild and uncontrolled landscapes of god and beast, and domestic landscapes subject to the regulating techniques of productive pastoralism, cropping, orcharding and gardening. French classicist Francois de

FIGURE 5.8 A grove of trees outside the ruins of Miletus on the coast of Anatolia, near the mouth of the Meander river. Photo by author.

Polignac argues persuasively that this is a threshold between open and closed, sacred and profane, and that it is marked typically by a sanctuary, or *temenos*, a site of worship and ritual (Polignac 1995). Many such sanctuary sites can still be seen from the ancient towns of Attica, Arkadia and the Peloponnese, as well as along the coast of Asia Minor (see Figure 5.8). Between town and temple in classical Greece, as now, there lies a landscape familiar to us through paintings and literature, a *locus classicus*, a "classical landscape with figures." Below the *polis* there is a plain through which a small river flows. In places it is uncultivated and supports a cover of deciduous and evergreen trees – oaks and olives, poplars, *Melia* – with understoreys of shrubs and grasses. The soil, largely eroded silt, is fertile and mostly put to agricultural, horticultural or pastoral purposes. There are swamps in gullies, with tree and shrub cover. The plain drifts into soft hills comprising marl, schist and volcanic lava. The low hills have been terraced in order to accommodate cultivation, until they become too steep and the oaks and olives take over. Above their range the conifer belt begins. Mountains rise out of this fringe; hard limestone which has lost its soils to erosion. There is no grazing here, not even for goats, only low alpine plants (Rackham 1990). Ridges run out from the low hills, and spurs from these on either side of the plain, which is quite enclosed by hills. The sanctuary is located on one of the foothills at the edge of the plain, looking down over it with a mountain range behind. There is a grove of trees there and a temple within them. This sanctuary is visible from many points but it is particularly conspicuous from the town on the plain where the inhabitants can see it every time they look east (Polignac 1995). The town seems to be embedded in the rural landscape that surrounds it. It is girt by a wall through

FIGURE 5.9 The pathway to an ex-urban sanctuary would have looked much like this, the path to the top of the Acropolis in Athens. Photo by author.

which a number of roads and paths thread, including a pathway to the temple on the ridge (see Figure 5.9).

Such a landscape is typical of any one of a number of different *demes* in the geometric/archaic era. The lowlands and plains of the Peloponnese, Boetia and Attica became thoroughly humanized during this period; ploughed, planted, organized. Temples on the edges of territories – ex-urban sanctuaries – mark the outer limit of the advance of agrarian civilization and set it in opposition to the wild domain of the mountains, the forests and the sea. The city of Argos, for instance, had a temple of Hera on the opposite side of the plain of Argos. In Arkadia, Mantinea's ex-urban sanctuary was the temple of Poseidon in the foothills to the north of the city. Korinth had a Heraion across the bay on the tip of the peninsula of Mt. Gerania that juts into the Bay of Korinth.[10] Polignac's theory that most cities were bipolar (having a temple in town and a major *temenos* on the margins of its territory) seems to have gained acceptance among archaeologists and classical scholars.[11] Certainly a network of paths stretched between sanctuaries and shrines across the archaic landscape of Greece. Deep wide paths linked city and shrine, where festivals passed back and forth on ritual days. The *deme* consisted of the urban center, the landscape that supported it, and the ex-urban shrine that marked its edge. The cult site established a community's solidarity by indicating the boundary of the relevant space. According to Polignac, social space was structured around the two poles, the inhabited area and the sanctuary, with the ceremonial pathway linking the two. Great processions took place on this axis at regular intervals, from the profane to the sacred and back again.

What are sacred groves?

Archeologist Darice Birge's definition of a sacred grove is simple and clear: it is "a stand of trees in a religious context, with or without associated structures such as altars or temples" (Birge 1982: 1). Such groves "are differentiated from their surrounding territories by visible boundaries and/or special regulations." The landscapes that are commonly referred to as sacred groves – *alsos* in Greek – had been part of the wider cultural landscape of Greece for a long time prior to the geometric/archaic period. On several occasions in the *Odyssey* and the *Iliad* Homer refers to such groves as well known and loved cultural landscapes; and the archeobotanical record shows that groves were planted in sacred precincts in Egypt in the previous millennium (Carroll 2003), whence the practice probably made its way to what is now known as mainland Greece. There are two main types of wooded area associated with terrain devoted to ritual, worship and sacrifice, that is to say, areas associated with god cults such as those of Zeus, Apollo and Artemis. The first kind is the *naturally* wooded tract of land which is understood as propitious or holy, and which attracts cult activities, perhaps an altar, and finally a temple. The second is the grove that is deliberately planted to accompany a temple or altar, and which defines and embodies a sacred precinct or *temenos*. This is a designed landscape.

While the archeological evidence for designed sacred groves is still relatively meager, particularly when compared to the evidence for temples, the disciplines of landscape archeology and archeobotany have over the last twenty years developed considerably. Researchers such as Osbourne, Rackham and Carroll-Spillecke have described and analyzed physical aspects of ancient Greek landscapes that have been uncovered by archeological projects. A picture of these landscapes can be built up from their work, and matched against literary references. Excavated groves include the cypresses at the temple of Zeus at Nemea; the sanctuary of Hephaestus above the agora in Athens where tree pits have been found arranged in lines either side of the temple; the gymnasia of Plato, Aristotle and Epicurus in Athens, some of which have been replanted; and the temple of Apollo Hylates in Cyprus, where excavated channels and pits show clearly where trees were planted. The archeological record demonstrates that trees were in many cases planted in a regular, ordered fashion. This accords with evidence from temple sites in Egypt, a number of which have been uncovered to reveal tree pits arranged in grids and lines. If the archeological evidence for sacred groves is currently scarce, classical literature abounds with references. There are several well-known allusions to sacred groves in Homer, for instance. In the *Iliad* (Bk IV) a hero plants a grove of trees to commemorate a fallen comrade. Book VIII refers to a site in Gargarus where a "fragrant" altar is accompanied by a grove of trees. In Book XVII a fountain is mentioned which has "an altar to the nymphs" and "a grove of water-loving poplars planted in a circle all round it."[12] Homer also mentions cases of sacred groves associated with specific gods. In the *Iliad* he refers to "Holy Onchestis with its famous groves of Neptune" (Bk II), and in the *Odyssey* "[a] sacred grove of Athena" where Odysseus sat and prayed to "the mighty daughter of Zeus" (Bk IV). A famous reference in the *Odyssey* is to the grove outside the city of Phaecia, which Odysseus passes on his way to meet Alcinous. "You'll find black poplars around a meadow and a fountain," Nausicaa tells him when she gives him directions to her father's city, "all dedicated to Athena" (Bk VI).

One of the most completely described holy sites in classical literature is that evoked by Euripides in his play *Hippolytus*. Hippolytus says to Artemis through the medium of her statue in the grove,

I have brought you this green crown,
goddess, fresh from the scene
where I spliced its flowers together,
a meadow as virginal as you are,
where no shepherd would think it wise
to pasture his animals, a perfect field
no iron blade has yet cut down.
Only the bees looking for flowers in spring
go freely through its cool grass.
Its water flows from the goddess
Restraint, who not only
leads in the rivers herself
but keeps the place a special preserve
for those whom modesty enters at birth,
the instinctively good –
these may pick what they will,
but the vulgar are barred from the meadow.
Now, blest lady, take this, embellish
Your gold hair – it comes from a faithful hand.
(Euripides 1994: ll. 112–130)

In this famous passage Euripides describes a place sacred to Artemis, "an inviolate and inviolable virgin," and "[a] feeling for virgin nature with meadows, groves and mountains, which is as yet barely articulated elsewhere, begins to find form" (Burkert 1985: 150). Euripides does not use the Greek words *temenos* or *alsos* to refer to this place however, although translators have recourse to the resonances set in motion by these words. Instead the term *leimotos* (meadow) is employed. The above translation, by Bagg, describes this meadow as "a perfect field," and "a special preserve," indicating its declination from the quotidian.[13] This is a place that iron, denoting the destructive touch of humankind, and a fallen state (the age of iron) has never sullied. Sacred plants, if taken at all, were cut with a bronze sickle (Hamilton 1982). Euripides also uses the word *kepeyei*, to "tend like a garden" (Hamilton 1982), a reference that Warner makes explicit in his translation:

> And native Shame waters the ground with river dew, and from his garden only those may pick the flowers who were elect from birth by a wise purity in all things, and never had to learn it.
>
> (Euripides 1958: 80)

The image of Aidos (variously translated as Restraint, Shame or Modesty, although there seems to be no real English equivalent) as a gardener complexifies the analogy between the idea of a "virgin meadow" and the virgin goddess, since the notion of virginity alludes to its innocence of any interference at all, even that of a reverent and chaste intermediary. Euripides' intention must be to establish the sanctity of the place through its declension from the natural order by means of gardening, a curious blending of *kepos* and *alsos*. Greve's translation suggests this reading. He uses the phrase "inviolate meadow" and then countermands this appellation with the figure of the gardener:

only the bees in springtime haunt the inviolate meadow.
Its gardener is the spirit Reverence who
Refreshes it with water from the river.

(Euripides 1960: 50)

There seems little doubt that Euripides intends this special place to be understood as a sacred grove (Ferguson, indeed, calls it just that) tended by Aidos as a garden. It is untouched not by all mortals but by those whose purity is unlearned, genetic. By way of the exquisite motif of the sacred meadow Euripides brings together *alsos* and *kepos* to suggest *temenos*, even though he does not use this word, in an intensification of the otherness that the notion of the sacred requires. It is at perhaps just this kind of holy ground that an altar might be placed, and eventually a temple dedicated. But there are many other sources. Pindar, for instance, describes Herakles as the first to plant a sacred *temenos* with trees at Olympia. The groves of the fourth and fifth century BCE *gymnasia* outside Athens – Academy, Lykeion, Kynosarges – are described in a number of classical texts (Herakleides, Aristophanes, Plato, Pliny, Strabo, Diogenes Laertes, for instance).[14] These groves are "shady, well-watered spots immediately adjacent to ancient cult places." Throughout the literary record we find references to shrines, irrigation systems and productive gardens. There is no doubt that trees were deliberately planted – in fact groves were often developed by statesmen, such as Lykourgos who financed the planting of the Lykeion gymnasium, and Cimon, who installed an irrigation system and planted trees at the Academy. Among other things these benefactors were funding the design and implementation of sacred groves (Carroll-Spillecke 1992). Perhaps the most useful descriptions of sacred groves come from the first century Roman writer Pausanius, who traveled through Greece and wrote extensively about the landscapes he encountered. His references to numerous sacred groves (still extant in 1 CE) confirm their association with specific deities such as Apollo, Artemis and Pan, and with springs and caves. His descriptions sometimes explain why certain sites were chosen, or became sacred. For instance, Pausanius describes a prophecy that Artemis would show a homeless people where to settle. When they make landfall a hare appears. They follow it and it disappears under a myrtle tree, which is where they decide to build their city. "They still worship that very tree, and give Artemis the title of Saviour" (Pausanius 1971). Artemis is also associated with water (she is often known as Artemis of the Lake) and her temples are often by springs – as are those of many other gods. Southwest of Phygalia the sanctuary of Demeter Melaina consists of an isolated mountain cave: "a sacred grove of oak trees surrounds the cave and cold water springs from the soil" (Pausanius 1971: Bk III, 22, 11–12). Quite apart from direct references to terrain planted with trees for sacred purposes, classical literature features many allusions to a wide range of tree species and other plants, and the Greek myths are woven throughout with arboreal references. There are specific places, such as the Garden of the Hesperides, where the goddess Hera planted the tree of immortality, laden with golden apples, and artifacts such as in the tradition of the gardens of Adonis, terracotta plant pots filled with annual species of flowers and herbs. There are also numerous pictorial representations of plants and trees from the classical era, particularly on funeral stelae and on black and red figure vases. From the archeological, literary and material evidence, then, it appears indubitable that there are sacred groves in Greece and the Greek colonies in the period 800 BCE to 200 CE.

The sacred

Classics scholar Walter Burkert analyses the Indo-European word for sacred, *hagnos*, as opposite to *miasma,* defilement, which signifies a disruption or dislocation of normal life. Typically, "disturbances of this kind are sexual intercourse, birth, death and especially murder" (Burkert 1985: 78). Like Burkert, mythographer Mircea Eliade contrasts the sacred to the more generalized profane. He argues that "the first possible definition of the sacred is that it is the opposite of the profane" (Eliade 1959: 10). Anthropologist Emile Durkheim also puts this difference between the sacred and the profane at the center of his analysis of religious life. "In all the history of human thought there exists no other example of two categories of things so profoundly differentiated or so radically opposed to one another" (Durkheim 1915: 53). Differentiation, it would seem, is at the heart of the relation between sacred and profane for, as Durkheim says, "the sacred thing is par excellence that which the profane should not touch, and cannot touch with impunity" (Durkheim 1915: 53). But, since the sacred flows into everything that does touch it, it must be kept within its own bounds. It is only any good if it is kept at a distance, just because it is a source of instability. Philosopher René Girard, writing later than Durkheim, Burkert and Eliade, pushes the notion of the sacred as a principle of differentiation much further. He sees the sacred as both the source and deliverer of instability, disruption and disorder – hagnos *and* miasma. For him it is the fundamental principle of differentiation in human society (Girard 1977). The radical, absolute difference between the sacred and the profane is a result of the violent expulsion of difference through the victim of sacrifice. If the difference between the sacred and the profane is absolute and irreducible, it follows that, apart from that differentiation, there is only unnameable, unspeakable undifferentiation. The victim is perceived as the embodiment of the *undifferentiated,* a substitute for human society, offered up by that society itself in order to protect the community from its own violence. In every act of sacral killing, sacrament and sacrilege merge (Burkert 1983). The sacred, according to Girard's account, is a purity, but a purity which generates *miasma*, and is free of it even though it creates it.

Hagnos is also not completely to be distinguished from *physis*, nor wholly identified with it. The *mysterium tremendum* or *mysterium fascinans* that classicist Walter Otto (1978: 1) calls "the feeling of terror" before the sacred, is a much more numinous experience than that of nature, for the sacred is *ganz andere*, "wholly other." "It is like nothing human or cosmic" (Eliade 1959: 10). It always manifests itself, Eliade goes on to say, "as a reality of a wholly different order from 'natural' realities." And yet, Burkert observes, since time immemorial much that we call natural has been known as divine (Burkert 1985). For instance Sun and Moon are major gods in the pantheon of the ancient Near East, and the Greeks had river gods and nymphs of spring and forest; there was Oceanus, and Helios and Gaia. These gods, however, represent a modality of the sacred that, in reinforcing the profound separation of mortals from immortals, places human being strictly within its ambivalent condition of being part of nature and yet separate from it. Humans share with plants and animals their "closedness from the open" (to use Agamben's words) and are differentiated from these species by their ability to suspend their animality and open a "free and empty zone" which is designated as sacred, an ability predicated on their separation from both nature and the sacred (Agamben 2003). Separation and difference would seem, then, to be the grand themes of the sacred.

Sacrifice

"Where there was no agriculture," (that is, in the wilderness) "there could be no sacrifice. Sacrifice thus signals man's control of the wild world, and marks the separation of man from beast" (Burkert 1985: 2). The order of social life, Burkert writes, is constituted through sacrificial killing which, as the basic experience of the sacred, enables *Homo religiosus* to act, and attain awareness as *Homo necans*. The distinction between wild and domestic is essential to civilized life. When civilized norms are transgressed as with Orpheus – who blurs the distinction between wild and domestic by charming wild animals, rocks and trees to follow his music, preaching that the sacrifice of animals is wrong (he was the inventor of vegetarianism) – death must follow. Orpheus is torn to pieces by maenads "reveling in the Bacchanalian rout" (Virgil 1907: IV, l. 626). It is sacrifice, the most important form of action in Greek religion, that explains the sacred groves of ancient Greece. In this vital ritual a domesticated animal is led to the altar, water and grain are sprinkled on its head to force it to "nod" assent, and then it is killed. Parts of the animal – the thighbones, usually – are burnt as the gods' portion, and the rest of the meat is distributed among those present. Connection to the gods is primarily by way of this act. If an individual is able to draw near to the gods, says Burkert, "he can do so because he has 'burnt many thigh-pieces of bulls,' for this is the act of piety: bloodshed, slaughter – and eating" (Burkert 1983: 2). It is in the deadly blow of the axe, the gush of blood and the burning of thigh-bones that the worshipper experiences the god most powerfully. "The realm of the gods is sacred, but the 'sacred' act done at the 'sacred' place by the 'consecrating' actor consists of slaughtering animals." Sacrificial killing, then, is the primary experience of the sacred in Greek life.

> In the experience of killing one perceives the sacredness of life; it is nourished and perpetuated by death. . . . Whatever is to endure and be effective must pass through a sacrifice which opens and reseals the abyss of annihilation. . . . Sacrifice transforms us. By going through the irreversible "act" we reach a new plane.
>
> (Burkert 1983: 40)

Husbandry could wrest order from chaos, but it could not mitigate this pure event. By labor humans could provide the material for sacrifice, but they could never do away with the requirement for sacrifice. When the women "raise their wavering cry" as the axe falls on Nestor's heifer they are marking dramatically the action that is at the heart of agricultural activity, on which rests the lives of human beings. The carcass sinks down:

> and they disjointed shoulder and thigh-bone, wrapping them in fat,
> Two layers, folded, with strips of flesh.
> These offerings Nestor burned on the split-wood fire
> And moistened with red wine. His sons took up
> Five-tined forks in their hands, while the altar flame
> Ate through the bones, and bits of tripe went round.
>
> (Homer 1965: ll. 362–434)

Athene is propitiated and the meal begins. It is evident that the whole cultural landscape of ancient Greece lies behind this dynamic of expenditure and consumption. In order to provide the sacrificial animal Nestor must farm. For cities such as Pylos to engage in sacrifice there must be agriculture (but not, significantly, horticulture, which plays little part in this complex of myths). Which is why Homer's stories of Odysseus in the lands of Circe and Calypso rigorously exclude anything to do with working the land, or with arable land insofar as it is worked. For it is labor that separates man from the immortals. When Odysseus leaves the realm of the fantastic to return to normality and accept the human condition, he goes to work. The productive farm participates in the order of things. Sacrifice in turn disrupts this order and restores the divine, the real.

Bataille and the gift

The connection between sacrifice and sacred groves can be elucidated by a consideration of George Bataille's conception of the gift, a theory of transgression he formulated on the basis of the role of potlatch in archaic societies. Writing before Prigogine, whose theory of dissipative structures he did not know, Bataille's theory of the gift almost exactly rehearses the role of waste in open systems. Whenever "there is an excess of resources over needs," he explains, "this excess is not always consumed to no purpose" (Bataille 1988: 22). Expenditure, or waste, as exemplified in the potlatch, or gift from one group to another, is actually a way of acquiring power, for giving is "glorious" (Bataille 1988: 69–71). It demonstrates superiority by making waste itself an object of acquisition. The most complete form of gifting, or the squandering of resources, is sacrifice, which "withdraws useful products from profane circulation" and destines them as "the accursed share" for "violent consumption" (Bataille 1988: 76). The sacrificial victim is torn away from the order of things and "restored to the truth of the intimate world." The notion of the gift emblematizes a transgressive, sometimes violent act. Bataille argues that this transgression is necessary in order to introduce into the reified world of things "the illuminations of sacrality" (1988: 55–61). Bataille's formulation of the gift is connected to his notion of individual sovereignty. To be sovereign is to not let oneself be reduced to the condition of an object. In a "reified" world, that is, a world dominated by "things" – the objects of production – sovereignty is the condition of freedom from the bondage of these objects. According to Bataille pure sovereignty is only to be won in moments of excess. How so? The potlatch, or gift, is an example of unproductive consumption ("excess") where the "sovereign" disdain of the chief or ruler for use-values is expressed in the outpouring of abundance towards his enemy or rival in order to challenge or humiliate them. Of course this is not a selfless act, it is driven by a calculated acquisitiveness for power. Sovereignty expresses itself in acts of waste precisely because it accrues power.[15] Bataille's work on the gift supports the proposition that sacred groves mark a boundary between the acculturated and the wild, between the profane, that is, and a realm variously described – by, for instance, Agamben (2003), Vidler (1992) and Eliade (1959) respectively – as open, uncanny or sacred. Their function as portals to this turbulent and unpredictable realm is established by sacrifice, a practice which, centered on a stone altar and requiring a ritual procession that breaks into violence during the ritual act, occasions the passage of the sacred into the everyday life of the people. As the site of violent sacrifice, the sacred grove makes a link to the profound chaos of nature. How is this connection between order and chaos to be understood? In geometric and archaic Greece, wilderness is turned into productive land by immense labor. Through sacrifice the products of this labor – grain, fruits, animals – are turned into a gift. This gift exemplifies the "endless and profitless consumption" of the divine order.

It re-establishes the intimacy with untrammeled nature that has been lost through the work, utility and "thinghood" that animal and plant husbandry requires. A double declension occurs in the grove, whereby the ordered system of the human-centered agrarian landscape is disrupted by a rearrangement of its essential, productive elements, just as the orchards, fields and pastures of the Greek rural economy are, in their turn, themselves disruptions of the order of nature. This declension – the differentiation between the systems of grove on the one hand, and farm and wilderness on the other – is often marked by boundary stones and herms. Within the grove an intimacy with the divine is vouchsafed by its separation from the orders both of the agrarian and the natural worlds. Bataille argues that the intimacy re-established by the gift is, "never truly cleared of external elements, without which it could not be justified" (Bataille 1988: 72). Just so, the terrain within the ancient sacred grove is subject to a set of operations that at once diverts it radically from the conditions that exist outside and yet at the same time includes them.

What, then, is the reason for this declension from the natural order? And how does nature continue to participate in the organization of an acculturated terrain that seeks to exclude its organizational codes? If the grove is a form through which human consciousness sustains and transmits its relation to a separate and extraordinary domain designated as sacred, then this occurs, it is argued here, by means of the transgression of ordered systems. For the ancient Greeks, the open and indivisible realm of the sacred, which cannot be grasped, stated, known or held, a time which is immanent rather than teleological, a domain which cannot be comprehended, is knowable only in flashes through the overcoming of limit conditions. The grove designates the terrain where such violent practices of transgression and excess may take place. The sacred and everyday life are two co-existing, mutually necessary economies whose boundaries can be crossed only by a force that can rupture closed systems.[16] Transgression renegotiates the boundary conditions of everyday life, and interrogates the limits imposed on life by closed structures of exchange, legislation, hierarchy, rationality. According to Bataille, human nature is subordinated to prohibitions that remove the violence of this nature from everyday life. The binding power of these limit conditions (or social norms) comes from the authority of something that is sacred. The ancient Greeks approached this sacred with the ambivalence of repulsion and attraction – the Dionysian terror and ecstasy. Since the transgression of social norms was enticing, because forbidden, the validity of these prohibitions was actually founded on their transgression. For the Greeks, then, there is a profound intimacy between law and the transgression of law. The whole rational world is established within the frame of prohibitions. Importantly the prohibitions are not themselves laws of reason. Instead they open the door to the sacred.[17]

The grove as a site of transgression

As noted earlier, the countryside and the town in ancient Greece are not separate worlds, but there is a line drawn between the cultural landscape and wilderness, just as there is a – different – line drawn between the profane and the sacred. The sacred grove, outside the *polis*, but still part of it, establishes these thresholds. If, following Bataille, we accept that in order to participate in the multiple realms of sacrality, wilderness, culture and society, the grove must itself be subject to cultural operations, then we might reasonably expect it to be specially planted, and in specific ways with specific species. While it is true to say that many holy sites were holy prior to human intervention, they could not become portals to the open until a reordering of nature took place with the alignment of species according to religious, cult or ritual requirements. There was therefore a transition from unplanted (natural) groups of trees (which may have signified *hagnos*) to designed and managed groves which were the sanctuaries of specific gods. *Alsos* is not a synonym, however, for *temenos,* which could refer to wooded and unwooded sanctuaries (though temples are often associated with sacred groves). Nor is it a synonym for *kepos*, which almost exclusively refers to productive gardens. It is quite possible that treed precincts marked sites that were already sacred, but it seems also clear that through human intervention specially planted groves focused and distilled that wilderness sacrality. It should be noted, however, that some sacred tracts of wooded land found their focus not in the realignment of trees at all, but in the placement of an architectural element, such as the altar of Zeus high on a hill at Kuçukkuyu, above the Bay of Edremit in what is now Western Turkey. These altars often developed over time into temples.

There is a case for assuming movement from some kind of tree cult-based reverence of arboreal sites to the motivated arrangement of particular species in particular ways, prior to their development into large or small temple complexes.[18] The connection between ancient tree-worship and the development of the idea of sacred groves could lie in such simple associations as that of Zeus with the oak tree (*Quercus robur*), the famed Dodona *temenos* being a site that celebrates this affiliation. Frazer writes that "the oldest and certainly one of the most famous sanctuaries in Greece was . . . Dodona, where Zeus was revered in the oracular oak" (Frazer 1925: 159). The name of Zeus is forever associated with this species of tree, and it could be said that every oak tree in Greece is sacred to Zeus.[19] Dodona, in northwest Greece, was a prehistoric oracle (the oldest, according to Herodotus), and in fact could date to pre-Hellenic times, perhaps as early as the second millennium BCE. Priests and priestesses interpreted the rustling of the oak tree leaves to determine the future. By Homeric times there were still no buildings on the site and priests slept on the ground in the sacred grove that was already legendary. It is not possible to say for sure if this arrangement of trees was specifically planted. Certainly, it was not until the fourth century BCE that a small stone temple was added to the site.[20] By the fifth century the association of temple complexes with sacred groves had become commonplace, the temples being part of a larger holy terrain. The extent of many sacred precincts was marked out by the tree plantings that focused the sacrality of the place. "The tree," says Burkert, "is more important than the stone in marking the sanctuary" (Burkert 1985: 85).

That sanctuaries were established at specific landscape locations already regarded as sacred is now generally recognized. Architectural historian Vincent Scully, for instance, says of the sites occupied by Greek sacred architecture that "the place itself is holy and, before the temple was built upon it, embodied the whole of the deity as a recognised natural force" (Scully 1979). Certain landscapes, he says, "are described in the *Homeric Hymns* and many other places as appropriate to or expressive of various gods" (Scully 1979: 1, 2). The Hymn to Apollo, for example, associates this god with the mountain of Kynthos and, more especially, the island of Delos which most "delights his heart" (Richardson 2003: 35, l. 146). Indeed it is clearly the case that combinations of landscape elements such as mountains, caves and springs are quite characteristic of Greek holy places. Most sanctuaries are associated with springs (and sometimes caves, which became grottoes and shrines) – Pausanius reports various sources of water in or near almost half the shrines associated with sacred groves (though this does not mean that those shrines that were not accompanied by sacred groves did not have water sources nearby) (Pausanius 1971). It is not only the ancient Greeks who associate wilderness sites with sacrality. Michel shows that wilderness places have been regarded as holy in many other European and non-European cultures (Michel 1973). Again, the question is the extent to which Greeks prior to the archaic period (when permanent temples began to be constructed) marked sites already considered holy by deliberate plantings of specific tree species. Burkert says that "often a tract of woodland belongs to the sanctuary, a grove, *alsos*, either constituting the sanctuary itself or lying immediately adjacent" (1985: 86). What was the extent to which sacred groves were created prior to the embodiment of the gods in architectural constructions? What were the criteria by means of which trees were selected for planting, and how were they arranged?

Scully argues that each Greek *temenos* "necessarily" differs from all others because it is in a different place from all others; both the form of the temple and its relationship to the landscape are derived from this difference. It also differs from all others as a consequence of the sacred qualities of the god it housed: "Apollo at Delos, for example, was not exactly Apollo at Delphi, nor Hera at Paestum Hera at Olympia" (Scully 1979: 4). It stands to reason that if sacred groves were planted as holy site markers prior to the permanent construction of temples then they must have been planned and organized in specific ways for specific purposes. If they continued to be planted in association with sanctuaries long after temples had been constructed on the sites, these plantings also would have had specific characteristics. The selection and arrangement of trees at sacred places would have been based on both religious and social considerations as well as horticultural conditions. That multiple species were associated with certain gods is clear from classical literature.[21] For instance not only oak, but cypress (*Cupressus sempervirens*) was associated with Zeus. The sanctuary at Nemea, where the sacred trees have been replanted, was famous for its cypress grove. A number of European visitors to Nemea in the nineteenth century remark on the species, its sacrality and, following Pausanius, the organization of the grove, suggesting that the cypresses that comprise it encircled or at least enclosed the site.[22] However, while archeologists have found plant pits to the east of the temple (Miller 1995), and these pits have been replanted with cypresses, their location does not suggest an enclosing ring of sacred trees as they are clustered between the temple and the *xenon* (although there is a sense of enclosure within the smaller area delineated by the trees) (see Figure 5.10).[23]

There are many other examples of tree–god relationships. Zeus' son Apollo, brother of Artemis, is associated with the laurel and is often depicted in vase paintings with a laurel wreath on his brow. In an important Attic work an image on a white *kylix* found at Delphi shows a laurel-crowned Apollo seated on a stool holding his lyre.[24] One of Athena's attributes is the olive branch – she planted twelve olive trees on the Athenian acropolis after her victory over Poseidon. Dionysus bears a *thyrsis* with a pine cone – his other botanical attributes are the vine and the ivy. A logical consequence of these associations is that sacred groves dedicated to particular deities might at least include, if not comprise solely, the trees that express their qualities. Therefore the sites at which specific gods were principally worshipped such as Apollo at Delphi and Corinth, Athena at Athens and Tegea, and Dionysus on Parnassus, can be expected to have been planted with laurels, olives and pine trees respectively. It is almost impossible to demonstrate this, for three reasons: first, the archeological investigation of sacred groves has barely begun; second, the tree species mentioned grow all over Southern Greece and are today found at each of the sites mentioned above, both as wildings and as designed plantings; and third, eyewitness accounts are unreliable since in even one lifetime tree plantings can change considerably. For instance, when Strabo went to the temple of

FIGURE 5.10 The cypress grove at the Temple of Zeus in Nemea stands amid the fallen columns of the temple. While tree pits have been excavated by archeologists the exact formation of the grove – for instance whether it was an enclosing circle – is uncertain. Photo by author.

Poseidon at Onchestos he expected to find a sacred grove, described by Homer as of wonderful beauty (praised in the *Iliad* and in the Hymn to Hermes). Instead he found a place bare of trees and greenery altogether. About 180 years later, however, Pausanius visited and discovered a grove of trees growing to a considerable height. He concluded that these were the same trees Homer had so lovingly described.[25] Pausanius also notes that most groves grew at shrines dedicated to the Olympian gods, some of which were simple in structure (an altar and a few low-key dedications) and some very complex (with altars, statues, stoa and more than one temple).

The testimony of Pausanius supports Polignac's theory of ex-urban sanctuaries. Pausanius alludes a number of times to the fact that sacred groves were planted on the margins of cities. For instance he mentions a sanctuary of Pan at Lykaion with a grove of trees around it (Pausanius 1971: Bk VIII, 37.10), the sanctuary of Demeter at Phygalia where "(t)here is a sacred grove of oaks around the cave, where cold water springs out of the ground (Bk VIII, 42.11–12)";[26] he describes a sanctuary of the Mistress ("the god the Arkadians worship most – daughter of Poseidon and Demeter") where "(a)bove the Hall grows the Mistress' sacred wood, surrounded with a stone barrier; among the trees inside it there are true olive and wild olive growing from the same root, and this is not due to clever cultivation" (Bk VIII, 37.10). He also visits a sacred grove of Karneian Apollo, with a water-spring in it a little way out of Pharai, and near the site of the ancient city of Trikolonoi, north of Megalopolis, "there is still a sanctuary of Poseidon left on a hill with a square statue, and a

FIGURE 5.11 The oaks and pines that surround the ruins of the ancient city of Priene show how important the groves were for providing seclusion, shelter, enclosure and, especially, a threshold to a divine realm beyond the travails of city life. Photo by author.

sacred wood growing around the sanctuary" (Bk VIII, 34.6). Pausanius' testimony provides some substantiation for the possibility that the *temenos* on the ridge or tucked in the bowl of hills is a place where two worlds meet, where it is possible to cross over from one to the other. A site of doubling, of passage, it opens in two directions at once. The sacred grove, carefully and deliberately planted – let us assume – with particular species – palm, pine, cypress, oak – acts as a gateway between a realm where natural forces are manipulated and controlled, and an infinite and multiple domain where nature simply takes its course: the wild country of the Dionysiac reveller is uncontrolled (see Figure 5.11).

The sacred grove offers intimacy through its association with the gods, primarily the Olympians Zeus, Hera, Apollo, Demeter, Poseidon and Artemis.[27] Planted trees mark it out as a threshold space where ritual enables a passage from the everyday reality of the world to the multiple, wild, state of nature. It is this ritual that distinguishes the sacred grove from the garden. Whereas the pleasure-garden provides a disturbance condition through its declension from the order of things as an artifact that exists through and for itself and not for an object-related purpose, the sacred grove provides a site where "the constant problem of being human without being a thing and of escaping the limits of things without returning to animal slumbers" receives what Bataille calls the "limited solution" of ritual (Bataille 1992: 53). Limited, because the ritual is not a complete return to immanence, but rather a fraught recognition of the impossibility of such a return, which would require the conjunction of knowledge and death. The human journey towards reconciliation with its own condition begins in a landscape that, prior to temples, prior even to planted groves, has been recognized as a place of significance precisely because it is wild: the spring, the cave, the ancient tree. Humans focus and align this significance through care and propitiation. A reordering of natural materials takes place. Trees are planted as a gathering action, a revelation of an intimate order that is both terrible and necessary, and completely at odds with the reductive ordering that occurs on the fertile plains and along the terraces among the soft hills (see Figure 5.12). The planted grove both holds in check and permits the passage of the "prodigious effervescence" (Durkheim's phrase) of life that breaks into violence during the ritual act. The horticultural intervention is itself a minimal ordering, just sufficient to cool the flames of a glorious, divine consumption that can destroy the grove by consuming it in its unlimited fire. But the destruction that breaks out in the festival regulates and limits the annihilating fury of immanence. When the festivals and rituals no longer are conducted in the groves, the groves disappear and the sacred landscape returns to the profane world, the quotidian realm of humans.[28] This landscape shows the devotees what makes them human; being part of nature and yet separate from its holiness is precisely what forges their destiny as ontological travelers.

The sacred grove is a declension from the order of nature: its particular operations introduce a different kind of order into the human world. The grove leaves production behind and introduces the chaotic domain of becoming and emergence. Like sacrifice, it rearranges

FIGURE 5.12 The ex-urban location of many sacred groves focuses the sense of separation and difference that marks the categories of sacred and profane. Photo by author.

the natural world by transforming it. Thus, through transgression a link is made to a state of disorder. As transitional zones between the cultural landscape and wilderness, the ancient sacred groves of the Greeks are located at boundaries and edges, between countryside and the realm of the gods. Even in the town or city the grove marks an edge condition set aside from both wilderness and productive lands, a diversion from natural order. Achieving its transformative energy through transgression, the sacred grove is a portal to the domain of disorder, where intimacy occurs with that which is beyond limits.

Rococo seductions: the paintings of Watteau and Fragonard

> It is the part of a wise man, I say, both to refresh and to recreate himself in moderation with sweet food and drink as also with scents and the amenity of green plants, with adornments, music, the exertion of games, theatre, and with other recreations of this sort such as anyone can do without injury to others.
>
> (Spinoza 2002: IV, xlv, 5)

> High society life . . . is closer to the pack.
>
> (Deleuze and Guattari 1987: 34)

Pleasure gardens are terrains where lived experience is a mixture of different types of relationality and encounter. The possible relations into which people may enter in pleasure gardens are different from those that other landscapes enable. The confrontation with nature as a chaotic and creative power of life conditions human interaction, mediated by human intention. Moreover, there is a connection between the specific design operations by means of which, in pleasure gardens, nature is disrupted and organized, and the kinds of encounter people have in gardens. Pleasure gardens are marked out by design from the processes of separation and combination that are immanent in the natural world. They are diversions from natural processes. To make a garden is to swerve from the order of nature through the intentional manipulation of natural hydrological, horticultural and topographic processes (often to excess). Similarly, disruptions of social order are enabled through human–human encounter in gardens, where gatherings of people connect with the indeterminate multiplicities of nature. Let us say a garden is an interactive milieu where individual humans can develop and explore their relations with other individuals and processes, human and nonhuman. In art this exploration often involves, not transgression itself, but a subtle and incomplete movement towards transgression. This section examines the inchoate, decentered garden realms of the rococo as depicted in the paintings of Antoine Watteau (1684–1721) and Jean-Honoré Fragonard (1732–1806). In these works we find ephemeral assemblages of seventeenth-century aristos moving in and out of relationship with each other across subtly defined social edge conditions. The study develops an argument to the effect that Watteau and Fragonard placed their lovers in gardens because these disequilibrious natureculture systems undergo continuous disturbance, permitting the emergence of an open-ended complex of difference and affect. The rococo was a time, like ours, of intense scrutiny of the partial and the incomplete.

Watteau's *Pilgrimage to Cythera* (1717), while not itself a depiction of a pleasure garden, makes a link to the classical tradition of the *locus amoenus* and thus provides a bridge from the trope of the erotic pastoral to the rococo fascination with imprecision and ambivalence in human relations.[29] The island of Cythera, geographically located off the southeast tip of the Peloponnese, was in classical antiquity a center for the worship of Aphrodite, the Greek goddess of love. The myth of Cythera has deep roots in French and Italian culture, where journeying to it is depicted as a difficult quest. One of the most influential descriptions of the mythical Cythera appears in the *Hypnerotomachia Poliphili* where Poliphilo describes Cythera as "benign and pleasing to the senses . . . a delightful place . . . delectable and beautiful . . . excellent and voluptuous" (Colonna 1999: 292). The island is laid out as a vast theme-park-like collection of gardens, including vegetable gardens, an herbarium, an orchard, a pleasant arboretum and, most famously, a pleasure garden with fountains and rivulets, lawns and flowery meadows spread under a transparent and luminous sky. Venus, the goddess of nature, dwells in a temple in a theater at the center of the garden that has a fabulous fountain on the side of which, etched in Greek, are the words "Seduction is like a spark." In this arena of love, after his long pilgrimage, Poliphilo receives the sweet agony of nature's primitive sexual power. His breast is pierced by Eros' arrow, and streams with hot blood from a wound that will never be healed. In order to live in perpetual bliss, the goddess tells Poliphilo, he must first be purified, "cleansed of every plebeian and vulgar stain," and "suffused with [her] dew." Only then can he and his sweetheart Polia be received into the "sacred college" of the happy and festive nymphs who inhabit Cythera and become "servants of abundant Nature" (Colonna 1999: 367).

The Parisians who viewed Watteau's painting knew quite well that Cythera was an erotic realm where free love was practiced and revered. It was even associated with a real place, the park of Saint-Cloud, designed by Andre Le Notre from 1663, in a bend of the Seine west of Paris where the Grand Cascade impressed painters who often came to the park to paint in the pastoral mode (Borsch-Supan 2000). The connection between pastoral, erotic love and the vital energy concentrated in nature imbues Watteau's *Cythera* with an elegiac quality underwritten by a subtle but distinct awareness of social protocols that are being forgotten, or discarded, in the intoxication of expectation and abandonment. The painting shows a long, mobile swarm of lovers moving towards a vessel that will take them to the land of free love. We sense that these people are on the threshold of a complete renunciation of social mores. Yet we are not convinced that they consider such renunciation necessary, for the delicate configuration of gesture, glance and slightly reluctant leave-taking, particularly on the part of the women, conveys an uncertainty present even in the later *Embarkation for Cythera* (c.1718–19) where, surrounded by imploring *amoretti*, the couples' progression to love is tremulous and hesitant (see Figure 5.13).

A seeming ambivalence moves across the surface of nearly all of Watteau's paintings, and is particularly active in his works, later than *Cythera*, that feature pleasure gardens. His *fêtes galantes,* the paintings that depict high-born groups of people lounging amorously in garden settings, flirt vigorously with undecidability. "We cannot always say what exactly the figures are doing, what gestures they are making, what emotions they feel . . . and . . . his works carry no obvious moral message" (Sheriff 2006: 23). Even when depicting large parties, Watteau groups his subjects into smaller trios and quartets, carefully separating them from the other groups, and at the same time relating them by the suggestion of gestures, glances and an unerring ability to conjure music as a barely heard, but distinctly felt, unifying element. Something "buried, or covered, inferred or concluded" is rising to the surface. The groups are supervenient: what is composed in one assemblage becomes a component of a new one. Whatever is discussed is not resolved. Under the cover of a courtly, puissant superficiality a

FIGURE 5.13 Antoine Watteau (c.1718–19) *The Embarkation for Cythera*. Courtesy of Staatliche Museen, Schloss Charlottenburg, Berlin.

dangerous secret is passed from group to group without any fixed order or sequence. Those who participate in these *fêtes galantes* are united only in a transitory way, like the gardens they inhabit, whose lanes and lines of trees never quite resolve into a legible design. The figures congregate in gardens which are themselves undecideable as to form, status, location, always on the point of slipping away, becoming some other kind of landscape. Located somewhere between park and garden, the settings have statues, walls, plinths, urns, steps and fountains, but they always open on to larger, treed landscapes, sometimes with fields or rivers framed by trees. Or do they turn away from these wider, open spaces, acquiring a meditative and comforting inwardness? Often paved underfoot, but just as likely to be grassed, the foreground plane tilts towards the viewer in such a way as to display the courtly, decorous, even-mannered – but not stilted – way in which the players in his scenes interact.

The rococo pleasure garden depicted in Watteau's works is asymmetrically configured, disordered in a deliberate or prearranged way, as if to encourage the pursuit of subtle, chancy, high-strung intermezzos: desire played out in discontinuous but intimate spaces among people the nature of whose longings we can only guess. A long, high note of seduction, however, sings faintly throughout these woods and riverine terraces. Beneath it, driving the delicate melody of desire, sounds a bass note of thudding eroticism. But Watteau's gorgeously arrayed, worldly aristocracts are not the shepherds of Theocritus' idylls. The *fêtes galantes* evoke the forbidden satisfaction of a country life that permits much more than that which is depicted in the scenes we see. Art historian Nicholas Mirzoeff describes seduction in the eighteenth century as "a violation of what were perceived as the normal, restrained relationships between the genders, in which it was natural for a man to desire a woman, and equally natural for her to resist his advances" (Mirzoeff 2006: 125). In early-eighteenth-century France seduction was, along with valor and honor, a central preoccupation of the aristocratic spheres. It was also a serious crime. By that time, seduction, and the violence and intimacy it designates, had ceased to belong to the order of nature, having transferred to the realm of artifice, to signs and rituals. To the order of law.

Is this the secret that is passed around the groups in Watteau's gardens? That seduction represents "the transubstantiation of sex into signs" (Baudrillard 1990: 13). That seduction is not energetic, is not "like a spark"? One has only to observe the declensional postures (couples moving off, singles standing aside, looking the other way) to infer operations of deception and guile, modes of circulation that are themselves secretive and ritualistic, intuitions that play by their own rules. Not sex then, but sexuality. Watteau transmutes the passion of crime into a light-headed, light-hearted commitment to the greater pleasure: the cultivation of ongoing sexual emergence, a refined, frivolous interpretation of eroticism that abandons the simplicity of heroic and celebratory sex for complicated desire. Made all the more complicated by the cautious gentility of a people who are aware they are play-acting – and by a dismissal of the sovereignty of complete immersion. *The Pleasures of Love* (1719) depicts a couple drifting away from a rhythmic, almost fluvial, swarm of lovers and *galantes* (see Figure 5.14). The couple looks back as they move off. Longing and acceptance mingle in their expressions.

FIGURE 5.14 Antoine Watteau (1719) *The Pleasures of Love*. Courtesy of Gemaldegalerie, Dresden.

The collective is the site of reason, and the couple cannot cross its border. If passion sets reason aside, if passion is "capable of carrying both man and woman to excess, beyond the limits of reason" (Mirzoeff 2006: 125), then what we see in Watteau's configurations, is a diagram of life where passion, recklessness and expenditure ("natural impulses") are transmuted into representation. Seduction is a disorder to which these lovers can never succumb. Passion, in Watteau's delicate garden scenes, elicits a strategy of exclusion in the form of courtly decorum, or in the rendering of transgressive practices as "an impossible dream-life of gay picnics in fairy parks . . . where [as in *Fête in a Park* (c.1817)] the ladies are all beautiful and all lovers graceful . . . and where the life of the shepherds and shepherdesses seems to be a succession of minuets" (Gombrich 1972: 358). By experiencing ever so decorous a perturbation, and subsequently integrating it into their careful, careless activities, these young men and women who seem just to sit and dream, or play with flowers, as they weave their Cytherean spells of what-might-be, involve the viewer in a complex reinterpretation of their conditions of possibility. Watteau puts the relations between the inhabitants of these dreamscapes (told in glances and gestures and half-mocking smiles) at the very heart of the action, at the core of their doing and becoming, driving the brittle social values inherited from the seventeenth century to the brink of a more open embrace of the complexities and ambiguities of human interaction. In the 1700s gallantry required the observance of etiquette and the preservation of social distance. In his *fêtes galantes*, Watteau uses decorous reserve as an emblem of intoxicating passion. These lovers who gather along the edges of excess, trembling with deferred – shall we say sublimated? – bliss, cannot plunge into intimacy but must remain circulating through a twilight zone of rare, almost intellectual, disequilibrium where they seem content to remain (see Figure 5.15). For intimacy "is violence, and it is destruction . . . it is holy, sacred, and suffused with anguish" (Bataille 1992: 50–52).

The gardens in Watteau's paintings introduce a perturbation into the organized systems of society. Even so, Watteau sensed, perhaps, that humans can be wafted gently and craftily through these fields of multiplicity and ambivalence in such a way as to expand the possibilities of life. Spinoza (who died seven years before Watteau was born) spoke of the art of organizing encounters as an effort of uniting and combining useful and agreeable relations. What are most useful and agreeable to humans are, to the rationalist Spinoza, other human beings seeking to agreeably and usefully combine their relations (Spinoza 2002: 358–367). In Watteau's *fêtes galantes*, we find life being constructed in this way: individuals seeking what is agreeable not only for themselves, but for others as well and, despite the deferment and delay, achieving something more than the social order within which they must spend their lives. Seeking pleasure in the "contractual" connections between bodies and bodies, and bodies and the world, Watteau's lovers demonstrate an extraordinarily fraught becoming-human, an overcoming that is catalyzed by the semiotics of social existence in eighteenth-century France. Watteau flourished in the period just after the death of Louis XIV, whose garden at Versailles incorporated a vast number of liminal spaces where the kinds of interaction described above could occur. These places were out of the way of the mainstream of court life. Courtiers and princes lingered among the serial conifers, the symbolic fountains and the low vegetal planes that described the geometry of Versailles. But way down the end of the transverse axes there were ambiguous edges where the gardens merged into surrounding farmland. These were the settings for Watteau's *fêtes galantes*.

On the other hand, Jean-Honoré Fragonard, in many ways Watteau's successor, placed his lovers on the architectural terraces closer to the chateaux. Here, the social ambiguity was

FIGURE 5.15 Antoine Watteau (1719) *The Meeting in the Open Air*. Courtesy of Gemaldegalerie, Dresden.

pointed up by the dialog between the vegetal and the architectural. It is in his paintings depicting young men and women in garden settings that Fragonard provides a compelling meditation on the nature of human–human and human–environment relations – in particular in the series of four paintings known as *The Progress of Love* painted in the early 1770s for the Madame du Barry's pavilion at Louveciennes (designed by Ledoux and sited on a rise overlooking the Seine) (see Figure 5.16). These paintings, *The Lover Crowned*, *The Meeting*, *The Pursuit* and *The Love Letters*, were painted specifically for the salon, or drawing room, which let on to the garden (Dupuy-Vachey 2006: 106). They show, in each case, a young man and a young woman, sometimes with a third party, in an undefined romantic entanglement. In *The Pursuit* (1771–3) the young man surprises the object of his affections on a garden terrace; she recoils. *The Love Letters* (1771–3) shows the lovers framed by hollyhocks reading billets-doux. The young woman theatrically places a coronet upon the youth's head in *The Lover Crowned*, and in *The Meeting* he is caught in the moment of leaping onto the garden terrace. But the paintings, now hanging in the Frick Museum in New York, do not depict individuals. At least, they depict individuals only insofar as these individuals are components of an assemblage. At first glance each painting is inhabited by a small group of two or three. In each one, however, there are sculpted figures, statues that increase the members of the group and the matrix of human–nonhuman relations. The figures are interchangeable,

FIGURE 5.16A
Jean-Honoré Fragonard (1771–3) *The Lover Crowned*. Courtesy of The Frick Collection. New York.

FIGURE 5.16B
Jean-Honoré Fragonard (1771–3) *The Meeting*. Courtesy of The Frick Collection, New York.

FIGURE 5.16C
Jean-Honoré Fragonard (1771–3) *The Pursuit*. Courtesy of The Frick Collection, New York.

FIGURE 5.16D
Jean-Honoré Fragonard (1771–3) *The Love Letters*. Courtesy of The Frick Collection, New York.

from mineral to organic, from painting to painting, and the subject of each painting has no specific place in the order of the works. There is no sense that the paintings compose a narrative. The order in which they should be viewed is not fixed, despite the historian Sauerlander's argument that they were hung in the salon in such a way as to connote order and narrative (Sheriff 1990: 71–74). He attributes meaning to the works on this basis. However, the fact that the paintings were hung in one room does not require the narrative progress that the name by which they came collectively to be known implies. They were intended to make a kind of disjunctive synthesis, to be seen, that is, at once; but the separateness and discontinuity of the moments caught are based on patterns of opposition, similarity and irresolution. The network extends across the walls of the room at the Frick Museum, from painting to painting, forging new connections every time they are viewed. Relations change, context shifts.

The principles of combination in *The Progress of Love* are not to do with sequence, or causation, or linear narrative. Instead there are moments of complexity within an overall pattern of organization that emphasizes combinatorial and relational promiscuity. The series depicts a group of eleven individuals caught in a tangled network of connections that, despite the confinement of the pictorial space in which they are depicted, define a vast and constantly changing semantic space illuminated by themes of disturbance, transgression, uncertainty and something of the rococo era's blurred, frightened etiquette that replaces erotic immersion in things. If we discard sequence and narrative, and even the (mostly posthumous) titles of the works and consider the four paintings in their capacity to come in and out of relationships with each other, viewing them in any configuration of four, we find a very classy meditation on the way human beings relate to each other and how their environments enter into combination with these relations. These people are not depicted in "settings," or "against a backdrop" of garden or park. They are shown as gathered forces. Eighteenth-century distinctions between nature and culture are dissolved in arabesques of gesture and figuration that weave what it is to be human together with what it is to be social and what it is to be part of nature. Gardens are naturecultures; humans are naturecultures. Naturecultures are not continua, but matrices or fields. The pleasure garden, indeed, is revealed as a field in which intensities occur, or coagulate, solidify, as – bodies, yes – but also as affect and raw, powerful emotion. People, very young people, are seen in various states of combination with each other and with the turbulent, fluid environments in which these moments of complexity occur. Fragonard shows the seamlessness between the order/chaos structures of the gardens and the vivid affect associated with the human encounter occurring within them. A girl crowns her lover – or is she removing the crown? She hesitates. The crown – a ring of florets gathered from the shrubs that ring the scene – hovers over the head of the young man while another youth sketches the moment. "Crown me, crown me!" the young man seems to be saying, while the will she, won't she pose of his amoretta demonstrates her glorious power.

Fragonard's masterpieces eerily but dramatically draw up the indefinite and ambiguous condition of the rococo in a painterly tumult of color and movement. But the subjects of the paintings are very precise. In *The Pursuit*, an elegant lover surprises a young girl who romps with her friends, offering her a rose, the ancient symbol of love and eroticism. The garden in which this occurs is an emotional correlative of the unscripted encounter: it explodes, too, just as the girl, who dissimulates and dramatizes, throws herself into the situation with the clarity and knowingness of adolescence. In *The Love Letters* a young man fondles his girlfriend who sits archly on a pedestal reading his (or perhaps another's) words of love, enjoying the rapt attention of the boy. The scene is quiet, focusing inward, but it is charged with an erotic electricity that energizes the couple, the flowers, the trees and the sky, all depicted as one system, one condition, one congeries of affect. Fragonard operationalizes the eroticism of the pastoral in his depiction of the pleasure garden as a realm of disorder. The terrace is part of a larger garden in *The Meeting*. The garden is itself set within an encompassing landscape of large trees and unkempt plantings. The walls, plinths, urns, sculpture and flagstones that give the terrace its structure seem slowly to be sinking into a tide of horticultural excess. Vines and tendrils prize apart the stones; roses and other flowers engulf the architectural components of the scene in waves of color and the surrounding trees toss foliage onto the terrace like the spume of wind-driven seas. All is movement and wind.[30] But the real drama of the scene is focused in the electric tension of two lovers caught in a secret moment of transgression. A young man pauses above the balustrade over which he is about to leap, frozen on the threshold of an erotic realm in which a girl, awaiting his entry, but also transfixed by the disturbance, as it were, off stage, stretches her arm towards him. Initially a gesture of welcome, the outstretched hand is becoming a signal to stay, to retreat, to disappear. Both are looking wildly and helplessly towards the left of the scene. Every feature of the composition is angled to emphasize the source of the turbulence – the intruding element, the unwanted visitor – even the statue of Venus which towers over the terrace is turned toward the interloper, and the *putto* at her feet is drawing her attention to the disruption. A potential transgression is being held in check. Something, initially excluded from the event about to take place, is making an entrance into the configuration of secrecy, intimacy and illegitimacy that was about to unfold in the garden. A garden world saturated with nature, potentially with ecstasy and sin, is about to be returned to the order of things.

If "[t]he sacred is that prodigious effervescence of life that . . . the order of things holds in check" (Bataille 1992: 52), then it is in *The Meeting* that Fragonard most explicitly depicts the relationship between pleasure gardens and the sacred. In the rococo period western artists, scholars and natural philosophers, increasingly – in *their* minds – divided from nature by the advance of the modern, found in the motif of the garden crawling with life, implacable and unpredictable, the location of a figure that shows how something beyond the everyday, beyond control, like nature itself, is always with us. Like the unconscious, laughter, violence and love, carnival and crime, the uncanny, the sublime, the open, it emerges through the normative social and spatial practices from which it is excluded. Gardens are always separate. Bounded by walls or fences, they are sites of deviation where nature is diverted toward human purposes. It is by suspending the order of nature that the pleasure garden permits the entry of a transgressive element through which a link can be made to something beyond the everyday. The pleasure garden is a realm of emergence and difference, of disruption – interruption – and transformation.

The interrupted meeting is a common theme in eighteenth century painting. It is found, for instance, in Boucher's *The Surprised Lover* (1750), now lost, and de Troy's *Surprise* (1723, London: Victoria and Albert Museum) where a servant is shown entering the secluded corner of a garden to warn lovers of an intruder's approach (Sheriff 1990: 75). Strategies of exclusion develop in circumstances where a dependence forms around that which is excluded. The outstretched hand, a sign of absence in *The Meeting*, locates a Serrean parasite at the heart of the lovers' love. This love, founded on an exclusionary Enlightenment regime, rejects the social only to make the rational a necessary operation whose continual negotiation with effervescent nature denotes the parasite as the noise in the system. "[W]hite noise is the heart of being; parasitism is the heart of relation" (Serres 1982b: 52). This is the paradox that irradiates Fragonard's great sequence, *The Progress of Love*: the paradisal garden is constituted by the parasite. In the interests of love we try to expel this parasite, which returns as the condition of its constitution – the condition of the demon.

It has been argued here that emergence and difference have often featured in verbal and visual discourses with respect to gardens and other landscapes. The paintings of Watteau and Fragonard connect human and nonhuman nature in a landscape trope that goes back through Rubens' less inhibited *Garden of Love* (1633) to the imaginary medieval heterotopias that also bear that name. The unspoken rules that regulate the activities of the rococo high-born tribe are deeply buried. Like mycelia, they bring the groups together and keep them apart, ordering their relational possibilities and extending beyond the human to include cultures of the vegetal and the inanimate. Watteau's landscapes are full of trees. He exploits their shadowy ecologies for human habitation and use. Culturally, and spatially, the gardens are co-extensive with the micro-communities that Watteau distributes with minute precision. In *Les Champs-Elysées* (c.1717) groups of men and women grow like fungi in the shade of trees, their patterns of distribution seemingly nurtured by the conditions for their cultivation that the trees provide (see Figure 5.17). Like precious truffles or delicate mushrooms these magic circles of young women and their accompanying gallants and troubadours emerge from subterranean interactions.

FIGURE 5.17 Antoine Watteau (1717) *Les Champs Elysées*. Courtesy of The Wallace Collection, London.

While Watteau's bands of merrymakers never reach beyond a tentative, symbolic sensuality to seduction proper, Fragonard goes further. His oeuvre still hovers about the risqué, but he pushes his impulse to break through fevered sensuality to a condition of abandonment and transgression. The huge, extraordinary canvases of *The Pursuit of Love*, so full of rhythm and movement, are themselves seductive self- and other-mocking correlatives of the condition of abandonment. They pull the viewer into a turbulent, erotic landscape where lovers are on the very point of fusing with the natural world, and thus becoming disindividuated passion. Like Poliphio in the *Hypnerotomachia*, they must first be purified from the inside out, bled of the "vulgar stain" of signs before they can be admitted to the "sacred college" of unconditional consumption. This, of course, can never happen: it is the fate of the rococo to pass by the possibility. We have, however, the markers of this flirtation with sacrifice: Watteau's and Fragonard's lovers who swarm in gardens momentarily forever. Is that which moves among the swarm, animating and regulating its flows – the secret of seduction – the sign of it, rather than its energy or spark? Are not Watteau and Fragonard commenting on the sad impossibility of sovereignty in eighteenth-century French life? Perhaps, rather than reading these paintings as celebrations of passion and unruly natural impulse, we should see them as the grave of love.

Landscapes of noir

Landscape is everywhere. It permeates our thoughts and actions, and enters our bodies as we enter it. Cinema has understood this since its inception. The landscapes that we see on the screen are rarely simple settings, though they are often referred to as just this. Screen writers and directors treat the landscape as a space–time field from which dramatic – and subtle – events emerge, and have created filmic worlds in which humans' ontological entanglement with landscape is both astonishingly immediate and undeniably fundamental to human–world relations. Westerns, war movies, Italian neo-realist cinema, bollywood, Kurosawa, Jane Campion, space opera – any one of a dozen different genres or auteur collections could offer fascinating insights about emergence in landscape architecture. Here, an examination of modernist film noir is chosen, because it enables the development of a theory of human–world relations appropriate to urban landscape architecture. This is because noir relates the ongoing emergence of human subjectivity to a nonlinear urbanism that exemplifies difference, openness, disturbance and relationality. These attributes of modern lived experience are explored by noir through what Lefebvre calls the "problematic of space." The spatial construction of affect is beautifully delineated in the challenging depictions of public sphere activity that noir has made its own. By formulating and transmitting the socio-spatial dynamics of modern urbanism, these films demonstrate the ambivalent, dysfunctional topology of twentieth-century urban space (Dear 2000; Dimendberg 2004). In noir we discover the organizational figure most suited to the modern city that emerged in the nineteenth century is not the plaza, or square, with its insistence on point or place. It is the dynamic labyrinth, a space–time dialectic whose constituent elements can be perceived only from place to place and moment to moment, never in totality. This urban realm is not available to cartography, for it has no coordinates that can easily be fixed. Film noir shows how nature, banished from the city, comes back as a turbulent, disruptive force. The modern urban landscape, uncertain and dynamic, is a realm for which conflict, understood as disturbance and interruption, is the very condition of its becoming. In this section the city of modern life is a force field, a landscape of social, aesthetic and cultural intensities that construct spatial orders and encode, distribute and organize the flows of human desire throughout these orders. The labyrinth, a place of violent oppositions, and found throughout film noir, symbolizes this city of modernism – decentered, multiple, indeterminate, open-ended and capricious.

Modernist architectural and urban theorists Siegfried Giedion and Lewis Mumford both highlight the indeterminacy of meaning to be found in modern urban space.[31] They write just as the labyrinthine dark city of film noir is being created in American cinema. Prior to this period the transparency that supported the establishment of classical space in the modern city had been carried through into the narrative space of mainstream cinema: a normal backdrop, invisible, the presumed condition of modernity (Easthope 1997). In the genres of the musical, comedy, drama and even gangster movies such as *Scarface* and *Public Enemy Number One*, the city is simply unmarked. It is just *there*. Normative representation of the city during the modern period, particularly as found in planning and design documents, acts as a filtering device that orders the urban according to cartographic codes of orientation, framing, delineation, scaling and so on – the classical modes described by Lefebvre. Urban planning and design methodologies of the period have the properties of a representational system based on a model of linear progression, of movement from research to analysis to design solution via synthesis (Sasaki 2002). They enshrine a hierarchy of dos and don'ts, of valid and invalid moves which seek to rationalize urban space and clarify its polyvalent configurations. When interference is encountered in an urban planning system formulated in this way the system simply expels it and moves on. When it is encountered in film noir, however, the apparent unity of the city is destabilized, an epistemological uncertainty is disclosed, and the urban realm is plunged into an intangibility of tone, mood, meaning and purpose.

According to most accounts of film noir, it almost programmatically casts nature as other, a dark continent that is gradually opened up for exploration, enabling triumphant return with trophies that signify a primitive, animal life, the existence of beings whose ontological status is lower than that of ordinary citizens: crooks, hoodlums, molls – necessary but inferior modes of human being. The protagonist of a noir movie, detective or otherwise, explores (willingly or unwillingly) a labyrinthine zone that lies, uncharted, inside the normative social geography of economic ratios, trade indexes, insurance companies and local body elections. The detective follows a path which leads from light into darkness, and finds himself in unknown terrain, exploring the criminal's natural habitat, a counter-world. This is the criminal underworld of urban fantasy, a geography valorized in the spectacularization of turn-of-the-century London's East End where aristocrats went on organized nocturnal tours (introducing the verb "slumming" into the English language). When the noir protagonist goes down into the labyrinth to solve the murder, or at least discover the murderer, he is going down to the moment of the primal scene, the epic hero's slaying of the monster, that is, of an inhuman nature, a monster who is nature, the double of the acculturated hero whose necessary violence is re-enacted as *nomos*, the condition of becoming human. Part demand for ritual purity, part desire to meet the transcendent goals of justice for which the murder of nature seeks atonement, the noir protagonist is a twentieth-century scapegoat. The noir discourse of the urban underworld merges ancient tropes for nature – night, woman, death – into one modality, where the nocturnal sex-murder forms the quintessential act. In nature, stochastic phenomena are rain, sunshine, ice, steam, fire. In film noir they are murder, violence, sexual intimidation, transgression. The criminal act provides the passage between the worlds of domestic inflation and economic crisis and that of the grim, violent haunt of the urban savage. It occurs, typically, in a street, an urban form that has been constructed in this discourse as a space of uncertainty and transgression. Film noir therefore provides a particular kind of access to the spatial configurations of modernity, both in cinematic form and in writings about that form. It reveals multiple spatialities as well as multiple temporalities. This revelation occurs by way of a disruptive movement. The civic construction of urban space as a seamless elaboration of democratic virtues is radically displaced by a violent event, which in turn leads to the disruption of social structures hitherto rendered transparent or unproblematic. If sex, murder and darkness are modernist tropes of nature, they enter the spatial framework of the noir film as natural parasites of the quotidian. Violence and intrigue become the noise that, through nature's sudden presence and intermittent, equally unpredictable absence, produces a new oscillation of cultural flows. Criminality becomes an immanent condition of urban space construction. Powerful social forces attempt to expel the parasite, in the interests of the communication of the effects of that power. These forces, however, cannot overcome the parasite's own command which, because it is not fixed or essential, but relational and multiple, always upsets the equilibrium of the system. Harmonious urban space is an abstract ideal that can never be reached. The parasite can never be excluded. It is necessary. Nature is always present, in all fields, at all levels. The many inhabits the one.

Operations of noir

By what operations, peculiar to film noir, are the social relations of the city destabilized? Three techniques are identified. First, the complex space–time structures of *D.O.A.* (director Rudolph Maté 1950) are related to the contingent, inchoate structures of feeling that imbue noir with its existential fragility. Second, the idea of lived experience is traced through four films, *D.O.A.*, *Double Indemnity* (1944, directed by Billy Wilder), *Detour* (1945, directed by Edgar G. Ulmer) and *Kiss Me Deadly* (director Robert Aldrich, 1955). Third, the relationship between the closed system of the film and the open system of the city in *The Big Sleep* (director Howard Hawks, 1946) is discussed. The techniques of destabilization examined here, then, are the re-spatialization of the city, the illumination of lived experience and the recovery of the open whole.

The respatialization of the city

On one level, the modern city was developed to deliver specific economic and legislative outcomes through the deployment of technological functions and informational transmissions. Modernist manifestos such as Le Corbusier's *The City of Tomorrow and Its Planning* (1929), Mumford's *The Culture of Cities* (1938), and the Goodmans' "A City of Efficient Consumption" (in their *Communitas*, 1947) are, at bottom, motivated by the notion that if we get the rules right the efficient delivery of these objectives will follow. Such is the city of what Lefebvre calls "abstract space," from which he regards the new modern city as emerging (Lefebvre 1991). Opposed to the "concrete" space of everyday users (recognized in various ways by the authors cited above), abstract space is the spatial framework for an economic mode of life in which the "reproducibility, repetition, and reproduction of social relationships" attempt to attain precedence over nature (Lefebvre 1995: 120). This is the space deconstructed by film noir as a brittle terrain punctuated by immeasurable abysses, and crossed by narratives of fear, loss, brutality and transgression. At the same time it is reconstructed as a locus of chance and contingency, the characteristics of a new urban space that, according to Lefebvre, is "not fragmented, but differential" (Lefebvre 1991: 202). Noir constructs abstract space as an affective field in which we find, like the lovers in medieval gardens, a totally human sweetness, and an infinite thrill, in an inherently paradoxical and violent difference. Noir also co-opts (and critiques) what came to be called *terrain vague*, the physical product of the construction of abstract space, the bits left over from the processes of urbanization fueled by economic ordering practices. Los Angeles has lots of it. Bodies are found in *terrain vague*. Elizabeth Ann Short, known as the Black Dahlia, was found near Olmsted Avenue in James Elroy's noir novel of the same name.

In noir classic *D.O.A.*, the protagonist Frank Bigelow is an accountant visiting Los Angeles on business, from his practice in small-town California. For reasons unknown he is poisoned in a bar with iridium, associated with the production of atomic power. Feeling unwell, Bigelow is told by a doctor that he has twenty-four hours to live. The remainder of the movie follows Bigelow's attempts to catch the person who has murdered him and find out why he has been poisoned. There are lots of shoot-outs in the film, as Bigelow blunders his way through the underworld of international crime. They all take place in some kind of compromised space. In some cases these are internal locations where a sudden confrontation between abstract space and its other occurs. In many cases the setting is a marginalized urban landscape, a *terrain vague* where anything is permitted. In the final showdown, Frank Bigelow kills his own murderer in a symptomatic abstract space, the Bradbury Building in Los Angeles, an edifice created by capital. Decorated in the style of the French *fin de siecle*, its wrought iron gallery and its glass skylight suggest the morphology of the Parisian arcade with its facades facing inwards to an interior lit dramatically by angular shafts of sunlight. Bigelow approaches the Bradbury Building from the street, at an angle, its quoin jutting out like a prow, external facade lit like the *Passage Jouffrey*. Entering its darkness, he runs down a corridor, slips over, picks himself up and rides up to the office of the import–export company of Halliday and Phillips in an ornate elevator. As Bigelow steps out on to the second-floor gallery that surrounds the central court, Halliday leaves his office. He turns to see Bigelow. This is the emotional peak of the film. They face each other in a network of shadows thrown through the ironwork balustrade by lamps on the banisters and stairs. The building's interior is presented as an organic space, like a cathedral – not sinister, but holy. Bigelow guns his killer down on the landing of an urban landscape dedicated to the holy twins of modernity: international commerce and rationality.

But there is more at work in this sequence than the sublimation of rationality through aestheticized violence. The internal courtyard of the Bradbury Building turns a perforated wall to the city outside. It is a landscape in which distinctions between inside and outside are blurred, as are those between public and private, and object and subject. A displacement occurs when Bigelow enters, signified by his stumble in the entry passage. Now distinctions between good and evil, hunter and hunted, life and death become blurred too, especially when in daylight the highly crafted ironwork catches the inexorable movement of the sun across the sky, and at night the dynamic and unpredictable flashes of street lamps, advertising neons and hotel signs. Bigelow's own inexorable search for his murderer ends in a garden of good and evil, signified by the agonal intersection of light and dark, with a confrontation between himself, dying, passionate, driven, and Halliday the smooth operator who would take Bigelow's life for the good of his business. The spatial closure of the Bradbury Building is perforated by the natural conditions of human life. The membrane between life and death, like that of the building's facades, is porous, and matter-energy moves back and forth across this permeable divide reconstituting existence, shuttling between reality and the Real. The scene of the shoot-out signals the advent of a nongeometric, interpermeable spatiality that invades the geometrized structures of the urban grid. Bigelow is hit and the whole system shifts to a higher order of complexity.

In a further, different, reformulation of modern spatiality, *D.O.A.* undercuts the romanticization of *terrain vague* as a site of freedom and possibility in its insistence on this realm as a space that can be wrested from the standardization of the city only by violent means. The moment it is incorporated into a narrative of functionality it ceases to be *vague*. Lefebvre, and after him Jameson, Harvey, Sola-Morales and many others, have argued that the modern city is characterized by a kind of interstitial space which is formed in the residue of dominant spatial structures (Harvey 1990; Jameson 1992; Sola-Morales Rubio 2002). While these dominant structures are effective in achieving the goals of power regimes, it is argued, they produce an excess created by what Manfredo Tafuri calls the battle for physical space construction which, he says, "leaves borders, remains, residues" (Tafuri 1990: 8). This excessive terrain is a no-man's-land which results, according to Kenneth Frampton, from the "wholesale dissolution of society," and is brought about by the decentering of capital and the rise of the multinational network (Frampton 2008). The connection here between the international forces of capital and Sola-Morales' *terrain vague* is a little underdetermined, as is its characterization by Bell and Leong as "post-urban infinite space" (1998). But the co-option, in noir, of marginalized urban landscapes as markers of ontological difference is consistently pointed. Bigelow's murderer gives him the slip in a degraded industrial wasteland near Venice Beach. Two gay photographers have their studio here. They offer Bigelow vital clues, but only after he has paid for the information – clearly this space is contaminated with vice and immorality. The space of modernity is not smooth, then, but crossed with space–time ruptures – wells or chimneys – which in certain sensitive conditions provide a passage to the sublime, that pertains always, as with the photographs of Lewis Baltz, beyond the frame. Bigelow discovers that it is not possible to inscribe freedom in this marginal terrain because the power regimes that bind him and play with him have no specific locus. Finally he realizes that his quest is not for a particular place. His question – who murdered me? – will not be answered by finding where the murderer is – even though he does. Neither resistance nor transgression are a question of place, so much as of a particular kind of practice, a social practice (or unsocial practice) that relies on a braided subjectivity that finds its meaning in its swerve away from the flow of normative civil society. The wasteland apotheosis is a vision of freedom, but not freedom itself. Not the sacred, but a sign of it.

In film noir the urban landscape is depicted as a "space of transformation" (Serres 1982f) in which systems pushed far-from-equilibrium spontaneously change state. Transformation occurs in time. We are continually aware of Bigelow's immersion in time. He arrives in San Francisco in the evening, goes that night to The Fisherman, a club on Fisherman's Wharf, sleeps, wakes the next morning, is diagnosed as having been poisoned and then spends the rest of the morning in panic. He flies to Los Angeles and in the afternoon talks in turn to Halliday, Mrs Phillips, Miss Foster, Stanley Halliday, Marla Rakubian, the photographers and Majak. That night he escapes from Majak, kills Halliday and goes to the Police Department where he tells his story. The whole spatially extensive episode has a duration of approximately twenty-four hours, of which we are repeatedly reminded, as Bigelow replays the doctor's curse, "You have one day to live." In a single swallow Frank Bigelow's ontological state transforms radically, from healthy all-American male to living dead man. His physical condition does not go into noticeable decline when he vividly confronts his mortality, but in an archetypal noir moment of transformation there is a sudden redistribution of his values and priorities. His realization of his condition of complete inconsequentiality is unredeemed except by the astonishing fact that this awful, unprovoked event connects him to a global system of transmissions enabled by the market price of iridium. His world is one world in the universe of worlds, whose constant spinning is lubricated by desire. Noir's respatialization confounds time with space. Placed irrevocably within a constantly moving reality, Bigelow's experience advances with it and we with him. And yet, despite this immersion we are never really sure what time it is – we can tell only when the narrative takes him outside and we can see whether it is dark or light. Otherwise the logic of forward movement is not of the "race against time" variety normally associated with the cinematic short deadline plot, but of a drifting, heaving temporality that repudiates clock time and replaces it with a Bergsonian lived duration. When Frank is galloping down the streets in the film he is not moving forward, but running on the spot. The sensation is one of an extraordinary evaporation of time occurring simultaneously with an inertial temporal depth. This is the far-from-equilibrium space-time of modernism, interactive and irreversible at the same time; the chronology of evolution, of magma and interstellar space.

The primary space of urban emergence, however, is public space. Twenty-six clock minutes into *D.O.A.* Frank Bigelow emerges into the urban landscape for the first time, out under the canopy of the hotel into the daytime street. He jumps onto a passing tram. Immediately we know this is San Francisco. The tram heads up one of the famous hills and there are views of the bay with the Golden Gate Bridge. On Sacramento Street he alights and enters a medical center. In a few minutes he will know he has been murdered. The next time we see Bigelow outside, he's rushing, almost falling down the monumental steps of a hospital. It's been confirmed: he's going to die, perhaps in one day. He swings into the street and begins a headlong flight from the truth. For two film minutes Frank runs manically through the dense streets, pushing against the crowd which flows in the other direction, out into the middle of the main road and across to the blank floor of the Embarcadero where he finally rests in front of a newsstand proclaiming *Life* on multiple covers of the famous magazine (see Figure 5.18). This long running sequence firmly establishes Frank's isolation from the crowd, and that the drama of his slow extinction will be played out in public space. As he weaves his way through the streets, driven by blind fear, Frank's plight is dramatized by the unknowing and unheeding crowd. The urban perspectives of street and park are framed and lit in such a way as to intensify the compulsive behavior of city dwellers. Volumes, angles, lines of sight construct these scenes, but we, like Bigelow, are unable to perceive or conceive the multiple pathways and complex spaces that define this reductive quotidian. We too are blind to the reality of the urban. Its intricate, rich *topos* has been reduced to Lefebvre's industrial web of "blind fields" (Lefebvre 2003). Lefebvre's use of the term field refers not only to successive or superimposed layers of facts or phenomena, but also to modes of thought, action and life. The industrial field of the city, he says, replaces the spatial grid of the rural field, clearing away "natural particularities" with a "methodical and systematically imposed homogeneity, done in the name of reason, law, authority, technology, the state, the class that holds hegemonic power" (2003: 34). Frank Bigelow comes from that ordered rural world. His profession, accountancy, is a paradigm of

FIGURE 5.18 *D.O.A.* (1950). Director: Rudolph Maté. Courtesy of MGM.

rationality, clarity and fact. He lives in Banning, a small town whose very name insists on its rectitude. Through no fault of his own, simply chosen by blind fate, Bigelow becomes caught up in an equation of disorder, relativity and surprise, a space-time that is differential. The people he encounters exist within a whole of which he can have no understanding. They connect to remote and unintelligible forces, which, as far as he sees, comprise a network of oppositional and contradictory behaviors, testimonies and structures. The urban space in which Frank finds himself asserts its topological qualities in contrast to the hometown with which he is pathologically connected by way of telephone, and to the stable, knowable Paula, his girlfriend. In the spaces of modern urbanism Frank whirls round and round, walking, running, driving a rental car, being driven, by car and by bus, from place to place, each one more unsettling than the last. He cannot even find sanctuary in his hotel room. It is always being invaded, parasitically, by reveling neighbors, by the incessant phone calls from Paula, by the psychopathic Chester and his hoodlums.

The theater of Bigelow's death is a succession of engorged streets, empty urban squares and dimly lit interiors. As he follows the central inquiry of the film – why have I been poisoned? – through to its violent conclusion on the stairs of the internal courtyard of the Bradbury Building in LA, we follow a movement from public to private space, a phenomenal shift that registers the increasing privatization of public space. The site of Frank's final apotheosis, where he declares his love to Paula, now the only certainty in his existence, is a street outside the Allison Hotel in Los Angeles. Here, the most intimate of performative speech-acts, delivered with the authority of privileged knowledge, is carried out in a space constructed by modernism as its most unpredictable and chaotic milieu. Between Paula and Frank, between the terms even of redemptive love itself, a greater force intrudes, the relentless machinic noise of modernity. In order for their love to achieve the purity Paula craves, and Frank has feared, the noise of modernity must be drowned out. It cannot be.

Frank's bloody engagement with his unseen enemies in the *terrain vague* of Venice, LA, is a fantastical negotiation, dramatized and simplified, of conflicting world-views played out in a marginalized public realm which, by its very nature, will never be stable. Filmic public space plays a critical role as the location for social representation and reconstruction, operations that occur on different planes and in different time-frames. It is impossible to understand this whole; Frank will never see the big picture. "The horror of the noir city," says Krutnik, "is firmly tied to the pervasive indeterminacy of meaning (Krutnik 1997: 99)." Film noir does not demonize criminals. They may be illegitimate, or even evil, but they are a necessary part of society (and know it). Their partiality is crucial. Public space is exactly where this partiality becomes legible. Nature irrupts into the public realm as rapes, knife fights, murders. These transgressions intrude between civil society and its official spatial representations in the landscapes of cities. Civil society's attempts to produce an impartial citizenship are an effort to escape from public space rather than enter it. Public space, says Deutsche, "is associated with openness, contingency, incompletion" – in other words, with partiality (Deutsche 1998: 308).

D.O.A. discovers and idealizes the potential of the street as a chaotic realm that holds out a promise of consummated desire, but can never deliver it. Sennett argues that such spaces, characterized by interruption and disturbance, are necessary in order to prompt human beings "to turn toward each other, to experience the Other" (Sennett 1994: 374). In film noir this is often an experience of conflict. "In the arena of the noir city protagonists must confront the strangeness of others and the strange otherness within" (Krutnick 1997: 89). This means that the spatial framework on which social representations are inscribed must stay open to alteration and revision. "Conflict," says Deutsche, "far from the ruin of democratic public space, is the condition of its existence" (Deutsche 1998: xi, xiii). Is this not why in film noir, settings tend to be in the public world rather than the domestic? Eating, for instance, occurs in public places such as the lunch counter. Crucial personal encounters take place not in the home but on a train, or in a carpark, in a hotel lobby, or in a seedy café. "The noir city is symbolically homeless, its denizens act out their lives in public spaces: in diners, bars, night-clubs . . . streets, alleyways . . . hotel suites . . . [p]rotagonists often find themselves in deserted industrial or commercial locations" (Spicer 2002: 67). Even Frank's hotel room is semi-public. "The noir city," Krutnik asserts, "is the principle stage upon which [the] drama of the obliteration of 'home' is played out" (Krutnick 1997: 89). Bigelow is caught up in this effacement of origins – connected to Banning and Paula only by telephone, he tries even to sever that link. ("Where have you been?" Paula asks nervously on the phone, "I've been trying to reach you for hours.") In these uncertain public spaces film noir constructs a filmic realm in which it is possible to distort the relationship between the seen and the unseen, between civil society and the people who participate in that society, between the Cartesian geometry of the modernist city and the uncanny, between the sayable and the unspeakable, between communication and the uncommunicable, between the solid artifactual architecture of urban design and the teeming unpredictable diversity of the urban fabric. And indeed the protagonist of the noir film, often but not always a detective, serves as a kind of weaver, a "proto-worker of space" as Serres calls him, who untangles, twists, interlaces and reassembles, passing above and below, rejoining the rational and the irrational, renegotiating the spatial geography of the city, turning the restricted economy of the metropolis towards a rupture in spatiality, through which the uncanny may emerge. In the case of *D.O.A.* it is everyman, the small-business owner Frank Bigelow who, hurled into a maelstrom and sacrificing his own life, makes possible an alienating *anomie* that exposes the disjunction between conflicting aspects of the modernist notion of public space.

The illumination of lived experience

"As the particular representative of modernity," writes Robert Lapsey, "the city evokes and calls into play the structures and processes of the subject with special intensity" (Lapsey 1997: 192). What, then, is the relationship between subjectivity and the city? The situation of the city-dweller, immersed in a field of shifting spaces, signifiers, forms and relationships, has often been depicted in twentieth-century art production. Poetry, such as that of modernists Pound and Eliot, highlighted the alienation of the subject faced with the secular, commercial and vain ennui of the modern city. Modern novels, like Joyce's *Ulysses* and Kafka's *The Trial*, mapped the disillusionment of human consciousness with the brutality of urban life, and visual artists – L.S. Lowry, George Grosz, Edward Hopper – represented the inherent loneliness and loss of autonomy associated with the vast, impersonal processes of urbanism. Philosophers and theorists have also probed the problematic of the "modern subject," a subject typically characterized as "decentered" and associated nearly always with urbanism. Sartre, Camus, Merleau-Ponty, Foucault, De Certeau, Bourdieu, Lacan and Badiou have addressed being, becoming and subjectivity from distinctly urban perspectives, Heidegger being an exception. These symptomatically twentieth-century inquiries are regarded here as investigations into "human nature." In the following it is suggested that this *human* mode of nature is of critical importance to an understanding of the role of urban landscapes in urban life. Existentialism is the frame. A modernist philosophical movement that explored the self-consciousness of living in a broken, dislocated world, a world into which the human being is "thrown" and "condemned" and yet "abandoned" and "free," existentialism was an urban way of thought par excellence. Film noir was its cinematic expression.

The action in *Double Indemnity* (director Billy Wilder, 1944) moves from the socially empty, but horticulturally lush, landscapes of residential LA where protagonist Walter Neff lives, to the physically empty railroad landscape of the city edge where he drapes the vamp's husband's body on the railway line. Understanding the urban landscape of Los Angeles as a field in which intensifications occur, permits a reading of the lived experience of the noir subject in which rather than a passive *dérive*, his filmic line of flight is that of an active agent. The urban field is the complex, shifting site of this agency, which is always reactive as well as proactive, and its flows permeate the consciousness of the subject. The protagonist is faced with the task of "putting right what can never be put right" (Lapsey 1997: 186). This is precisely the task of Walter Neff in *Double Indemnity*. Master of his own misfortune, ambiguously responsible for the dangerous mess he is in, Neff faces ruination, and within this hopelessness he finds, blurred and flimsy, a project – the reversal of this experience of ruin. And here we find the existential premise of the film. Neff put the body of a man he had just murdered on a railway track to make it look like he had fallen off, and been killed by, a train. Neff throws himself into the situation in order to take over, in his thrownness, the self that is thrown and "go right under the eyes of Death" (Heidegger 1962: 434). The implication is that Neff's life up to this point had been taken from him. The justification for his existence had gone. He was alone and in his solitude had discovered that life had been given him "for nothing." Hence, the "existential" decision. But *Double Indemnity* does not leave Neff without a possibility for – and this is not redemption – the excruciating experience of a totally *lived* self-reality. Right out there on the railway track, alive, aware, the inauthentic subject of the modern city lays his old self down to destruction. It is dark night. Filmed from below so that he looms threateningly against the pitch sky, Neff drags the body onto the line – it is another quintessential noir situation. Badiou, remember, describes the situation as:

> an ordinary multiple, a multiple that is obviously infinite because all situations in reality are infinite. It can be a historical, political, artistic, or mathematic situation; it can even be a subjective situation. I take situation in an exceptionally open sense, and to capture that openness I say it's a multiplicity.
>
> (Badiou 2000: 46n6)

Badiou argues that not every human being is always a subject, but some human beings become subjects; those who act in fidelity to a chance encounter with an event which disrupts the situation in which they find themselves (Feltham and Clemens 2003). In most noir films an event occurs in the first ten minutes and the rest of the film is a chain of reactions to it. The actor creates out of a situation a prolonged investigation of the consequences of the inaugurating event. It is this inquiry that the noir film chronicles and presents. It is an urban investigation, which traces the intersecting trajectories of the "they," the subjects that are not the subject of the protagonist, through a world that holds the situation in specifically urban modes of articulation. The noir film premise that you don't have a subject until you have a project, exemplified in *Double Indemnity*, and in *Scarlet Street* (Fritz Lang, 1944), *D.O.A.* and *Kiss Me Deadly*, places the viewer within the lived experience of the subject of the film. In this way we live his experience of the city through his experience of his own contingent and quite destabilized subjectivity. It is precisely noir's emphasis on lived experience that makes its urban field so real and present. It *creates* the city by discovering the material of the urban world and in so doing redeems the urban from its dormant state of virtual non-existence, making its reality physical. In *D.O.A.*, for instance, a number of similar scenes are shot in the streets of San Francisco and Los Angeles. In both night and daytime scenes the streets are depicted as frantic with motion, traffic weaving – cars, trucks, buses, trams – and the sidewalks full of pedestrians all seemingly determined to get somewhere important quickly. Bigelow's predicament propels him through the city at an even faster rate than these seemingly fanatical urbanites, and against the general flow of their movement. We experience these pedestrians as the "they" on which Bigelow has ordinarily depended for his self-affirmation. And we feel the jolt of realization that comes when we discover that despite the fact that Bigelow's quest is hopeless, and that he must die, he is nevertheless, in his *project*, impelled by a different order of desire than they, and that his living/dying is in fact a way of being in a closer relationship to something they can only discern vaguely through their compulsion.

"They" however, are in the city too. This is, after all, the site of their lives, and they can all encounter situations too. Some do, some don't. The probability of encounter is not great according to Badiou. Whether an event belongs to a situation or not is actually undecidable, as is whether a subject will recognize its occurrence. This undecidability is the mark of the urban. All noir city-dwellers are immersed in contingency and unpredictability, not just the protagonist of the film. Film noir evokes an urban *landschaft*, an environment in which people are plunged for reasons other than, and in ways that cannot promote, mere visual consumption. The landscape becomes a realm in which social obligations based on symbolic values reign. It is configured according to patterns of occupation, activity and appropriation. The urban landscape is the realm of the everyday in which urban dwellers find the redemptive value of minute, ongoing transgressions (De Certeau 1984). *Landschaft* betokens the forms and ideas that structure society in general, not to be identified with the abstract civic realm which controls by controlling representations of space, but one where there is no clear separation of self from scene, subject from object. A world of unconscious engagement, achieved most usually through the ordinary practice of daily labor, *landschaft* provides a connection to the material and symbolic qualities of the urban landscape. Through this connection, human–urban landscape relations are constructed with intimacy and care even though they are fleeting and disparate, just because the urban is a realm of multitudinous promises and scarce discoveries. In film noir, human nature, itself an open realm of multiplicities, resists and breaks down normative social and spatial practices. Specific sites are therefore the sites of specific performances of resistance and transgression. These sites, by their very nature, are impossible to design exactly because they are potential situations.

One of the ways Bigelow's subjective phenomenology is made apparent is through the adoption in the film of the narrative voice. *D.O.A.* does not employ the device of voice-over however, but of a story narrated to another, in this case the Captain of the Homicide Department. "Do you want to hear me out or don't you Captain? I don't have much time," asks Bigelow, about to expire on the Captain's floor. The Captain replies, "You tell it any way you like." Bigelow's subjective experience is translated by the use of this and other noir conventions such as the point-of-view camera angle and claustrophobic lighting to interpret events as they are filtered through his consciousness. Director Fritz Lang remarked: "I use my camera in such a way as to show things, wherever possible, from the viewpoint of the protagonist; in that way my audience identifies itself with the character on the screen and thinks with him" (Bogdanovich 1967: 85). This subject-centered approach, the noir film correlative of novelistic stream-of-consciousness, is specifically and self-consciously "modern." As early as 1946 French film critic Nino Frank identified it as a feature of noir movies:

> And so these "dark" films, these films noirs, no longer have anything in common with the ordinary run of detective movies . . . The step forward is a big one; after films like these, the characters of ordinary detective films will seem puppets. Today's spectator is sensitive to nothing more than this impression of real life, of lived experience, and – why not? – of certain disagreeable realities that do in truth exist, whose repression never served any purpose; the struggle of life is no invention of our own time.
>
> (Frank 1996: 8–9)

In the first decades of the twentieth century, Whitehead's philosophical project was to "modernize" the analyses of experience of Descartes, Locke and Hume, the so-called "rationalists" of the Enlightenment for whom, he says, subject and object were in a "bare relation" of knower and known. The whole point of his inquiry, he says, is to demonstrate those factors in the current stage of "Western civilization" which "jointly constitute a new element in the history of culture" (Whitehead 1961: 8). Whitehead argues that experiencing is a process of actualization of time and place (1961: 167). Crucially, the basis of experience is emotional. The subject, he argues, has a "concern" for the object (that is to say, the world). "And the 'concern' at once places the object as a component in the experience of the subject, with an affective tone drawn from this object and directed towards it" (1961: 167). Whitehead undertakes a qualitative meshing, or merging, of subject and object, where these become relative terms. This forms the basis of his theory of lived experience. "An occasion is a subject in respect to its special activity concerning an object; and anything is an object in respect to its provocation of some special activity within a subject" (1961: 167). Thus, the subject not only flows out, but the object flows in, the character of the dialectic being determined by the space–time causal nexus, taken together with the formal experience of the subject and the affective possibilities of the object. This "prehension" is therefore partially determined by the "affective tone" of the complex subject–object relationship. Bergson runs a similar line: individual experience is undivided; the subjective experience of an individual is at once an aggregate of separate fragments *and* a materially indivisible whole (Bergson 1998). For neither thinker is the substance of the world resolvable into pure or independent material forms. Rather, these latter shift and fluctuate in and out of various arrangements.

This is how many noir films analyze the city. The relationship between the protagonists and the urban matrix is constantly shifting from mode to mode, but all the time moving under the influence of an "affective tone." This is communicated by means of a range of techniques including a considered deployment of the *mise-en-scène*, *chiaroscuro* lighting, mobile camerawork and the use of fractured narratives. The urban nature of the filmic experience is delivered through the creation of a distinctive emotional milieu, the shiny rainswept streets, the neon-lit nightclubs and deep, poetic urban landscape perspectives. These are meant to convey an image of the city as the pre-eminent site of danger and desire. It is just at this point that Whitehead's analysis of lived experience is most relevant. What film noir shows us is the utter complicity of the experiencing subject in the construction of this image.

The male protagonist, typically confused, or divided, and often deeply introspective, creates a distinctly modern, affective experience of the city through an immediate transaction, shared intimately by the viewer, of symbolic elements that are presented as quintessentially urban. The capacity of noir to reduce the complex modalities of experiencing to the simple "fact" of immediate, lived experience represents the urban field as a dynamic transactive phenomenon, where the subject is not different from the act of experiencing, and the object is not either. This combination of affective tone and immediate experience presents the disordered and violent city as an absolute reality, to which we can only mutter our assent. Crime, sexuality and death are the parameters of our lived experience, played out as intensifications of the urban matrix, *specific points* where a hole opens up and the artificial, "civic," rational city is sucked into the natural cauldron of human desire whence it first emerged.

The lighting of film noir is an important aspect of its representation of the urban realm. Noir owes much of its repertoire of lighting effects to early German cinematic expressionism, known for the intensity of its interrogation of the personal (De Certeau 1984; Spicer 2002). Deleuze identifies in this Germanic cinema "an organic zig-zag line" which passes through things and people, and imbues them with a nervous energy (Bogue 2003: 59). In *Detour* (1945) – whose very name invokes the clinamen – we find an expressive visual register pitched so high that the path into darkness can barely be seen. Only the darkness itself is visible, a palpable signifier of grim necessity. The lighting of this film perfectly exemplifies Arnheim's description of "the artistic use of colour and of the absence of lighting" (Arnheim 1958: 62). Black and white film provides the perfect medium for the struggle of the One and the Many: "The primitive but always effective symbolism of light versus darkness, white purity versus black evil, the opposition between gloom and radiance, is inexhaustible" (Arnheim 1958: 62).

This merging of agonal movement and image in film, expressing the immanence of subject and object, forges an identity of matter and light. The image is movement as matter is light.

Light, therefore, is in things, it actualizes the thing for vision (Bogue 2003: 34). When something is not lit it cannot exist. The darkness of the noir landscape is the darkness of the nonexistent; subjects, elements move in and out of existence seemingly arbitrarily, but their contingency is always conditioned (not motivated). Both *Detour* and *Kiss Me Deadly* (1955) use a visual register characterized by a hard, impenetrable light and a deep, black dark to dislocate and disrupt meaning. Noir lighting techniques such as the variability of lighting positions, deep focus and low-key light placement serve in both films to illuminate a resistant, cold world. In *Kiss Me Deadly* this world is distorted by the use of the 3-D deep focus technique of *Citizen Kane* (directed by Orson Welles, 1941), so that an anamorphic surreality provides a counterpoint to the studied violence of Mike Hammer, "a serial killer with a detective's licence" (Davis 2001: 40) (see Figure 5.19). The lighting highlights the disjunctive nature both of Hammer's right-wing value system and the ambiguous morality that pervades the film, which is a "left-liberal critique of capitalism," an "allegory of a society that seems about to self-destruct" (Spicer 2002: 73). *Kiss Me Deadly* ends with an apocalyptic nuclear conflagration that is a *tour-de-force* of noir lighting. The radioactive substance that is the center of the film's narrative explodes like an atomic bomb. A Los Angeles beach house goes up with it, and the protagonists escape through a series of moving, interlocking planes of jet black and white light, as through a portal to another world. Flames arc against a deep void and the huge elongated shadows of the fleeing couple are flung diagonally along the beach, as foaming sea waves, irradiated by nuclear light, spin into the darkness. These two, it is clear, are escaping into nothingness, the nonexistent, the absent.

In *Detour*, immense black wings fan out behind the bad girl in the hotel room as she lifts the phone to tell the cops her hapless sucker has killed a man. All that exists (all that is framed on screen) is the image of the face: the girl who emasculates and loves it. Her black brows

FIGURE 5.19 *Kiss Me Deadly* (1955). Director: Robert Aldrich. Courtesy of MGM.

arch and her dark lips curl: holes in the face, the planes of the face lit against the pitch beyond. The modern primitive, forcing its way out of the darkness, facial holes streaming evil, a Gothic pre-organic vitalism spreading through the closed system of the room. This energetic force of light connects all the things it touches, and draws them into its web of changing and developing relations: her eyes, her earrings, teeth and lips, the bottle, telephone and lampshade, a pliant geometry which includes the angle of her brows in the same diagram as the plane of shadow on the wall behind. Here, light and shadow are separate infinite forces in perpetual conflict. It is this confrontation that generates the geometry of shades, and through the struggle of light and shadow the intensity of her will moves in a jagged pattern at the very edge of the visible. Now we know she will die. She is too luminescent to live. Too brilliant, too dazzling, too full of the terrible primitive fire that burns the world.

This blazing light force has its origins in the German expressionism of Murnau (*Sunrise* 1927), Von Sternberg (*The Blue Angel* 1930) and Lang (*M* 1931), and in the baroque. These film makers had tried to transcend nature and discover a spiritual destination in the dynamic sublime of Kant (Žizek 1993: 199–226). This attempt at transcendence can be aligned with the discovery of something else, described by Deleuze as the light of "a non-psychological life of the spirit which belongs neither to nature nor to our organic individuality, which is the divine part in us, the spiritual relation in which we are alone with God as light" (Deleuze 1992: 54). Night, woman, death. The image of the nocturnal city is an image of modernity, an image based on transgression, and the city as the geographical location of the most intense transgressions. In noir the dark city wears a halo of fire.

The recovery of the open whole

In *The Big Sleep* (1946), Philip Marlowe is hired to find a blackmailer and is drawn slowly into a complex modernist geography of crime. His search takes him back and forth across the urban landscape of Los Angeles, a landscape that has been crossed many times since, by geographers in particular, mostly also drawn, like Marlowe, by disciplinary curiosity to discover how the specific is related to the universal. Writers like Banham (1971), Davis (1990, 2001) and Soja (1980, 1989, 1996) have explored the spatial and temporal intersections of Los Angeles and all found in its labyrinthine spatiality a passage from the local and specific to the global and entire, from subset to open set. How are the relations between the Many and the One investigated by Marlowe, in many ways these urbanists' precursor, in *The Big Sleep*?

The film puts the Many on the screen. LA is depicted as the site of a coming together of a multitude of everyday things, but things whose everydayness is rooted in an arch-modernist formulation of discrete particles in motion, propelled by forces beyond their ken. The signs of this energetic multiplicity are everywhere; seven different firearms, mostly other peoples', pass through Marlowe's hands in the course of the movie; countless automobiles flash through the narrative with sleek modern efficiency; there are multiple phone calls, street signs, even many more trees than in other noir. People are always going through doors, from one side of the film to the other, from the known to the unknown, from closed and framed to out there, "out-of-field." The film traces the urban space of LA by foot and by car through a series of chronotypes: streets, gardens, bungalows, apartments, highways, nightclubs, carparks, restaurants, alleys and bars. These singular locations form a closed set of spatial elements that fills the screen space. They are cut out from a much larger set of elements and framed as a particular type of modern terrain – a *divided* terrain – through which the detective, an indeterminate center, a living image of marginality, moves and around whom the objects of the city swarm and part as he proceeds. The protagonist is locked, always, inside the frame, in the assemblage itself, with the moving elements that it is his fate to sort through, take apart and reassemble in a new configuration.

All the moving elements connect to the larger picture in some way however. Every distinct entity is part of a larger whole. Marlowe's task is the determination of that whole, a task that, ultimately he must fail to complete, because to understand the parts is not to approach the whole – and he can never really understand the parts because they participate in an indeterminate whole. That is why the characteristic Marlowe pose is quizzical/serene; he wants to be comfortable with the unknown for that is to be modern. But the narrative always moves off into the vast uncertain out-of-field, connecting galvanically with all the things in the world – the places where the cars come from, the economic universe of capital, the intimate gestures of love. Critically, by means of creative operations within the world, operations unique to cinema, there is a movement towards a beyond-the-world. "The creativity of the world," Whitehead writes, "is the flying dart, of which Lucretius speaks, hurled beyond the bounds of the world" (Whitehead 1961: 177). *The Big Sleep* presents LA as a world cut out of the world. It is a set hived off from an even larger collection that changes as its component elements change. But even this is not the set of all sets – this is not the whole. "The whole is . . . like thread which traverses sets and gives each one the possibility, which is necessarily realised, of communicating with another, to infinity" (Deleuze 1992: 16–17). A closed system is never absolutely closed, but is both connected in space to other systems by a thread, and integrated into a whole "which transmits duration to it along this thread" (Deleuze 1992: 17). The whole, therefore, is not beyond, but inside – and this is the peculiarly modern formulation of the One and the Many, a formulation that *The Big Sleep* serves admirably to illustrate. Elsewhere – and this Marlowe does know – is in here. The thread Deleuze speaks of is also the thread the detective follows, the line of clues he pursues through the labyrinth of the minotaur.

How does *The Big Sleep* present the open whole? The movie places itself as part of a larger set, a moving image of modernity that is revealed by the assemblage of shots and takes that release it from the flow of matter-energy, the "plastic mass," the stuff of which films are made, Los Angeles, the world. *The Big Sleep* invites us to consider the city as an open whole and itself as a single movement that expresses this openness by means of a distribution of its multiple elements through the various sets and subsets of the film. These systems are traversed by the characters of the film each in their singular way, each calling up the whole by means of their position within, and in relation to, the multiplicity of collectives. The multiple systems of the film are presented through a cinematic technique which shapes the turbulent flow of the world. The diverse modulations of the sensorial, the kinetic, intensive, affective, rhythmic, tonal and verbal are shaped by framing, shots and cuts into a specific configuration of image, light and movement that ignites the space–time field of urban LA.

In the first frame of the film Marlowe knocks at the door of the Sternwood mansion and is ushered (a better word might be "inducted") into the closed world of the old rich, symbolized by the decrepit figure of Sternwood himself, immobilized in a wheelchair in his temperature-controlled orchid house. He must remain here, he tells Marlowe, because as "a dull survival of a very gaudy life" he has no blood in his veins. Sternwood neatly connects the moral system of modernity to the vegetal life of the open world, a movement which is counterposed by the contrast of the physically closed system that his orchids require for life with the spatial freedom of his wayward daughters, and for that matter, of Marlowe. "Their flesh is too much like the flesh of men, and their perfume has the rotten sweetness of corruption," the epiphytic Sternwood complains of the orchids that share his life. Sternwood does not appear in the film again though his off screen presence regulates all its interactions, the still center of a constantly moving collage of characters whose vectors become so interwoven that it is virtually impossible to follow the plot.

In the next scene, Marlowe leaves this man, the source, ultimately of all the corruption and violence in the film, and the butler directs him to Vivienne Sternwood, who is to become the rabbit he pursues through the spatial labyrinth of the city. He meets her in her bedroom, the locus of her *femme fatale* power, and the chase begins. From the private, hermetic, hothouse of General Sternwood, Marlowe steps into the nonlinear space–time territory of the labyrinth, a spatial condition which engenders what might be called the "urban uncanny." This condition is one of slippage and ambiguity between states – reality and nonreality – but also, it could be argued, of movement along passages between the open realm of the sacred, the gift and the restricted, closed systems of the city. In *The Big Sleep* the uncanny is discovered, not by means of an ontological formulation like Vidler's, nor in a theory of perception, like film theorist Strathausen,[32] but in a relation between the closed and the open.

The Big Sleep deploys a changing set of scales, places and modes of spatial practice to present as clearly as possible the simultaneity of the urban uncanny, its embeddedness in everyday life. Benjamin drew attention to the crowds, heterogeneous spaces, scale shifts and corresponding response of the privileged point of view, "the endless quest of the detective tracking his clues through the apparent chaos of modern urban life" (Vidler 1992: 4). Fear, revulsion and horror, he wrote, were the emotions which the big city crowd aroused in those who first observed it (Benjamin 1973a: 131). *The Big Sleep* does not so much evoke as contribute to this destabilization of normative spatial relations, producing an anamorphic effect where "the individual is no longer the origin of the coordinates, but one point among others, and does not know where to place itself" (Caillois 1984: 70).

Adapting Roger Caillois, Vidler argues that the spatial paradigm of total control that generates the uncanny derives from a fear of the dark. He cites Boullée's "architecture of the uncanny" as an experiment in the projection of "dark space" which plays on the shadow/light ambiguities between absolute flatness and infinite depth, between his own shadow and the void. The high contrast cinematography of *The Big Sleep* dramatically manipulates light and shadow both to upset our understandings of spatial configuration and to imply an infinite field beyond/within the two-dimensional format of the image. It implies a metropolitan uncanny, at the center of the urban, in Lefebvre's words "a hole which is sacred and damned, inhabited by the forces of death and life, times dark with effort and ordeals, the world" (Lefebvre 1996: 88). The "fear of the dark" that inspires the uncanny, then, leads Marlowe right back to "the world" where occupying a social space generated by the interactivity of everyday life and the open, streaming multiplicity of the void, he faces the ultimate ambiguity of sex.

With *The Big Sleep* it is possible to pull anywhere on the thread of its densely woven space–time field and have the openness of the world unfurl. It is composed of innumerable instants, each one characterized by a specific site or locale where the solid – the hardware – breaks and disintegrates, while the software – the program – resists and goes on. These jagged moments are not so much privileged moments, essential, summative, as *instants quelconques*, any-instants-whatever.[33] The violence is inherent in every moment, and every moment is like every other: full. The privileged instants of *crisis* (Marlowe discovers the drugged and compromised Carmen at the house of her blackmailer Geiger), *paroxysm* (he encounters

Vivienne waiting for him in his office) and *intensity* (Marlowe chases the shooter Carrol) are such singular instants extracted from *instants quelconque*. Another such moment is when Marlowe and Agnes are in a parked car. Marlowe pays Agnes $200 to tell him where Eddie Mars' wife is. She describes the situation perfectly:

> Joe and I were out riding on Foothill Boulevard a couple of weeks ago. We passed a brown coupe and I saw the girl who was driving. She was Eddie Mars' wife. There was a fella with her, Canino – there's some people you don't forget even if you've only seen them once – so we got curious and Joe tailed 'em. You know where Rialito is? Well about ten miles east there's a side road. Just off the highway there's a two-bit garage and paint shop run by a guy named Art Huck – a hot car drop most likely – and a frame house right behind it. That's where Eddie Mars' wife is holed up.
>
> You're sure of that? asks Marlowe.
>
> Why should I lie?

The *mise-en-scène* is a subset of a subset. Marlowe and Agnes are head to head, virtually touching, on the bench seat of the car. The scene is composed of a series of over-the-shoulder close-ups from within the car. Each shot is framed by its curving surfaces; outside the windows another scene, a larger set, the night-time urban landscape of LA, flows by as they speak (see Figure 5.20). The visual content of the scene is divisible into parts which belong to various sets and the woman refers throughout to another even larger one outside the frame, the road to Rialito, on which a garage stands, behind which a house harbors Marlowe's quarry. Communication occurs from set to set, and to further frames outside the field of the scene. Spatially the landscape of the film is composed as a nested hierarchy of organic and mineral systems which are referred to back and forth throughout the *durée* of the film, protagonists passing from one to the other as the narrative draws them spatially from framework to framework, setting up the city as an interactive ecology itself giving rise to other unseen sets. This is a by-product of the spatial equation of frame-by-frame time sequencing, the temporality of modernity as understood by science and commerce which cuts duration into homogeneous units fixed in space. Bergson calls this the "cinematographic illusion" (Bergson 1998: 322). It is the noise that life rises above.

Noting that humans' most basic experience of the world is one of constant flux, and citing physics, which suggests there are no irreducible bodies in the world, simply "modifications, perturbations, changes in tension or energy," Bergson argues that duration and matter, time and space, are inseparable functions of one another. Matter, he says, forms a world of undivided flux, and life cuts matter up into living containers (Bergson 1998: 249). Life expresses the world of flux; life is a movement, life and matter are movement. What is expressed by the One in this movement remains enveloped in each entity of the Multiple, in a hidden, partially evident manner. The enveloped One is a sign that points beyond itself. Crucially, that which is enveloped may be unfolded. Through movement the objects of a closed set are related to duration, our basic psychological sense of time, the dynamic continuation of a past into a present and toward a future that is truly open and unforeseeable. When Marlowe and Agnes are sitting in the car (the closed set, the space container) their bodies are continually changing in terms of their relation to each other, and the whole is changing too: its spatial parameters are moved and the story follows their movement. Through this movement in space, and by means of the movement of the camera (which enables us to see movement

FIGURE 5.20 *The Big Sleep* (1946). Director: Howard Hawks. Courtesy of Warner Bros.

disengaged from bodies) the elements of the closed set are related to open duration. The whole divides itself in the elements and the elements are reunited in the whole, and between the two everything changes. This occurs in every scene, and it is an accurate account of a particularly twentieth-century urban spatiality in the form of the networked city. Imbricated practices of representing and inhabiting space show film noir itself to be a parasite of popular cultural manifestations of late modernity. The popular film genre, the musical, for instance *Singing in the Rain* (1952), presents the city as an unproblematic *setting* for film life, rather than as a condition of modern existence.

One of the most characteristic intensifications of the urban field in film noir is the act of violence. Murder, particularly sexually motivated murder, brings together a range of different taxonomies – geographical, social, political, legal – within a new set. Each new intensification of the field has a unique set of descriptors, not the least of which are the geographical coordinates of the murder scene. Whether it takes place in an alley or a hotel room, the murder scene is the point at which major taxonomies of noir (dark and light, life and death, crime and the law) intersect to introduce a modern spatiality that is constructed by pure transgression. In *Kiss Me Deadly* the protagonist Mike Hammer picks up a blonde hitch-hiker in a trench coat (and that's all) who spins a mysterious story (see Figure 5.19a). Eight minutes into the film she is dead, having been apprehended, tortured and murdered, and Hammer is thrown into a dizzying sequence of situations that culminate in a murder scene – the flaming house – of nuclear intensity (see Figure 5.19d). (Certainly nuclear conflagration is the closing metaphor in this film about the black market in radioactive materials.) The first crime in the film, however, even more tellingly makes the point. We do not see the actual murder of Christina. The camera is in a room with a chair, a table and a bed. In frame are the legs of the bed, and Christina's naked legs dangling from the table. There are two sets of men's legs – trousers and leather shoes – and also dangling into shot from one of the men's hands, a pair of long pliers. The woman is screaming. She goes limp.

> She's passed out. I'll bring her to.
>
> If you revive her, you know what that'll be? Resurrection, that's what it will be. And you know what resurrection means. It means raise the dead.

Something unspeakable has been done to her, something primitive, involving the pliers and her vagina. It is an event, so feral it is modern, that suddenly and shockingly stakes out the territory of the whole film as an investigation of the primitive modes of modernity. It constructs such an intense spatiality that it must be cordoned off, partitioned, bracketed: "CRIME SCENE DO NOT CROSS." On discovery, the crime scene becomes part of a legal system that constructs it anew by means of its own codes, techniques and practices, so that it cannot be separated from the organization of signs that surrounds it. This "intensity" is super-sensitive. Tiny, microscopic changes can quickly be magnified and brought up to macroscopic level. The shamus finds a glove, a blonde hair, a smear of lipstick on a cup and huge shifts take place in the relationships between elements of other systems. "It was either you or your sister, angel, which of you is going to own up?" Which makes a big difference. Specific local conditions and local knowledge move from level to level of the situation under investigation. The local even designates the site within the global at which specific multiplicities and the open whole conjoin. Movement from taxonomy to taxonomy is therefore significant. The local and the global reproduce each other so that at certain times universals (values, laws) appear to run through the crime scene system (i.e. pre-murder), at other times not, such as when the maverick shamus appears with his subversive agenda. The landscape of crime is multiple and diverse: each taxonomy relates a specific patch of urban terrain to others, but no taxonomy can bind them all. Thus does crime defamiliarize the city – or in noir, both aesthetically transfigure it and renature it – without social transcendence, from within, through the internal operations of its systems, which move continually from equilibrium to disorder, switching levels and continually becoming different.

The movement of bodies in space, and of relations between them, a movement which occurs in different time frames in different sets, which jumps spontaneously from level to level at points of intensification, this is the habitus of the modern noir subject. This subject is in the grip of larger forces, the "laws of fate," which are disturbed "without reason," by chance, by the clinamen. Through this encounter the subject begins an interrogation of a situation beyond his or her control. The real action quickly moves out of frame, leaving a hole through which the noir film finds a way to re-present the open whole from moment to moment, scene to scene, by slicing its material out of the movement of *durée* itself.

It is found already in the poems of Baudelaire, of whom Benjamin says "he indicated the price for which the sensation of the modern age may be had: the disintegration of the aura in the experience of shock" (Benjamin 1973a: 196). In *A Une Passant,* for example, Baudelaire describes the quintessential urban encounter in a "deafening street", a brief epiphany in which he glimpses "a fleeting beauty" in a "flash of lightning – then darkness" (Baudelaire 1961: 221). An unknown woman's face comes into the poet's field of vision, bringing to the city dweller a "figure that fascinates," and through this a tremor, a rapture of sexual shock with the street "screaming all about me." The important thing here is not so much the nature of the shock, however, but the wild current which propels the woman past him. Like Bigelow, and Neff, she is cut out of the stream of life, vividly, dramatically, but momentarily. Through this bewitching "moment of enchantment," the poet makes what Benjamin calls a "catastrophic" connection to *durée.* It is an "actualisation of the *durée* which rids man's soul of obsession with time" and places him within "the realm of ritual." Likewise, suspended outside time, the noir film slices from the uninterrupted stream of Becoming a passage to the uncanny, a modern, urban nature. It is the gift of nature to the denatured urban world.

Film noir, then, invites its viewers to consider the landscapes of the city potentially as sites of events that are at once extremely abstract and shudderingly immediate. Noir films' combinations of image, movement and light depict the flows of matter-energy through all their modifications by means of kinetic, sensorial, affective, tonal and rhythmic modulations. We find that these films are made of the same "plastic stuff" as us, but directors may shape their "signaletic" matter, through the vertical processes of framings, cuts, shots and editing, and through the horizontal processes of long, medium and close-up shots and images, whereas we experience fully the identity of consciousness and *durée*, as well as the freedom and openness of *durée*. All our thoughts and emotions are embodied, and connected to the physical world. When we view a noir film we gather the possibility of a further connection, to the uncanny field onto which the film mysteriously opens. It brings us a condition to which the fragmented industrial city might aspire, a condition which Jonathon Raban considers "a state of natural magic." The city points beyond itself to its other self, its twin, that which remains enveloped in each aspect of its multiplicity, its nature. "For most of their inhabitants, cities like New York and London *are* nature, and are as unpredictable, threatening, intermittently beautiful and benign, as a tropical rainforest" (Raban 1974: 157).

Noir shows how the modern city had to be respatialized in order for lived experience to confront the open whole. It clearly shows how deviation and transgression (the swerve) are essential for human nature to wrest the urban landscape away from its representation as a playground for docile bodies, in order for life to be lived "deeply" as Bataille says it must be. As an interconnecting set of charged sites the landscape provides a shifting, difficult but necessary milieu for the staging of human experience. This was already understood in the baroque period, but the necessary landscape modalities for human enrichment did not make the transference from pleasure-garden to city. Noir tells again how important they are. And they are simple: the nearness of human and natural infinitude; the sweet agon of the sacred; the necessity for transgression. Film noir spreads out before landscape architecture an understanding of urbanism that pushes human nature to the forefront of its depictions of the urban condition. Landscape architecture can call nature back to the city as a parasite of urbanism, in the same way that noir shows us the violent crime as a door opening on a parallel universe of values.

Writers such as Williams, Lefebvre and Serres separate the city from nature. They are explicit about this. Serres says the city is founded on the exclusion of nature. Thus when the garden does appear in the city it is as a sign of an absence. Like the garden, the essential operation of the city is that of sacrifice, the sacrifice of nature. For nature to appear in the city, then, it must become artifice. Olmsted underlines the artificial character of the park when he justifies Central Park as art. But nature is always already in the city. Natural process is the "infrastructure" of the city. Human beings, for instance, are natural, and if, as Lefebvre says, "there is a production of the city . . . it is a production of human beings by human beings" (Lefebvre 1996: 101). The city depends on that which it excludes. This dependence takes the form of a continual renegotiation of its relationship with nature. Nature therefore is the parasite of the city. It is the noise in the system. But because the city itself is naturalized, this nature cannot be seen. The city appears denatured even though it is permeated by representations of nature. Therefore it must be revealed. Urban landscape architecture is thus a matter of revelation. Landscape architecture, however, has given away the key to the revelation of nature under the guise of human accommodation. Is this not the same reason that, despite the existence of the parasite being the condition of existence of the city, we try to expel the parasite?

Public space plays a critical role as the location of social representation – it is in public space that the most powerful representations of nature are staged. A theory of multiple, competing and mutually exclusive public spheres provides a framework for urban space that opens it up to alteration and revision. That this can be achieved through disturbance is demonstrated by film noir. The noir film acts as a parasite that distorts the relationship between civil society and the people who participate in that society. It occurs *in* public space (the cinema), is generally *about* public space (the city) and helps *construct* public space (as uncertain and dynamic). It is a noise that disturbs the even communication of "citizenship values" from the state to the people, and this helps redirect the cultural politics involved in shaping civic spaces. The noir film *is* nature. What landscape architecture can, and must, offer the city is an openness for the city, a technique for "letting go." This requires work in several arenas – physical design, social codes, cultural values – all of which is directed toward the enabling of a parasitic nature, a turbulent, disturbing nature rather than the anodyne operations of urban design with its tropes of square, boulevard and park. Happily, transgression loves order.

Last Year at Marienbad

It is not only cinema that rejects the idea of landscape as a separate, physical correlative for emotional interiority. In the *nouveau roman*, a French literary movement of the 1950s and 1960s, landscapes and gardens feature prominently in the breaking down of boundaries between perception and the perceived. Rather than jump to novels, however, this section examines a film written by the most eminent practitioner of the *nouveau roman*. The sensibility is modern, the *mise-en-scène* is baroque. Two gardens that the Bavarian Elector Max Emanuel made at the time Watteau was painting his *fêtes galantes* feature in French film director Alain Resnais' *L'Année Dernière à Marienbad* (1961), known in English as *Last Year at Marienbad*. Resnais combined a number of Bavarian chateau gardens into the setting for his cinematic exploration of memory and desire, but the most prominent and recognizable in the film are those at Schleissheim and Nymphenburg. Construction on both began between 1684 and 1705 by Henrico Zucalli and was continued by Domenique Girard, whom Max Emanuel engaged from Versailles (Reinhardt 2000). The gardens were initially designed in the classical style exemplified by the work of Le Notre, but developed away from the hard clarity of central axis and subordinate cross-axes towards an organization of individual garden zones arranged parallel to each other.

> Asymmetrical layouts were preferred in which a deliberately planned disorder reigned, and numerous small, intimate garden spaces could reveal the development of a pure, unrestrained delight in play and entertainment. The gardens shut themselves off from the countryside and turned inwards. They were less easily seen as a whole, and no longer contained a single viewpoint from which the overall scene could be taken in. On the contrary, the various parts developed a strong individuality, and sometimes seemed to split off into autonomous areas, only loosely linked by axes.
>
> (Reinhardt 2000: 298)

The relationship between the baroque landscape multiplicities that Reinhardt describes and Resnais' modernist take on the emotional ambiguity of pleasure gardens come together in *Last Year at Marienbad*, scripted by *nouvelle roman* writer Alain Robbe-Grillet (1961).[34] The film is an investigation of the relationship between feelings and affections, and therefore of the very conditions of pleasure. Resnais seems to interpret human subjectivities as, in Reinhardt's words above, "asymmetrical layouts in which a deliberately planned disorder reigns." The thinking, feeling human cannot be seen "as a whole" and "no longer contains a single viewpoint" from which an overall sense of the interactive field of forces and movements in which they are entangled can be taken in. In his novels, such as *Jalousie* (1957), *Maison de Rendezvous* (1977) and *Topologie d'une Cité Fantôme* (1976), Robbe-Grillet famously, and controversially, elides subjectivity altogether.

Last Year at Marienbad presents a situation. An unnamed man X, and a woman, A, are brought together in a palace that has been converted into a hotel. X tries to convince A that they have met before, last year in Fredericksburg, or Marienbad, under similar circumstances, and that something critical happened between them; but the woman cannot or does not want to remember. It was a brief encounter, according to X. They met by chance in the corridors of the hotel, and in its garden where they conversed about everything and nothing (see Figure 5.21a). While X is very precise, there is nevertheless a general sense of vagueness, perhaps because of some impropriety, or simply because A is not playing this game of "We've met before . . ." Eventually she does begin to remember, or construct a memory of what happened. Something momentous: a declaration of love, a seduction, or a rape, a murder even. X maintains that A convinced him to wait a year before they should meet again, perhaps in the same place, and at which point, he says she said they would leave together. For much of the film A prevaricates but, finally, at midnight X and A do indeed leave the hotel but only, as the voice-over says, to "get lost forever" in its garden. The film depicts A's hesitation and X's insistence. For the viewer there is no way to decide whether X is telling the truth, or whether it is A that is dissembling. It asks, can we ever know, between two persons, which is the way to the truth, or are we always "lost forever" in unknowing?

The vast pleasure garden that the protagonists alternately view from the terrace or wander through as they speak, is not a mere backdrop to human frustration and bedevilment. Resnais' stunning black and white images, comprising deep dark blacks and crisp, vivid whites, involve the actors in the *mise-en-scène* as much as in the eighteenth century the new geometries of the baroque ordered the placement of trees on the open planes of the garden. The garden is dark foliage and bright paths. The characters are dressed in black and white evening dress throughout the film. The configuration of geometry and affect within the rational diagram of a baroque garden emphasizes the pre- or post-rational overlapping of temporalities in which they are depicted (see Figure 5.21b). X and A are caught up in a set of co-existing mutually necessary realms – the now, the then, the quotidian, the extraordinary, the spoken, the unspoken, the banal and the horrific. None of these layers of human experience can be separated one from the other. They are intensive, of course – how can you measure emotion? But Resnais reminds us that affect has an extensive dimension. This is one reason why the pleasure garden features in his film. More than this, the chateau gardens represent the movement of the multiple beyond its instantiation in "subject" and "object." Despite the pain and doubt that X and A suffer, Resnais seems to be saying that it is better that life is lived in a more

FIGURE 5.21 *Last Year at Marienbad* (1961). Director: Alain Resnais. Courtesy of Reel Productions.

open – albeit undecidable – economy of affect and disorder. By contrast, the fragile social framework within which the guests at the hotel are caught is a rule of law that permits only a limited response to an existence that is fraught with contradictions. The encounter of X and A dramatizes the dilemma: while at the same time depending on, and perpetuating this social order, they are acting in ways that transmit their own doubt of this stability, and which express an awareness of their vulnerability to prevarication, misprision, loneliness and loss (see Figure 5.21c).

It is no accident that Resnais and Robbe-Grillet chose seventeenth-century baroque gardens in which to explore the geometry of affect. The new ideas of nature and rapidly developing representational technologies of the seventeenth century were used by garden designers to lay out palace grounds through the application of procedures which permitted the introduction into garden space of an explicit connection to something beyond the material conditions of the garden itself. Seventeenth-century pleasure gardens expressed the new rules of nature by ordering landscape elements (visual axes, vegetal planes, lines of conifers), and then disrupting them by means of the introduction of the infinite, the anamorphic diagonal and the theatrical projection of the human body through space. Le Notre, for instance, organized the tensions produced by a precise marking out of terrain into a human confrontation with the baroque loosening of the sacred from religious life. This different order of reality – encountered on the flatbed of the baroque garden – becomes implicated in human life in a new way.

In his *De l'infinito universo e mondi* (1586) Giordorno Bruno (1458–1600) described as perfect, not the circle or sphere, but that "which consists of many parts, disposed in a fixed sequence and closely joined together." There is, he writes, "a single general space, a single vast immensity which we may freely call Void; in it are innumerable globes like this on which we live and grow; this space we declare to be infinite, since neither reason, convenience, sense-perception nor nature assign it to a limit" (Lovejoy 1964: 126). Bruno propounds an infinitely extended universe that has no need for a hierarchical structure, neither center nor boundary. The baroque garden as exemplified by Le Notre combined the rigid geometrical organizing capabilities of astronomers like Copernicus and Kepler with Bruno's rhizomatic spatiality. But these gardens' affective potential is only fully realized when the vital energy of courtiers and nobles courses across their mathematical fields. The sacrality of the ancient sense of the stable, immutable, metaphysical infinite of God is merged here with Bruno's fluid, unbounded network of innumerable objects in space. The result is a garden model that, like the universe, is atomistic, polycentric and autonomous, but also dynamic and open. It is this new garden spatiality – an innovation of the French formal garden – that permits a specific and unique interplay between the human body and the designed landscape. But Resnais' allegory of modern life needed more than this. It required a setting that generates difference by staging the ongoing, unending negotiation between rationality and lived experience.

This negotiation can be traced through the work of Pierre Gassendi (1592–1655), who postulated that atoms had materiality *and* a spark that moves them, an internal force (Debus 1977: 115). According to Gassendi, atoms were endowed at Creation with an internal principle of motion, a "natural impulse," or "force," which always maintains their motions. A force that could transfigure the human body (as in the sculptor Bernini's *St Teresa*) was an active, essential component of a substance, not an external *vis insita* or *vis impressa* (Debus 1977: 115). Gassendi's physics imply a baroque body ablaze with a concealed fire that at its highest pitch could convey the intensity of a physical force. If we combine this conceptualization of immanence with the fundamental insight of philosopher Gottfried Leibniz (1647–1716), we discover what it really is that drew Resnais to the baroque pleasure garden as a setting for his film. Leibniz saw forces as local properties within a dynamic system; individuated non-self-identical points, each of which is spatially related to other points and communicating with them through the godhead, a medium by means of which they stay connected (Leibniz 1973). Each of these "monads" expresses each other, registering through its internal law (calculated by God) every change that has occurred in other monads and responding to them. This formulation precisely describes the body that moves through the spatial matrix of the baroque garden. Versailles, for instance, was created to receive, display, divert and control absolutely the monadic subjects of the king of France. It is just such a subject, calculating, watchful, aware, that glides across the stage of the garden, moving in and out of infinitesimally fine relationships with other subjects, never motionless, always following delicate, carefully observed protocols, cautious never to transgress, and yet transgressing always. The trajectory of this body through the space of the baroque garden simultaneously constructs and destabilizes the ideals of proportion, closure and projection that inform both the garden and the courtly etiquette and ambition that propel the king's subjects across it. As Weiss argues, the courtiers' experience of the baroque garden is a continual negotiation between lived and mathematical space (Weiss 1998). Individual people enter this environmental calculus as beings detached from each other, or at least connected only by a kind of human *mathexis*, whose trajectories through the spatial field of the baroque garden can also be plotted and predicted.[35] The logic of the garden is synaesthetic, not purely visual (Weiss 1998: 49). It serves to activate the tension, complexity and ambiguity that characterize the intersection of the *mathexis* and the social. If, as suggested earlier in this book, and following Spinoza, a body is defined by the possible relations into which it may enter and, in fact, develops desires for relations, then the multiple relations (visual connections, power politics etc.) that criss-cross the spectacle-space of the garden[36] in *Last Year at Marienbad* can be delimiting of an individual's power of acting, or enhancing of it. What unfolds in *Last Year at Marienbad* is a vital process of differentiation that occurs between X and A, a movement that links an indefinite temporality (*Last Year . . .*) with a half-remembered spatiality (*. . . at Marienbad*) in a rigorously structured here and now. Robbe-Grillet says that the film occurs in "a perpetual present," it is "a world without a past, a world which is self-sufficient at every moment and which obliterates itself as it proceeds" (Robbe-Grillet 1965: 152–153).

The application to gardens of a theory of pleasure other than that of Spinoza's would necessarily yield a different interpretation of pleasure gardens. But Spinoza, if anything, is a philosopher of life. That gardens are places where human beings may discover and experience what human life can be, is one of the fundamental insights of Harrison's *Gardens: An Essay on the Human Condition*. In this treatise Harrison explores the ancient wisdom of the philosophers – Plato, Aristotle, Epicurus – who taught in gardens especially created for this purpose (Harrison 2008: 55–81). These are places, Harrison writes, which are "at once removed from and open to the world" (Harrison 2008: 70). It is just this openness to the world, a function of the garden's separation from it, that Spinoza's explication of human pleasure permits us to realize. Spinoza affirms the productive dynamic of the world and places humans within this dynamic as participants in the constitution of its life. If life, therefore, is constantly being constructed, then in pleasure gardens we find a process of world-composition that dissolves the difference between thinking, feeling and living.

Borders and assemblages

Borders have always been tricky. Westphalia, in Germany, for instance, is often associated with fluid boundary conditions. No exact definition of its borders can be given, because the name "Westphalia" was applied to several different entities in history. For this reason, specifications of the region's area and population vary greatly. Things are no different today. If anything borders, edges and boundaries have come under increasing pressure in recent years, politically, biologically, ecologically and culturally. What is a border now? A couple of examples will illustrate the issues. Since 9/11 the United States has revisited its border security protocols and is actively transforming – even ambiguating – the whole notion of what a border is and does. At the same time as it is tightening border security, the US (and its closest allies) want to make national borders effectively obsolete. Now not just thought of as porous or permeable, borders are ceasing to be considered as filters between states or *lines* of demarcation at all, emerging instead as "transnational, increasingly interoperable, systems or assemblages of control technologies strung out along the world's infrastructures, cities, circulations and bodies" (Graham 2011: 42). These "transitional zones," of course, are being established in order to more perfectly anticipate, channel and monitor movement so that acceptable flows can be demarcated from those that are not. As always, it depends often on which way you're facing. Back in the 1960s, in *The Ghost in the Machine*, Arthur Koestler invoked the Roman god Janus as performing the role of gatekeeper, looking both inwards and outwards from political and geographical systems (Koestler 1967: 48). These days Janus no longer patrols a barrier structure, but rather a shifting meshwork that blurs established categories and accepts the international truth that the enemy is always already "within the wire."

Closer to our own disciplinary borders, Kristina Hill declares that the ecological sciences have seen "a wholesale re-evaluation of boundaries" which present "special challenges for professions that propose site designs" (Hill 2005: 131). She argues that in ecology notions of boundary, border and edge are based on metaphors that have been conditioned by "the experience of living in a body that appears to be separated from its surroundings by a skin" (2005: 135). But when graphs replace maps to describe and represent the energy flows between different conditions, and biological relationships yield to explanation in terms of metrics rather than geographic dimensions, the idea of the site as a distinct, bounded entity loses its explanatory relevance. Furthermore the once precise human skin, considered in biological science as a significant boundary, is no longer so discrete. The body, itself once thought of as a geographically distinct site, is now regarded as functioning as a node in a system. "The notion," Hill says, "that humans are somehow separate from the flows of energy and organisms that define ecological systems is one of the most thoroughly reified metaphors in popular culture and has influenced the kinds of research undertaken in ecological science" (2005: 140). This has changed. There is plenty of recent evidence to show that the human body is permeable to many kinds of flows, from energy and materials to organisms that are necessary to maintain bodily health. The major gain here for ecology is that ecological sites, reconsidered as open systems, can be studied in the light of new conceptions of demarcation that depend on the density and intensity of biological interactions over time. It is just at this necessarily imprecise and indeterminate location that a new shifting boundary between the disciplines of ecology and landscape architecture can be theorized.

Hill speaks, perhaps using a rather too edge-informed metaphor, of a "new conceptual bridge" connecting "the contemporary paradigms of ecology" to the "theoretical and practical concerns of landscape architecture" (2005: 150). This is a reference to the new disequilibrium ecology on the one hand and the rise of emergence theory as a guiding paradigm in landscape architecture on the other. Particularly, but not solely, in landscape urbanism. Now that cities are being understood as open, nonlinear systems, enriched like ecosystems, by disturbance, and dependent on feedback loops to achieve greater levels of complexity and resilience, landscape architects working in the urban realm are harnessing the metaphors of flow and self-organization to develop new kinds of solutions to growing urban problems.

But the conceptualization of cities as systems is not itself unproblematic. Landscape architects face new dilemmas as they consider time-dependent, open-ended design strategies for clients and communities used to the installation of the finite, bounded equilibrium landscapes whose passing we are witnessing. All landscapes are vested in value systems: aesthetic, political, cultural. It is these values that are being tested. This section explores, quite modestly, the provocation of traditional values in the reconfiguration of boundaries between what is acceptable and what is unacceptable in the small university town of Auburn, Alabama. Boundaries are significant in the establishment of cultural formations about what belongs to a place and what does not belong; what is natural to a region and what is not; what is foreign and what is not. As Stephen Graham remarks, the new international security protocols construct boundary conditions that determine "the preemptive separation of mobile bodies into kinetic elites and kinetic underclasses" (Graham 2011: 45). Kinetic elites are those movers that are deemed risk-free, while the underclass comprises people and other types of circulation that are deemed risky or malign. The City of Auburn, like all terrain-based clusters of populations in interaction, has its fair share of issues to do with indigeneity, naturalness and exoticism. As always, the negotiation of these issues, particularly in public space, leads quickly into the more obviously value-laden concepts of place, identity, purity and contamination.

The act of drawing boundaries that inscribe an interior and an exterior dramatizes the kinds of movements that take place between them, even if the boundary is itself a zone rather than a line (see Figure 5.22). An element – a biotic agent, for instance – may cross a boundary from one direction, move across subsequent borders and become lodged within a condition, bounded like the others and separate from them. Let us call the directional movement lines *vectors*. A vector may consist of a prior condition that permits the migration along it of a certain agent or population that is in some way compatible with it. Inside the bounded zone there may be other bounded zones that exist in an overlapping or nested way. Each boundary defines what we have called an assemblage, which is composed of different entities that operate together interactively. Like the new international borders an assemblage is both bounded and borderless. An assemblage can also be called an ecosystem, though this term, while capturing the organizational coherence, does not get at the structural, let alone constructional, potential of assemblage. Let us also postulate some *populations*. Through processes of aggregation, disaggregation, appropriation and displacement, interaction and growth, these populations enter into and ultimately contribute to the formation of assemblages.

The assemblage this section explores is public open space in Auburn, AL. Recalling the discussion in Key Concepts, we find that it is possible to consider public space as a physical and discursive territory into which actants may bring their own belongings. Such a milieu is much more than a space. It involves a sense of belonging to and causing change within – like the concept of place, but without the essentialism. Public space grows, diminishes, ebbs, flows and moves around. In any adjacent or overlapping territories elements exchange position, traveling along vectors and establishing new populations and new assemblages.

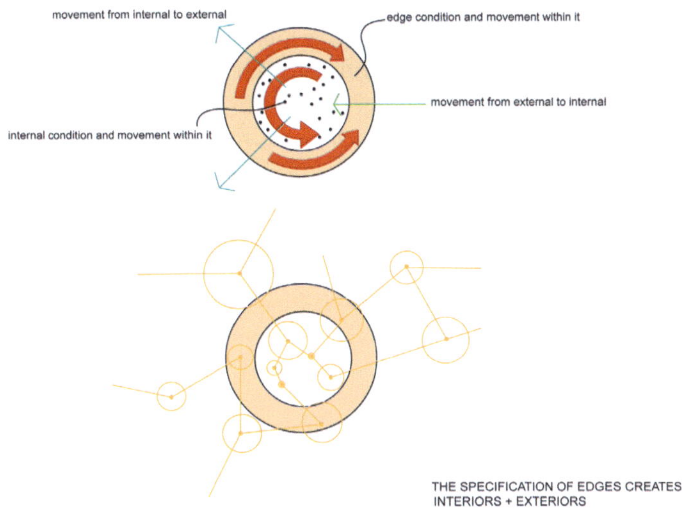

FIGURE 5.22 Diagram showing how the specification of edges creates interiors and exteriors on conditions that are intrinsically without borders. The inscription of borders influences movement patterns. Courtesy Qian Deng, Long Zhou and author.

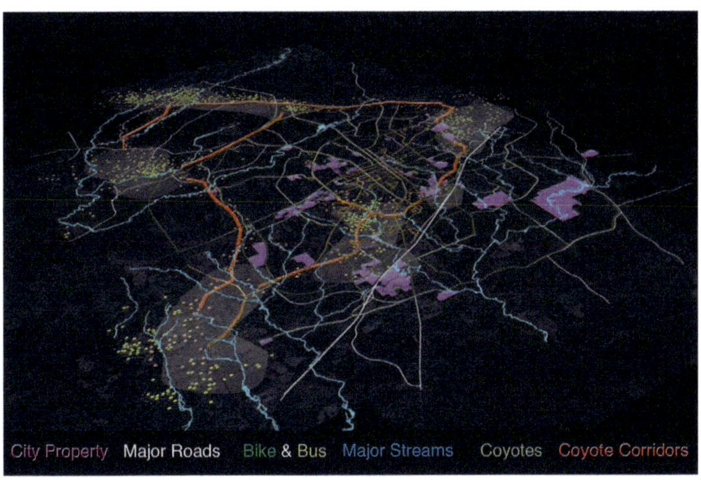

FIGURE 5.23 Researchers in the School of Forestry and Wildlife at Auburn University have tracked coyote movement through and around the city. Clusters and corridors have been identified. The map shows urban coyote movement patterns in Auburn. Courtesy Qian Deng and author.

Populations derive their ability to increase their inherent potencies from the environments they inhabit. Additionally, through their interactions with those environments they change them, and increase the ability of those environments to support and accept new organisms. This can occur at a range of levels. At the genetic level, biological populations' success rates depend on their ability to transmit genetic information across both generations and boundaries. This is as much the case for humans as it is for nonhuman species, though humans act as cultural replicators as well as genetic replicators in the sense that there is also a transmission of semantic information across generations, about the traditions and customs of a particular community, or about formal and informal rules defining structures (for example structures of authority) in a hierarchy. Any assemblage, then, consists of groups of concrete individual entities in relationship to each other and to the larger systems in which they participate. These systems themselves are made up of, for example, nucleotides, chromosomes and genes on the one hand or sounds, words and sentences on the other. When populations of replicators encounter a boundary, it acts as a filter or sorting device, depending on the factors discussed above (De Landa 2006: 44). The nature of the boundary itself is instrumental in how that encounter involves organisms and their environments in the degrees and kinds of adaptation that occur both in the assemblage and its components. Because assemblages are scale-free, new flows and novel structures may occur at multiple spatial scales simultaneously. In the discussion that follows, the City of Auburn is considered as an assemblage that comprises, among many other species, subspecies, circulations and structures the following populations: Creek Indians, kudzu vines and coyotes.

Kudzu, coyotes and creeks

Kudzu

Soil loss due to the exploitation of forest, plain and mountain resources, particularly the transformation of land primordially grazed by large animals into agricultural terrain, has been a constant threat to urban centers throughout history (Crosby 1986). This is because urban centers have a dependency on the agricultural land that surrounds them. Even though methods of soil erosion prevention and control have been known since the beginning of agriculture, many urban hierarchies have failed to implement such knowledge. Sometimes, however, conditions get so severe that potential solutions are applied before very much is known about them at all. When the once fertile plains of the Southwest United States succumbed to over-cropping followed by repeated drought, kudzu vine looked like manna from heaven.

A member of the pea family, kudzu (*Pueraria montana var. lobata*) is a climbing, coiling and trailing vine native to Southern Japan and Southeast China. Outside its countries of origin it is regarded as a weed because it climbs over trees and shrubs and grows so rapidly that it kills them by heavy shading and root competition. The plant is edible, but most often is treated as a pest and sprayed with herbicides. Introduced to the United States from southeast Asia at the 1876 Centennial Exposition in Philadephia (in the Japanese Pavilion), it astonished the public with its growth rate, soon becoming known as "mile-a-minute" (Hinman 2011: 39). Before long it was being planted en masse – with taxpayer funds – first as a cheap fodder crop and then, during the Depression as an agent in the fight against soil erosion. Kudzu was planted extensively in the 1930s on the orders of the Soil Erosion Service (later the Soil Conservation Service). According to a study at a Georgia research station the plant could stop 80 percent more water and 99 percent more soil from running off than cotton could. In 1935 the Soil Conservation Service launched a kudzu planting campaign across the South. Within a decade the amount of kudzu in cultivation went from 10,000 to 50,000 acres. In 1943 an entrepreneurial farmer, Channing Cope, inaugurated the Kudzu Club of America, pledging to grow eight million acres of the vine throughout the South. In 1949 he published *Front Porch Farmer* advising the planting of kudzu to hold and restore eroded farm soil. Kudzu spread through the South of the United States largely as a result of human intervention (Hinman 2011: 42). But kudzu has no natural predators in the US and in 1970 the Department of Agriculture officially declared it a weed. It is now in thirty-two states. The Forest Service has not been able to devise an herbicide that does not leach into surface water used for irrigation. Auburn University researchers are continuing with herbicide trials. Meanwhile Department of Agriculture scientists are exploring whether insect predators native to Asia can destroy kudzu without harming crops and livestock.

Although a vigorous, independent colonizer of exposed soils and interstitial terrains, kudzu would probably not have thrived if it had not been spread by conscious human effort. Some imported plants win their own propagation battles, but others' replication procedures occur as a combination of cultural and organic forces (De Landa 1997). In such cases, landscape transformation becomes institutionalized and carried forward through strategies driven by the market-led extraction of surpluses, by the protection of social investments, and by technologies of control. Eventually humans become caught in a biological net of adaptation and succession, and their cultural energies are captured by the organic replicator. In the case being discussed here, this is the kudzu gene, which now requires massive new technologies of control directed not at its expansion as in the 1930s, but at its eradication. Both organic and cultural change involve replicators moving across boundaries, and new assemblages arise by selective retention of variants, which is what researchers are working on in "the fight against kudzu." The interaction between social and organic agents demonstrates the discursive element in assemblage construction. This powerful semiotic component in the evolution of new landscapes can, as the story of kudzu shows, develop dramatically in favor of the organic replicator when cultural and social institutions such as market capitalism reinforce growth by self-selecting what counts as the conditions for growth. The Dustbowl is an example of cultural adaptations (trying to farm prairie) that had disastrous – let's say maladaptive – organic consequences. In the case of the Dustbowl it was the manipulation of land tenure and land use policies by corporations and their state and federal agents for export agriculture that imposed maladaptive conditions on small farmers, leading directly to the success of the kudzu genetic replicator in the South. Botanically speaking, "weeds" are plants that spread rapidly and outcompete others on disturbed soil (Crosby 1986: 149). They are specialists at colonizing soil that has been stripped by landslides, floods, fires – and poor farming practices. Durham cites the imposition from above of habits and customs that are maladaptive as a major example of opposition between genetic and cultural replicators. If the discursive element of an assemblage refers to the way in which the assemblage enters into discourse – how it is described, framed and allocated meaning – then here it was the application of policies and procedures from above that contributed to the rearrangement of connections between farmers, soils, plants and federal agents.

We see with the arrival of kudzu into the Southwest and the consequent reformation of the assemblage, a classic example of an open system (which includes humans and nonhumans) being subject to disturbance and social and biological conditions adjusting accordingly. The social replication of values associated with the new entrant to the assemblage in the first instance caused acceptance of kudzu as a friend, followed by the gradual conversion to its current status as enemy and concomitant measures to achieve its eradication. *Pueraria lobata* never changed, *Homo sapiens* did. The result of this negotiation is that an extremely efficient genetic replicator from Japan has taken over vast tracts of disturbed terrain in Auburn, AL and can consequently be seen all through the suburbs, along highways where it smothers tall trees and is cursed by travelers, and is spreading horizontally across public open space where it is mown periodically by contractors.

Coyotes

In the United States the pre-European distribution of the coyote (*Canis latrans*) was confined to the arid regions of the Midwest, from Alberta, Canada in the north to Mexico in the south. Between 1900 and 1939 coyotes began to extend their range eastward and northward, migrating into the Mississippi Basin and moving north to Alaska. After the 1940s they penetrated into the eastern states, moving into Ontario first, then down into New York, New Hampshire and Massachusetts and finally through into Alabama, Georgia and Florida. The migration of *Canis latrans* genetic material into the east was directly related to the eradication of the native wolf population in these states and the creation of more and more grassland habitats where mice, voles and hares thrive (Grady 1994: 28). Because of their extremely efficient genetic replicability, rivaling rats – and humans – in their ability to survive under almost any conditions, coyotes have become the most intensely studied predator in North America.

Auburn University's Department of Forestry and Wildlife researchers have been tracking the coyote population of Lee County, AL for many years.[37] According to their most recent data, there are approximately 600 coyotes in Lee County.[38] These animals drift across the boundaries between urban, semi-urban and rural zones in packs of six or seven and in pairs, defending and extending their home ranges and core areas, and building territories through negotiation with other populations. The home range of a pack consists of the entire area it patrols, which depends on such factors as population density and the availability of food. A home range can be anywhere between three or four square miles and forty square miles. Within the home range is a smaller area that is defended, though not strenuously. This is the territory. A coyote that wanders into a pack's territory will be watched and perhaps escorted out again (Grady 1994: 40). The territory is the pack's regular hunting area and its topography – and geography – are intimately known by each member of the pack, even young pups.

They are the largest omnivorous wild animals living in Auburn. Like many other megafauna they are semi-nomadic.[39] In this town they move from place to place (settling down briefly in January to have their litter), through a forested environmental field that largely ignores the human structures of the city even as it coincides with them. What kind of ecological relationship do coyotes have with humans? While coyotes do not prey on humans, the reverse is often true, so the human–coyote relationship could be said to be that of a predator species to its prey. Even though humans do kill coyotes, they are not dependent on them for food. Ecologically, therefore, their relationship might be seen as commensal, where one species, the coyote, benefits from the habitat created by urbanization, while the other, the human, is not affected in any significant way. But it is probably more accurate to say that the two species are equally involved in what is sometimes called a network mutualism. This term describes a relationship in which species have an indirect effect on each other, in this case by means of a food chain that extends from insects through birds to small mammals – all of which contribute to the biotic health and vigor of the urban system – and beyond these to the forage that coyotes find in rubbish bins and pets' bowls. As a predator of rodents and insects the coyote benefits humans by removing pests and limiting the spread of their populations. A recent study of Auburn's coyote population shows that around 50 percent of their diet consists of rabbits, rodents and deer (the latter being mainly roadkill) and 15 percent anthropogenic (what humans inadvertently provide), with the rest being made up primarily of persimmon, native fruits and grasses, exotic fruits and the occasional birds and insects (Santana

2010). The number of North American native birds killed by coyotes is less than 0.1 percent of the two billion killed annually by domestic cats. The number of North American human infants killed annually by coyotes registers as zero.

In terms of "lifestyle" the coyote population enjoys a relative autonomy and a direction of its own. Coyotes are both migrants and nomads, for these are two different things. The migrant goes principally from one point to another, even if the second point is uncertain, unforeseen or not well localized. But the nomad goes from point to point for a reason or out of necessity. For nomads points are relays along a trajectory. These attributes, in coyotes, are mixed, as a coyote population seems always to be in a migrant condition, flinging off individuals who move to new habitat structures to create new possibilities, at the same time as the main body of the population, in the case of Auburn, moves continually through the gradient of the city, eating and drinking (see Figure 5.23). What the coyote population of Auburn draws our attention to is its continual construction, or inhabitation, of a parallel landscape, a layered landscape whose spaces are not closed and marked out into territories and milieux according to fixed rules, but which are open and indefinite and always negotiable. The coyote population distributes itself throughout a space without borders or enclosure. "The life of the nomad," say Deleuze and Guattari, "is the intermezzo" (Deleuze and Guattari 1987: 380). It is a movement between other movements, an interior movement. Organized by the rule of law, human Auburn is a static space defined by walls, enclosures and paths between enclosures, a division that nonhuman Auburn is continually redefining as a smooth continuum of "habitats," "ecotones," and "biotic communities." As the largest representative of wild Auburn the coyotes adumbrate the possibility of a landscape that has no points, no lines or even land. As a deterritorialized agent whose movements through the city are directed by the logic of the pack, the coyote makes of the whole urban landscape an edge condition. Coyotes move through the urban forest as barracuda through the swells of the sea, making the landscape just as they are made by it.

Like the kudzu vine, coyotes rearrange the assemblages with which they come into contact. While their co-existence with Auburn humans may be commensal or even mutualistic, nevertheless these descriptors are ecological in the restricted sense. They do not take into account the human values that systems ecologists such as Kay (2008) and Lister (2010) insist are productive elements of ecological succession. The presence in Auburn of powerful genetic replicators has caused a reaction; while coyotes have acclimated to urban conditions urbanites have not acclimated to them. There are reasons for this. Small cats and dogs are a food source for coyotes, albeit not a regular one (Gehrt 2007: 308). The Auburn packs are getting bigger, dispersing greater numbers of juveniles into the urban woodlands all the time.[40] Pet food goes missing from the porch more often. Being opportunistic replicators, coyotes sometimes breed with female dogs, producing a larger specimen known around Auburn as the "coydog." Coyotes have a dozen or more vocalizations so there often seem to be more "out there" than there really are. Their howls are common at night, especially on the fringes of town. The Auburn Wildlife Services Office and the Department of Conservation periodically trap coyotes, when the complaints get too numerous to ignore. The trapped coyotes are put down with sodium penthobarbytol, rather than relocated and released, as they would simply come back. Eight coyotes were trapped in Town Creek Park in 2010. In November 2011 trapping recommenced and went though until the spring. However, as the Wildlife Officer said at a public meeting, as soon as a bunch of coyotes are moved out, new ones will come in. "They will always be here. Auburn is a green space city – coyotes love it."[41] The eastward movement of these highly successful genetic replicators in the twentieth century reflects in reverse the westward migration in the eighteenth and nineteenth centuries of another highly successful population of replicators, like coyotes opportunistic, omnivorous, precociously adaptive, the American pioneers.

While the coyotes of Auburn, AL are classified as urban, ex-urban and rural populations, this human taxonomy has little to do with their lives and movements. Coyotes are swift, determined vectors whose trajectory has taken them across a species boundary into a condition of continuous encounter with humans. The primary realm of this encounter is public open space, be it urban stream corridor, woods on the fringe of a park or an urban cemetery. Camilla Fox of the Animal Protection Institute in Sacramento notes that, while most people are unaware that there are coyotes living in their midst, increasing human–coyote interaction is an urgent issue facing urban administrators and wildlife managers alike. In the Chicago metropolitan area most large green spaces are occupied by coyote packs of up to twelve individuals (Fox 2006: 287). In many communities the first response to a sighting is fear and a knee-jerk reaction to trap and kill whatever animals may be living in the area (Fox 2006: 288). In many states, such as Alabama, the coyote is classified as a non-game "predator" or "furbearer species" and may be killed year round in unlimited numbers. In Alabama there is no closed trapping season and no bag limit. The question facing landscape architects is, how do we make a world in which coyotes and humans may co-exist? Recent developments in the sociology of science suggest a way forward, providing direction for the design of public open space.

In *The Politics of Nature* (2004) Bruno Latour develops a comprehensive attack on the division of the social, traditionally regarded as concerned with matters of value, from the realm of science, which has always considered itself concerned exclusively with facts. In the world of science there were objects (nonhumans like coyotes, rocks and micro-organisms); in the world of society there were subjects (social beings). Planet Earth was neatly divided into two types of being, two types of knowledge and two kinds of reality. The problem of science was how to represent nature to social beings – of which humans were the only class (Latour 2004). Landscape architecture joined in the game: its problem was how to construct a world for humans from the realm of objects (nonhumans were not regarded as part of the collective for which landscape architects worked; they were regarded as a separate constituency). Part of the problem, well known among landscape designers, was how to develop a visual language adequate to their complicated subject matter, in other words, how to represent nature to social beings. Latour argues that things have changed. We realize that it is dangerous and misleading to think in terms of two assemblies, science (fact) and society (value). We must, says Latour, get rid of the notion that there will always be two blocs, nature as it is and the representations we make of it. He suggests that the distinction between nature and society should be blurred, and that we should think of the inhabitants of the Earth as a collective of humans and nonhumans that are not considered epistemologically or ontologically distinct (Latour 2004: 41). This means abandoning the notion of nature and replacing it with the idea of naturecultures, an assemblage of humans and nonhumans. In doing this we still leave intact the two elements that matter most to us: "the multiplicity of nonhumans and the enigma of their association" (Latour 2004: 41). Accepting this idea as a working proposition, landscape architects can concern themselves with where this collective comes together, and how the collective becomes gathered into a whole. It implies a search for what makes up the common world. Instead of oppositional approaches to the gathering of humans and nonhumans, we can foreground the webs of cohabitation and encounter that landscapes comprise, permit and sustain. We can work on an account of human–environment relations that emphasizes the affective nature of human relations

FIGURE 5.24 The author has designed a number of public open spaces where humans, kudzu and coyotes may encounter each other. This design, near Town Creek park, shows a retaining wall with coyote dens let into it, and a kudzu pergola that measures the daily growth of the vine. Courtesy Qian Deng and author.

FIGURE 5.25 Another design at a Water Services facility, develops a "critter orchard" nearby a known coyote corridor. The orchard provides seasonal food for the creatures that attract coyotes, and shade, shelter and fruit for the people that work at the facility. Courtesy Qian Deng and author.

with the nonhuman and with other humans, in the context of constructed landscapes. We can consider, further, that landscapes encourage a subtle but passionate confrontation with what it is to be human in a constructed world. They do this by enabling specific kinds of encounter with other beings and species that share this world – such as urban coyotes.

While there is more than a single site for the collective encounter of humans and nonhumans in the constructed world, the most charged terrain is that to be found in the public realm. Public space may well, in fact, be the primary site of this encounter, if we extend the notion of the public to include nonhumans (which Latour does) (see Figures 5.24 and 5.25).

An important consideration in the theoretical construction of public space as the primary site of human–nonhuman encounter is its operation as an assemblage, an open system, a multiplicity. Public open space is peculiarly sensitive to changing conditions, and itself changes along with these as it is continually constructed by its users. When we say that people construct space we can see through that very possibility that space is an active component of the social assemblage. Through the construction of space this assemblage can itself be constructed, or maintained, or challenged. And when we refer to open space we refer not only to the spaces of the city but all the spaces of our lives.

Creek Indians

The eighteenth-century ancestors of the Creek Indians who lived among the woods and streams in the area now occupied by Auburn, Alabama were themselves descendants of the peoples called by archeologists the Mississippians. These peoples developed a complex culture, living in fortified towns characterized by large earthen mounds; they maintained trade networks, engaged in warfare and shared symbols and rituals. The Mississippian culture emerged around the confluence of the Mississippi and Ohio Rivers during the period 800–1000 CE (Blitz 2008: 38). Archeologists and historians estimate the population of Native Americans in the whole Mississippian world as approximately 1.5 million, with a 90–95 percent collapse after contact, due to the contamination of their genetic spectrum by new replicators. Prior to the arrival of the Spanish, followed by the English and French, the Indian polities most associated with the South did not exist. The Cherokees, Choctaws, Chickasaws, Seminoles, Catawbas and the Creeks came into existence some time in the late seventeenth and early eighteenth centuries, in response to contact between the Old and New Worlds (Ethridge 2003: 22). Although already accustomed to movement both cultural and geographical, Native Americans entered into a period of migration (not nomadism, which characterized their early geographical dynamics) that required of them a considerable openness to ecological adaptability. Two massive migrations of genetic material occurred. The first was the dispersal of the Mississippi culture, which lead to the inhabitations of the Coosa region in North Georgia between 1000 and 1400 CE. Cultural replicators flowed south from the area of the confluence of the Mississippi and Ohio Rivers and hardened into the rituals, practices and material cultures of the Maskoki peoples. The second, precipitated by the Spanish entrada of the sixteenth century, was a migration from the Coosa province south and east, which led to the development of the four "power towns" of Coosa, Apica, Coweta and Kasihta (Wickman 1999). It is possible to understand this, in effect, self-replication as a kind of traveling assemblage or moving ecosystem comprising a whole array of species that migrated together, entering into new relationships with the ecologies of new regions – humans and their, as Crosby puts it, "servant organisms" (Crosby 1986: 25) (see Figure 5.26).

The new country of the Maskoki (Creek) peoples encompassed five significant provinces and four corresponding ecosystems: the Conifer Forest ecoregion of the Coastal Plain, the transitional Mixed Forest ecoregion of the Piedmont and Fall Line, and the Mesophytic Forest of, respectively, the Ridge and Valley and the Appalachian Low Plateaus physiographic provinces. All four regions lie within the Temperate Broadleaf and Mixed Forest Biome of eastern North America (Ethridge 2003: 37). Each forest system was different, and each had its own mosaic of topographical and vegetative features, coalescing in both forest and non-forest ecosystems. The seventy-three Creek towns that by the eighteenth century comprised

FIGURE 5.26 The migration of the Creek Indians from Mississipian terrain near the confluence of the Ohio and Mississippi Rivers to North Alabama, and from there, after contact with the Spanish, to the lower territories of Alabama and Georgia, near Auburn. Courtesy Qian Deng and author.

FIGURE 5.27 A map overlays the nineteenth-century tribal territories of Creek Indians in and around Auburn, with the plan of the city as it is now. Courtesy Qian Deng and Sudha Thapaliya.

the Creek Confederacy (itself a construction) were mostly located within the ecotone of the Fall Line, which is the transitional ecoregion between the upland Mesophytic Forest and the Coastal Plain. Although in this large area of approximately 62,000 square miles, the flora species are quite variable, depending on the underlying substrata, soils and elevation, it was dominated at the time of the Creek inhabitation by an oak-hickory-pine forest (Ethridge 2003: 37).[42] The socio-cultural geography of the Creeks, like the physical region with which it interacted, was not a single bounded entity. Its borders were amorphous and its population was culturally diverse and socially heterogeneous. At least five separate languages and their dialect families were in use throughout the Southeast in the sixteenth century, making it impossible to assign specific geocultural borders to these families (Wickman 1999). Most researchers indentify the rise of Mississippianism with the introduction and spread of sedentary agriculture, particularly maize. Mound-building itself is a sedentary occupation (Blitz 2008: 39; Wickman 1999: 26). The geographic migrations that occurred as a result of introduced pathogens fragmented this culture and its core elements relocated south and southeast during the period 1540–1763 (Wickman 1999: 40).

For about 300 years, then, Creek Indians interacted with the generally complex adaptive ecosystems of the Piedmont Fall Line ecotone. These ecologies were primarily disturbance-driven: fire played a considerable role and determined much of the species composition. Both the flora and fauna have co-evolved with the forest and have adapted to the periodic forest fires. "The fauna in a fire-dependent ecosystem must be able to tolerate and/or find refuge from low-level burns. Ground-nesting birds, for instance, are relatively few in these forests, compared with canopy nesters and cavity-nesters" (Ethridge 2003: 38). Species such as reptiles and amphibians that live in or that can escape to water are plentiful, and small mammal species that live high in trees or burrows under ground are also numerous. Large mammals – white-tailed deer, beaver, otter, rabbit, bear, bobcat, mountain lion, wolf, human – have to be able to run fast, from each other as much as from fire. But the Neolithic hunter-gatherer Creeks were also agriculturalists, and it was the raising of maize that occupied their days. Creek towns were largely farming communities comprising family-worked lots in a community field. The women, who were the farmers, practiced swidden, or slash-and-burn, agriculture. The fields were often located in the flood plains of the rivers with the houses situated on the flood-free terraces above them. Slash-and-burn agriculture is usually regarded as inimical to ecosystems, since it requires the continuous burning of woodland as planted areas become increasingly infertile. The Creeks prepared land by ring-barking trees to open the forest canopy to the sun, and then burning off the undergrowth, leaving the trees gradually to fall and rot. Every few paces they planted three or four kernels of maize together between the charred trunks, piling soil up around the maize stalks a few weeks later and planting beans to climb the stalks. Since the Creeks had not domesticated beasts of burden and did not have the plow, this method, which could even be carried out by children, was the most efficient and productive that they could devise (Crosby 1994: 172). In the winter most people were involved in the hunting of the aforementioned deer, beaver, muskrats, red fox, raccoon, bears and other species "under the pelt." This activity caused a seasonal migration cycle that had the hunting parties returning to the towns in the warmer months. However, the number of people "in the woods" during the hunting season left the towns quite empty. Archeologist Chad Braley estimates that 300 hunting camps could be made in one season (Braley 1994: 49). The hunting was done with bows and arrows and fires were often set to entrap deer or run animals out of hiding (Braund 1993: 65–66).

Euroamericans arrived along the Saugahatchee River, in the area around what is now east Alabama, in the middle of the eighteenth century. By the end of the 1700s a frontier exchange economy was well in place, with a great variety of goods, services and labor moving through the Creek territory as people attempted to balance their commercial interests with their daily needs. European goods flowed into the system through a larger exchange network. Three interdependent economies emerge as a Creek Indian social ecology: hunting and gathering in the rich ecotone, farming on the fertile flood plains and trading with the seemingly affluent

Euroamericans. By the turn of the nineteenth century a hunter-gatherer/farming/trading assemblage had been developed by the Creek Indians of what is now known as Alabama. By the turn of the twentieth century, Maskoki speakers were all but non-existent in the state, providing another example of genetic material on the move (see Figure 5.27).

At the beginning of this section, the region of Westphalia in Germany was invoked as a case of a geographical entity with fluid boundary conditions. The borders of European countries have shifted around for hundreds of years. After World War I, the provisions of the Treaty of Versailles redrew the maps of Europe, ostensibly for one final time. As subsequent events have shown, reconfiguring the boundaries of European states did not end conflict between communist, democratic, fascist, colonial and nationalist forces, and in fact exacerbated religious and ethnic migration. The evolution of even a small geography, the landscape of Auburn, AL, also involves processes of migration, displacement and adaptation. Diagrams of these processes, like the cartographies of European powers, are compositions of vectors and boundaries. They show evolution, disturbance, recombination and what Maturana and Varela call autopoeisis, the propulsive flow of energy through a living system that responds to turbulent fluctuations by becoming more varied and complex, richer, more adaptive and ultimately, perhaps like Westphalia, more resilient (Maturana and Varela 1980: 78–79). Certain species attract discourses of condemnation, denunciation and blame, which often result in marginalization, relocation and sometimes elimination. These discourses can be considered as sites of struggle between cultural and biological replicators. Landscape architecture can bring together migratory and/or nomadic human and nonhuman species in public open space design in such a way as to draw attention to the role of cultural values in discourses around indigeneity and exoticism. The rationale for this is the simple fact that public open space is often considered as a site of contestation and negotiation, a condition regarded as a hallmark of democracy. If we extend this civic concept to ecological democracy we find that the rights of nonhuman species must be considered along with those of humans. Latour has argued that the distinction between nature and society must be blurred in order to do this. We should consider the inhabitants of the Earth as a collective of humans and nonhumans that are not considered epistemologically or ontologically distinct. This means abandoning the notion of nature and replacing it with the idea of naturecultures, an assemblage of humans and nonhumans (Latour 2004: 41). Can such an assemblage come together in public open space to create a common world? All this section does is trace the barest outline of an approach, and perhaps suggest, as always, further – much further – study.

06

PROPOSITIONS

This chapter discusses how a landscape emergentist approach to public space design might address some of the issues landscape architects face in the creation of a public realm that is open and democratic, but deeply influenced by what it calls the "social machine" – the social economic forces that are beyond our control. It proposes a revaluation of the notions of space and place, and suggests that a shift from an equilibrium point of view to a disequilibrium perspective may help in the development of a collective imaginary.

In "Do Landscapes Learn: Ecology's 'New Paradigm' and Design in Landscape Architecture" (2000) plant ecologist Robert E. Cook contrasts what we might call the doctrine of spatiality prevalent in landscape architecture with the time-based perspectives of ecologists. Lacking ecological literacy, landscape architects, he suggests, are out of step with the new paradigm of ecology, which sees nature as "dynamic, contingent, and full of change and uncertainty." Following Mozingo, who had argued for a new ecological aesthetic in landscape architecture, Cook, balanced on the cusp of change, calls for "a renewed acknowledgement of temporality in landscape design" (Cook 2000: 130). He echoes Mozingo's earlier charge that landscape architects were not seeing change as "a vital, imaginative force," and that they needed to move beyond a "fixed vision of the landscape" in order to create an aesthetic that is adequate to the flux of the world in which they worked (Mozingo 1997: 52). Would a "temporal aesthetic, one that is informed by the real life of landscapes," Cook wondered, "give new meaning to 'ecological design'?" (Cook 2000: 131). This repudiation – or at least augmentation – of spatiality as an aesthetic ordering principle by the long-view ecologist and the accompanying call for a "temporal aesthetic" has now been absorbed into landscape architecture. The result, unsurprisingly, has been a renewed reflection on the intertwining of space and time in constructed landscapes. This is seen in the development of programs on the part both of practitioners responding to complex urban conditions, and of urban administrators, academics and others who write competition briefs. But Cook's formulation of landscape architecture as primarily spatial, and of ecology as suggestive of a temporal corrective, is inadequate as an indicator of the changes that have occurred, and are occurring, in landscape architecture as a result of the advent of emergence theory. Emergence has, in landscape architecture, ventilated a number of domains of practice and inquiry in ways that are as yet barely discernible. One way to consider these domains is through the hybridization of landscape and urbanism.

In the 1970s and 1980s – the era, in landscape architecture at least, of postmodernism – it became clear that the problems besetting critical risk landscapes such as evacuated urban cores, coastal dune systems, riverside flood plains, deindustrialized terrains and mass housing developments, were structural and organizational. The urban forms of these conditions could not be addressed by the traditional supply of architectural objects that was the main form of amelioration of difficult conditions (Waldheim 2002: 12). It was also apparent that many of the engineering solutions to widespread urban environmental deterioration – revetments, canals, levées and dams – were not able to contain the "forces of nature." It was during the years through which postmodern design waned, while in landscape architecture something else as yet unidentifiable began to grow, that the various planning-based solutions to urban problems started to take hold. Responses to problems associated with sprawl and community impoverishment, for instance, involved traditional neighborhood development, smart growth, transit-oriented development and new urbanism. These initiatives provoked a dilemma in landscape architecture as the rise of new urbanism occurred at the same time as two important developments in disciplines associated with landscape architecture: the new paradigm in ecology was starting to take root (*pace* Cook), and cultural and social geographers were incorporating insights and procedures from post-structuralist philosophy (Cresswell 2010; Daniels 2010). Planning-based modes of urbanistic solution-finding were very successful with the urban planning and design departments of city councils. The models were exported from Europe and America to many other countries, including Australia, New Zealand, Canada and Mexico. The global spread of retro-urbanism sharpened debate about social geographies that are central to landscape architecture, and especially significant in the evolution of emergence in this discipline. The concerns focused on issues to do with spatiality, place, consumption, heritage and identity. A significant development was the growing problematization of the urban/suburban field as a quintessentially physical, authentic and essentialist *site* of civic transformations. Just as a gradual rejection of the reification of public and private space was occurring, the procedures of urban transformation were narrowed rather than opened up by the development of planned communities. These were based on the assumption of a professionally produced social matrix rather than one that ultimately is both deeply discursive and collectively produced. The connections between concepts developed in social geography and the growing requirement for landscape architecture to provide ecologically productive and environmentally sound urban landscapes became more clear. These wide-ranging areas of concern coalesced in the critical discourse of systems theory (see, for example, Sheffield 2009).

Subsequently, it is possible to separate the operations of emergence in landscape architecture into two broad themes: those that have to do with social systems and those that have to do with natural systems. The challenge for landscape architecture as a discipline is how to bring these together. With respect to the former, a growing body of contemporary work, founded on relatively early experiments (for instance the work of Walter Hood), attests to a developing interest in the relationship between landscape design and management and how people do or could live. There has for many decades been a relatively easily identified connection between environmental conditions on the one hand and quality of life on the other. This is the focus of the environmental justice movement (see, for instance, Pellow and Brulle 2005). With the arrival in cities, in suburbs and along their edges, of huge, highly visible and unplanned agglomerations of poor, migrant, indigenous, populations, there has been a compulsion – among landscape architecture professionals – to explore how the modification of territories can help to change these conditions. This has taken the form of an inflection towards the environmental justice movement, whose vision has been summarized as "the development of a holistic, community-based, participatory, and integrative paradigm for achieving healthy and sustainable communities for all peoples" (Lee 2005: 219). When landscape architecture combines its disciplinary objectives of creating meaningful, healthy, safe and provocative urban landscapes with this vision, the result is a powerful social and environmental force. The process of implementing environmental justice goals in any community occurs over time. It builds on existing community capacity to enhance the social capital that the community has invested in its own resources. A central premise of social capital is that communities have networks, and that these social networks have value. Social capital is distributed through multiple channels, as flows of information, norms of reciprocity (mutual aid), collective action, and broad senses of identity and solidarity (Putnam 2000). Communities are rich repositories of relationships and networks, which can be called upon to address problems. It is a question of asset-building rather than service provision; learning from experience and doing, and discovering new concepts and information in the process – the defining operations of emergent systems. Novel approaches to the problems associated with particular communities are derived from within the community, not from some outside condition in the form of aid and advice, for instance (Lee 2005: 219). Understanding communities as emergent systems, landscape planning and design can incorporate strategies that encourage and empower peoples to become involved in the development of small, incremental interventions that are meaningful and useful to them. These strategies, based on the establishment of appropriate initial conditions to guide and regulate self-organization, can propagate multiple options for growth and development.

Cook cites the work of ecologist Michael Fastie on succession in Glacier Bay, Alaska. Fastie's description of the colonization of post-glacial sites by trees serves as an excellent succession model for social systems.

> The existence of qualitatively distinct pathways at similar sites at Glacier Bay demonstrates that no single sequence of species replacements and no single mechanistic model of plant community change is mandatory. . . . Multiple pathways of compositional change at Glacier Bay appear to be a function of landscape context, which, in conjunction with . . . dispersal capabilities and generation time, affects seed rain to newly deglaciated surfaces and thereby alters the arrival sequence of species. . . .
>
> The single species differences early in succession at otherwise similar sites at Glacier Bay therefore could have substantial consequences for successional pathway and ecosystem function for many centuries.
>
> (quoted in Cook 2000: 124)

"Multiple pathways of compositional change" which "appear to be a function of landscape context" aptly describes the interactive character of human communities and the landscapes that they inhabit. Just how humans respond to, or participate in, the landscapes that designers construct is evoked in Fasties' description of downward causation as having "substantial consequences for successional pathway and ecosystem function for many centuries."

What might be the hallmarks of public space design that took seriously the fundamental propositions of emergence?

In the first instance we would look for a multitude of connections between environmental design and social outcomes. The nature of this connectivity will be such that the emergent social phenomena would have an autonomous existence, and yet depend on the designed structures to enable and support them. Landscape architects are agents of change, says Hood, but once people are encouraged actually to see the familiar environment around them, they are empowered to become involved in its ongoing transformation.[1] The higher-level properties of the designed landscape, such as its functionality, aesthetics, intelligibility, affectuality and affordances, will supervene on the organizational conditions it provisions. These will include formulations of spatial elements, adaptability, accessibility and openness, and the articulation of material components such as surfaces, plants, water, sunlight, shelter, seating as well as the relations between these. In Auckland, New Zealand, the redesign of the primary public open space in the heart of the city, carried out by Ted Smyth and Associates, was based primarily on arranging the familiar components of floor, seating and trees. A carefully considered strategy of reconfiguration multiplied the possibilities for opportunistic occupation of the space by a much wider range of social groups than the previous layout had afforded. New types of occupation emerged almost immediately. Not only are there now multiple micro-sites for simultaneous individual and small group use, but these informal gatherings can occur alongside major formal events such as dramatic performances and political rallies.

Second, a landscape architectural commitment to emergence will be a commitment to strong emergence. That is to say, it will claim that higher-level functioning in some way influences the lower-level functioning. Emergence in landscape architecture takes place in, and partly enables, multiple forms of existence. Non-material features such as concepts, information and desires will have causal effects in the material world of forces and particles, fish and insects, which means these nonmaterial events are accorded an ontological reality. Such immaterial features include human mental structures and events like conceptual schemes, plans, intentions and emotions, as well as socially constructed elements such as games and commodity prices. In effect, landscape architecture requires an emergentist pluralism. Many levels of reality emerge in the world and the various objects and processes at each level have their own types of reality. Ecologist Ronald Trosper makes a distinction between cultural emergence and material emergence. If the components of the system are ideas, concepts and values, we have cultural emergence. If they are material we have material emergence (Trosper 2005). Landscape systems mix these up. Human and nonhuman structures are assimilated to each other. The logical relationships between ideas intermingle with the exchange of materials, physical connections and processual participation of nonhuman structures. As a result of this autocatalytic operationality, social and cultural responses would be able to cause new organizational and material arrangements, which in turn would permit a third feature of landscape architectural emergence: the appearance of novel social properties. A co-evolutionary relationship would develop. These new organizational patterns and programs would have a propulsive algorithmic function, and themselves drive material change. Adaptive landscapes empower people to change them. Moreover, since quite complex behaviors can come from very simple interactions of simple components (the butterfly effect), people's ability to make small changes to their environmental conditions can spiral upwards into dramatic revisions of social relationships. The planting of fruit trees in a street by neighbors can do this over time, but the sudden availability of a low-rent retail premise for the use of a community organization has an immediate effect. Here the catalyst may be the redevelopment of an empty lot into a playspace, or a constructed wetland plaza that permits humans and nonhumans to interact in a collectively realized realm. The human reception of such a plaza will not be determined by the underlying ecological structure of the design, but by the affective power of the situation it affords.

A fourth component of emergence in civic landscape design will be the assumption of local control over local conditions, and the ongoing empowerment of local stakeholders, requiring a refocusing of attention on the emergence of small-scale shifts in governance and management, and a commitment to currency and relevance rather than a final result. One of the most significant factors will be a reliance on short-term increments of transformation, a system of visible small changes that occur within and against a background of groundswell long durée transformation. It is impossible to start from a "genetic code" – a local tradition, a community's or an extended family's habitus – and predict the structures that are created from that code. Crucially, the supervenience relation puts the responsibility for unpredicted outcomes on the designer of the initial conditions – the landscape architect. In any given situation there are several possible states that have a conditional equivalence. Which state the system occupies at any given time is a function of its history (which includes the landscape architect). There is not a preferred state for the system.

The landscape architect will establish a set of initial conditions, based on a description of the situation that generates a process of realization. New situations evolve from the initial conditions, their components connecting and disconnecting, reconfiguring in particular ways, moving, doing something, creating, enabling, empowering, changing, autocatalyzing, wandering between possibilities. Urban landscapes are strange attractors. As "regimes" they develop according to specific histories that could go any way. "The slightest perturbation can push the system from one regime into a very different one" (Stengers 1997: 67). The degree to which a landscape can respond to fluctuations in its physical and social contexts and its internal conditions is a measure of the openness of the systems of which it is comprised. Different landscape systems respond in different ways to the same and different fluctuations. A natural wetland responds to a 100-year flood in one way, a levee system in another. Openness is a declaration of nonresistance. It is a reliance on system resilience that is brought about by assimilation and adaptivity. Landscape architects make urban landscapes that are resilient due to their structural ability to assimilate and adapt. The landscape architect designs landscapes that are far-from-equilibrium. The fluctuations of far-from-equilibrium landscapes cease to be what Serres calls noise or turbulence (Serres 1982b). Instead they play a role in changing the macroscopic regime of the landscape system.

Fifth, there will always be at least three types of simultaneously applicable explanation possible for any designed landscape. The wetland plaza mentioned above – a combination of social space and ecological performance – will be explicable in terms of upward causation (bottom up), same level explanation and downward causation (top down).

a) *In bottom-up terms*
   Because access and egress, spatial organization, sun/shade factors, seating opportunities and aesthetic qualities work together to create a flexible, attractive, functional and inhabitable space that connects to its context and reflects its cultural history.

b) *In terms of same-level explanation*
Because people of all ages and backgrounds are occupying it in a variety of ways.

c) *In terms of a top-down explanation*
Because it is planned, designed and constructed to be vital and energetic by a team of highly trained and experienced designers and planners working in cooperation with an enlightened administration.

These explanations are all simultaneously applicable. The higher-level explanations, which involve a mix of inherent social, urban and natural processes, rely on the existence of lower-level explanations, but they are quite different from them and are neither reducible to them nor dependent on them. The structure of a wetland plaza (for instance) actually represents a maximum of what we might call social improbability, since the human reception or response to the design is not determined by the underlying hydrological conditions or how these have been harnessed and regulated. Whether the plaza can function well as a social environment is a matter of how high it is in, among other things, affective content relative to the background conditions of its formation. The affective capacity of the plaza emerges from the interaction of systemic discursive elements with other processes, including what are at least perceived to be natural processes.

If, somewhat simplistically, we say that in its capacity as a situation an urban landscape is at once a social system and an ecosystem, then a sixth aspect of emergence in landscape architecture will be that the landscape architect is responsible for a description of that situation. This description, while always contingent and incomplete, will form the basis of the special involvement of that designer with that system. The description will emerge from an involvement of the designer with the situation that includes a participation *in* the situation sufficient to develop the terms *of* the situation, and to enable design to be regulated *by* the situation. An emergentist landscape architect will combine techniques of reconnaissance and imaginative forensics with what Meyer calls situated "forms of notation." These will be based on modes of expressive drawing, writing and photography that are adequate to the multiplicities that comprise the field in which he or she is immersed (Meyer 2008). The important consideration here is that the landscape architect is part of the system that is being designed, with a concern for the world and for the creative harvesting of its relationalities.

So, how does landscape architecture respond to these requirements? While emergence theory has an identifiable prehistory in landscape architecture, architecture and urban planning (see Waldheim 2002; Almy 2007; Meyer 2008) the answer to this question will be confined to an examination of contemporary practice. The most obvious place to look for emergence in contemporary public space design is in the design practices and writings of landscape architects who align themselves with landscape urbanism. This is not to say that many landscape architects who do not consider themselves landscape urbanists do not also at least partially or episodically utilize landscape operations that promote emergent conditions. However, landscape urbanists have both formally and explicitly invoked the operations of emergence, and adopted or devised a language that speaks directly to many of the emergentist protocols mentioned above.

Landscape urbanism

Across a range of texts Waldheim has identified the components of landscape urbanism:

1. Basis in postmodern critique of architectural urbanism
2. Recognition of the indifference of city development to the stylistic oscillations of architecture
3. Landscape as a singular medium in its own right
4. Development of landscape architectural practice in the wake of modern urban decay (brownfields and industrial sites)
5. The demise of physical design in the discipline of planning
6. Integration of infrastructure and public space design
7. Assumption of the operational methodologies of field ecology
8. Rise of the event/program for public urban terrain (and identification of programmatic indeterminacy)
9. Replacement of plan by strategy
10. Orchestration of diverse contents and elements
11. Introduction of phasing
12. Equivalence of urban systems and natural systems

(Waldheim 2002, 2006a, 2006b, 2007)

The last six are directly connected to emergence. Some important aspects of landscape architectural ambition are missing from Waldheim's list, notably its passion for social transformation and a burgeoning realization that the discipline is deficient with respect to its responsibility as a social force for political critique. Richard Weller's "defining characteristics" of landscape urbanism begin to address these omissions:

1. Aligns itself with contemporary scientific paradigms of nature as a self-organizing system
2. Conceptualizes, interprets and engages the city as a hybrid ecology
3. Emphasizes the creative and time-developmental agency of ecology in the formation of urban life as opposed to envisaging an ideal equilibrium between culture and nature
4. Includes within the purview of design all that is within the landscape – infrastructure and buildings etc. – and does this at scales which bridge the divide between landscape design, landscape ecology and landscape planning
5. Experiments creatively with computer-driven methods of mapping social and ecological forces which affect a given site so as to get closer to the complex dynamics of the landscape
6. Aims for structural efficiency and instrumentality by design and apprehends both site and program as creative subjects and opportunities
7. Generally privileges a rational understanding of site forces, not the designer's subjectivity
8. Foregrounds the landscape as the ultimate system to which all goes and from which all comes, a template for urbanism.

Weller also says more critical discussion of landscape urbanism is required in the following areas:

- What about aesthetics?
- Is operational performance a type of neo-functionalism?

- Are landscape urbanism's data and diagrams trustworthy or productive?
- There is a need to better articulate what is meant by complexity and self-organization.
- LA discipline still needs to acquire "a genuine reconnection with aesthetics, poetics, philosophy and representation."

(Weller 2006/2007)

Another professed landscape urbanist, Chris Reed, identifies an issue that complicates matters considerably – the scale of work at which many landscape professions are required (or have ambitions) to work. He lays out an agenda for landscape urbanism that places the landscape architect firmly in the position of large-scale infrastructure designer – "an urbanistic system-builder, whose interests now encompass the research, framing, design, and implementation of expansive new public works and civic infrastructures" (Reed 2007). Here we find a problem that besets landscape urbanism. The site has expanded to include everything. Professional work routinely includes strategies for CBD densification, the construction of networked brownfield remediation projects, redevelopment of railway easements as mixed use communities, integration of postindustrial zones into commercial, retail and residential urban fabric, three to five storey residential subdivisions accommodating 5,000 people in sensitive coastal dune systems, vast, denatured urban parks. These infrastructural scales of work are reflected in many schools' landscape architecture design studios. At the author's school, students have recently designed, among other things, a vast constructed wetland system with huge mixed-use developments inserted within it in Mobile, AL, an open systems-based redevelopment of five miles of foreshore in Baltimore and a six-mile greenway system that intends to reformulate the landscapes of 18,000 tax delinquent properties in Birmingham, AL. It might be thought that marshaling and reformulating these vast systems requires a logistics of control (and a large dollop of hubris) that belies the protocols of emergence. Christopher Gray states the problem:

> Considering one of the consistent critical shifts which prompted landscape urbanism theory is the move from the representational to the operative, there is a definite contradiction that arises when such dynamic systems are formally represented and finitely defined in pieces of fixed architectural infrastructure . . . systems and flows of energy that have been sensitively mapped and analysed are all too often translated into massive elements of built form in the rush toward an architecture that is in some way recognizable. Riparian patterns are literally transformed into floor plans; land-use analyses determine the elevational data points for monumental grading exercises; and habitat surveys determine grids for space-frames that cover acres of ground.
>
> (Gray 2007: 99)

One significant consideration, raised by Bart Lootsma, is that the ecological aspect of a complex site is only really represented in the organization of the end product. During the process of designing with vast, interconnected systems of transit and commerce, any organic development in the diagrams or buildings is subject to a "brutal break" with the move to final architectural and built form. "This approach in the end produces only a simulacrum of life, not life itself in the original 'ecology'" (Lootsma quoted in Gray 2007: 99). Ecosystems are too unstable for the kind of final product outcomes in which even small changes can be ruinous. Obviously this is a key issue, and one from which landscape urbanists should not shrink if they wish not only to embrace indeterminacy because it is trendy, but because it is effectively the single most appropriate design paradigm for urban landscape architecture. The whole concept of emergence defies finality, or any positing of an end product.

If, as most landscape architects say, the twenty-first century project for landscape architecture is the project of urbanism, then how this task should be addressed is a significant question. What is missing in action from most of the tracts of landscape urbanism are three critical dimensions for the future of urbanism: social transformation, political critique and a productive ecologism – in Lootsma's sense, and all understood as part of a process that never ends. There have been some early signs of the combining of environmental justice, the intransigent problematics of consumer society, and spatial politics with the promise of an ecological landscape architectural urbanism. For instance Mitchell and Van Deusen praise the Field Operations design for the Downsview competition as highly encouraging of publicness in a political sense, and castigate the same firm's Highline as developed in the service of social segregation and real estate objectives. However, most of the urbanistic discussion of spatial politics, the role of the market in the development of civic space, the complexities of community participation, the paucity of designed democratic space, the problems of privately owned public space and so on, have come from cultural and social geographers rather than landscape architects. It is therefore significant that Mostafavi and Doherty (2010) situates ecological urbanism within a philosophical position that incorporates a powerful vision of social transformation based on design as a political practice. While he does not specifically demonstrate how emergence theory can help find a way through the very difficult problems designers face (from a range of directions) when designing public space, he does place many of the components of emergence at the heart of the *modus operandi* he advocates. Like the landscape urbanist writers cited above, Mostafavi lays out what he considers to be the key aspects of ecological urbanism, as follows:

> *Transdisciplinarity*
> the collaboration of planning, design social sciences and sciences
>
> *Scale and scope*
> multiscalar modes of practice and borderless sites
>
> *Adaptivity*
> an adaptive urbanism that is flexible, reversible and provisional
>
> [And he adds] an awareness of patterns of legal, political and economic activity [the conditions of governance].

Ecological urbanism, Mostafavi writes, is concerned with "the articulation of the interface . . . between the urban and the political." It "recognizes the importance of the urban as the necessary site of conflictual relations" and promotes an agonistic pluralism in order "to address the central question for democratic politics" (Mostafavi and Doherty: 2010: 48). Unfortunately, he does not actually say what this question is. However, he does acknowledge "the role of the urban as the provider of spaces of difference and disagreement" and that the purpose of ecological urbanism is to engender "greater opportunities for social and spatial democracy." Importantly, the fact that Mostafavi founds ecological urbanism on Guattari's concept of an ethico-political reformulation of the subject, means that the role of the human actor in the construction of urban life is recognized and, to a certain extent, incorporated into his concept of "the ethical performance" of the city. This is really the only nonecological text devoted to an emergentist approach to the design of the urban condition that realizes the critical importance of the subject in the process of self-organization. It is strange that at the beginning of his introduction to *Ecological Urbanism,* Mostafavi states that the speculative design innovation that his new approach augurs is a response to the paradox or oxymoron, as he calls it, of urban ecology, since ecology "invariably excludes human intervention." Strange, because the inclusion of the human subject in the ecosystem being studied is itself one of the innovations that ecosystem ecology brings to urbanism. At any rate, at the end of the essay Mostafavi endorses Guattari's conception of "an ethics of the ecological" as an "inherently political project with a commitment to countering the global dominance of capitalism."

This is a very significant alignment, as the design of cities is itself deeply political and always involves political processes. One of – if not *the* – primary failures of landscape architecture has been to address the centrality of the political in the prosecution of its environmental and urbanistic goals. One practice that has foregrounded the political nature of design is Stoss LU. Landscape urbanist theory, director Chris Reed argues, must include bureaucratic and administrative strategies to cope with "the dynamic nature of politics and public adminstration." Reed turns the "decreasing control" of the designer into a positive condition for practice by reconfiguring the designer's role into one of coordination. The designer "becomes a catalyst for a directed but ultimately uncontrolled and uncontrollable sequence of successional changes and adaptations to future conditions." This operational mode, he says, allows for "future inputs and adaptations" that lead to "transformations of form, program and role within the city" (Reed 2007: 227). Reed emphasizes the political nature of this position and develops diagrams that lay out potential pathways that user and interest groups might take, depending on how they expand and network, combine with each other and activate different initiatives and "individual sub-projects." "(T)he more diverse the program of the project, the more potential for active stakeholders and participants to pressure political entities for change and action" (Reed 2007: 227).

While landscape urbanism has placed emergence clearly at the heart of its project, there has as yet been little description and explanation of how emergence has actually been used in design process. Reed is an exception. In his "Landscape Urbanism Practices: Precis and Case," he first aligns his practice with Forman's theory of environmental change through ecological succession, and with Cook's "new ecology," and then attempts to show by means of a case study how the kind of "responsive, adaptive, open-ended and unpredictable landscape" (Reed 2007: 225) that they describe in nature may (or may not be) achieved through design. Interestingly, the Papago Trail-Metropolitan Phoenix project operationalizes four "typologies": nodes, gardens, conduits and fields. Each of these typologies conjures up ideas of impermanent networks and flows, and Reed's description of the strategies used to instigate these "wildly divergent forces" (Reed 2007: 228) that play across the landscape reveals an intention to engage with them in imaginative and sensitive ways. He uses a protocol of letting go, establishing a framework where various programs "might each find a place for occupation, orient themselves, and physically interact to build a series of rich and even imaginative overlays and juxtapositions between radically different uses" (Reed 2007: 236). Strategies are developed for accommodating uncertain future uses within a landscape formed by landform manipulations that encourage colonization and succession (by plants, people and programs) on the basis of how water rises and gathers within the forms. Airport parking, civic festivals, markets, agricultural production and informal social appropriation are specific functions that may or may not eventuate. What happens if they do not is unclear, however, as Reed defers to the interventions of government institutions and the uncertainty of funding as "unique opportunities for collaboration, cross-fertilization and identity-making" (Reed 2007: 237) and avoids speculating on how the divergent futures might actualize differentially depending on circumstances. Within this field of unpredictability a range of adaptive landscape interventions is arrayed, each using the dynamic nature of existing conditions as a basis for form-finding. For instance, a buried channel is tapped into in order to "release flood waters, creating wetness and dryness regimes along the conduit," (Reed 2007: 228) landform fields are laid out to accommodate foreseeable uses, a "submerged linear garden" (Reed 2007: 229) connects two separate areas of the site. Combining the top-down and bottom-up techniques that Mostafavi regards as essential to the provision of emancipatory infrastructure, the proposal also exemplifies the new curatorial practice of territorialization.

The writers mentioned (Waldheim, Weller, Mostafavi, Reed) are just a few among many whose promotion of, and ideas about, landscape architectural practices of urbanism incorporate emergentist strategies. The question this chapter asks is, how do we turn these undeniable material and infrastructural gains to the creation of our shared goal of truly democratic public space?

Landscape architecture and public space

The advent of emergence theory in landscape architecture has been characterized by calls for a reformulation of the entire imaginary of landscape design. In "Terra Fluxus" Corner calls the imaginary the "fourth theme of landscape urbanism," seeing it as an antidote to the "absolute impoverishment of the imaginary" which is the cause of the failure of twentieth-century planning. "Public spaces," he says, "are firstly the containers of collective memory and desire, and secondly they are places for geographic and social imagination to extend new relationships and sets of possibility." Corner states that "political change through practices of place construction owes as much to the representational and symbolic realms as to material activities" (Corner 2006: 32). In an earlier essay he outlines the creative agency of ecology, particularly its potential to "inform more meaningful and imaginative cultural practices than the merely ameliorative, compensatory, aesthetic, or commodity oriented" (Corner 1997: 82). Similarly, Richard Weller makes what he calls "the relative impotence" of landscape architecture to lead to a "reshaping of the world" the basis of an argument for ecological design. It is in ecology, he says, that "we find a new conceptual imaging of landscape, one which landscape urbanist sensibilities apprehend as a hybridization of natural and cultural systems on a globally interconnected scale" (Weller 2006/2007: 73). Weller had previously wondered whether landscape urbanism "underrates aesthetics and imagination and overrates operational performance," perhaps in a rush to "avoid ornamentation." This leads him to suggest that "(i)n nearly all landscape urbanist work to date . . . there is great realism but a distinct lack of intellectual and aesthetic adventure" (Weller 2006/2007: 70). Landscape architecture, Weller insists, still requires "reconnection with aesthetics, poetics, philosophy and representation" (Weller 2006/2007: 70). There is a sense, then, that it is within the combinatory and adaptive protocols of ecological thought that a recovery of the lost poetry of landscape architecture is to be found (much contemporary discussion is motivated by the ubiquitous allegory of loss and recovery). Mostafavi's argument that in the "territory of ecological design" we might find "a new sensitivity" to the project of urbanism emphasizes the city's "relationship with the qualitative value of things" and contrasts the novelty of this value with current urbanist practices that do not evince such a sensitivity. He uses Guattari's ecosophy with its three registers of environment, social relations and human subjectivity to begin a rethinking of the values of production. This reconsideration of the values of (material and immaterial) production are no more apparent than in the recent establishment of the James Rose Center's competition called *Suburbia Transformed: Exploring the Aesthetics of Landscape Experience in the Age of Sustainability*. The goal of this international design competition is "to promote and celebrate residential designs that go beyond 'green' by explicitly using sustainable strategies, tactics and technologies to enrich the aesthetic spatial experience of people."[2]

The social production of public space

The idea of the imaginary has percolated through discussions of the urban, the public and the social and is significant for its insistence on the collective making of a shared world. Sartre introduced the imaginary into modern philosophical discourse in his *L'Imaginaire* of 1940 where he analyzed the role of the imagination as a necessary component of human becoming. Because we can imagine, he says, we are ontologically free (Sartre 2004). A consciousness that could not imagine would be incapable of the perception of unrealized possibilities and therefore of freedom of thought or choice. All our engagements with the world have the potential to activate the imaginary process, and because the imaginary process is intentional, the world is constituted not from the outside but by human beings based on their intentions toward it. In recent years the imaginary has been folded into social thinking. Sociologists John B. Thompson and Charles Taylor invoke the social imaginary in order to explain the creative and symbolic dimensions of the social world. For Thompson, the social imaginary is the dimension through which humans create their ways of living together and their ways of representing their collective life (Thompson 1984). Crucially, the imaginary is not synonymous with the fictional or the irreal, but capable of very real effects. It is also not a set of ideas, rather it is what enables practices of a society (Taylor 2004). It has an instrumental dimension.

In contemporary western cities the social conditions for the formation of public space work to filter and mediate the affective and the imaginary. This is a difficult phenomenon to trace, but it is central to the defense of public space as an emergent realm in whose ongoing construction landscape architects are uniquely positioned to assist. When we talk about the formation of public space there are two important distinctions to keep in mind. First, the design of public open space is only one aspect of its formation, since generative spatial practices are part of the repertoires of all the populations that participate in the urban assemblage. Second, the idea of public space involves the circulation of material and discursive articulations throughout a vast social territory that includes movements such as Occupy Wall Street and debates about citizenship, identity, planning, health and ethical conduct, many of which bear upon practical legislation that affects public space design, such as the role of the homeless in the development of urban terrains. Intrinsic to these considerations is the fact that public space formation is influenced by powerful social and political forces beyond the control either of landscape and urban designers or of the public organizations that purport to organize the urban realm on behalf of their constituencies. For instance, in contemporary society the discursive production of space is hugely influenced by what Adorno called "the culture industry." In the 1930s the Frankfurt School criticized the industrialization and corporatization of culture. They analyzed the ways in which people could be manipulated by advertising and other means to create a "false state of reconciliation" brought about by the belief that the social world is rational, conducive to human freedom and happiness, and unalterable. Society, Frankfurt School theoreticians such as Horkheimer and Adorno argued, is in fact deeply irrational. It is an obstacle to human freedom and happiness and very mutable (Adorno 1991). The culture industry, according to Adorno, imposes its contents and its products on the people. When the customary practices of communities – those practices which produce their loves, hates, rebellions and resistances – are mediated by institutionalized communities of cultural production such as mainstream cinema, advertising, popular music, even the self-organizing communities who use Facebook and Twitter, these customary practices, or folk cultures as the Germanic Adorno called them, become something else – commodities. Adorno regards it as absurd to call something popular because a lot of it has been sold. To be genuinely popular, culture must spring from the lives of those who produce it, not be dispensed by an industry (Adorno 1991). Adorno has been regarded as mistaken in considering that people express their culture through the *use* of things, rather than producing culture through the *making* of them (the position of cultural materialism; see Harris 1979; Jackson 1989). Contemporary cultural geographers consider citizen-subjects to be much more active in the making of public space than does Adorno. Writers such as Wylie and Rose suggest that the making of subjects is no longer only a matter of the operation of disciplinary power upon relatively benumbed and docile bodies, "it is also an exercise by the self *on* the self, 'taking care' of the self, a self-fashioning, a cultivation of one's own body and mind through regimes of practice," reading and exercising, bicycling and camping for instance (Wylie 2007: 211–215; Rose 2006: 547). But Adorno's insights, influential in the development of cultural geography, carry a message for the designers of contemporary public landscapes.

The relationship between popular culture and the production of public space is so strong that we notice when it is absent, and realize how pervasive it is. Many landscape architects work hard to create zones of collective inhabitation in which the influence of the commercial and industrial dispensations of contemporary life are radically diminished. These undertakings are successful most obviously in the design of large parks, where space can be manipulated with greater freedom than in intensive urban precincts, and yet even in terrain that is seemingly only trees, grasses, walks and wide open space, the social forces of stratification are at work. What David Matless calls "cultures of landscape" are influenced by "styles of moving and looking, forms of land settlement and management, aesthetic ideals, exploratory discourse, scientific visual practices" which set the parameters for the design even of landscapes seemingly removed from urban spatial practices (Matless 1998: 1). In such "urban natures" as Buro-Kiefer's Hans-Baluschek-Park in Berlin, and Busk Meadow by Gross.Max, carefully controlled spatial organization and the sophisticated formal elaboration of paths, embankments, rows of trees and clusters of productive gardens draw attention to the ways in which landscapes are produced. Sports fields, community gardens and park management regimes all play their part in the construction of space and the regulation of spatial practices. These landscapes are all *produced* in one way or another. They do not arise spontaneously or autonomously from the interaction and labor of ordinary citizens; nor is their regulation and management a matter simply of community decision-making processes. There are few landscapes that are not actually physically created by means of sets of social and economic relations. Parks such as Hans-Baluschek-Park gain even their aesthetic values through the circulation of ideas of nature, and conventions that create forms of appreciation and reception. Ordinary people *are* involved in the construction of public space but their involvement is woven through with protocols of moral and aesthetic discrimination that realize patterns of landscape-making and spatial practice. If landscape is not so much the setting for human activity as the outcome of it, then what other generative activities are we talking about, and what motivates these? Often landscapes are created and presented to the public by political groups motivated by political conditions. These landscapes – such as parks, squares and waterfronts – are the production of types of discourse within which specific terrains are selected for development depending on wider networks of funding, influence and political power. Private commercial interests, public organizations and community groups all claim discursive authority over these landscapes at the same time, and the formulation of collective terrains in the public realm is a matter of how these disparate actants plait their interests, codes, practices and metaphors together to produce the landscapes that serve as public space. When landscape architects work with such entangled vectors of influence, inevitably they are themselves inside the frame of the assemblage whose ever-shifting alignments they are attempting to make potent and productive through design.

We can readily see how the affective and the imaginary are often filtered in the development of public space. If landscape architecture can work through the reasons for this, it can promote the realization of affective and imaginative spatialities that, while civically produced, are not as hierarchically constructed as many public landscapes tend to be. A useful way to introduce the problem associated with the inclusion of ordinary imaginaries in the development of public open space is through Žizek's analysis of the contradictory formulations of expression and representation. Žizek contrasts the material processes of socialist culture with the "shadowy" political processes of representative democracy, both of which coexist in the contemporary world (Žizek 2008: 366–370). Put simply, communities are encouraged to *express* their singularity at the same time as they forfeit this singularity through the processes of democratic *representation*. They are encouraged to exercise their right to evolve as long as their futures can be controlled by regimes of institutional maintenance and regulation. This is the famous trade-off of participatory democracy, and it has its corollary in the formation of urban conditions. The design of cities, neighborhoods, community landscapes and civic or public space is subject to participatory measures in order for community involvement to occur. The question becomes, how do people participate in the development of a public space that is both a material and symbolic realm? How are we all to be enabled to take our places in the extension of Corner's "new relationships and sets of possibility"? This is not just a question of the capacity for humans to construct places of great geographic and social imagination, but of how all beings are involved.

Richard Sennett has put his finger on one part of the problem; Hannah Arendt provides a swerve away from it. Public participation actually eventuates, Sennett says, in a double resistance to the imaginary because "the people," who base their understanding of political action on a personal construction of value, can only find collective agreement through the application of a rationalism based on a personalization of public space (Sennett 1994). Sennett therefore sees a danger in the investment of new conditions of democracy in popular community. He suggests that an "ideology of intimacy" has arisen, which assumes that social relationships are real, believable and authentic the more they approach the inner psychological concerns of each person. A direct expression of emotion through community participation in the construction of the public realm transmutes political categories into psychological ones (Villa 2001: 259). This causes the public realm to develop as a macro-social correlative of the church pulpit, the entertainment industry, the kitchen table and the forbidden chambers of sex and desire. Political considerations become revalued as issues of personal morality and the making of public space becomes imbued with the ebbs and flows of a family feud, rather than the operation of *communitas*. How, then, can the expression of a life in common with others, actualized through legitimate and legitimating contention, be grounded in such a way as to permit the full expression of the "aesthetics, poetics, philosophy and representation" which, Weller argues, landscape architecture needs to make possible?

Political philosophers John Rawls and Jürgen Habermas both see the answer to the problem of public space formation in democratic systems in the application of rationalism (Rawls 1971; Habermas 1989). Habermas, one of the most influential theorists of public space formation, regards the prospects for a democratic public realm as hinging on a "re-rationalization" of the public sphere. If public space has dissolved into a terrain in which the commercial, the anecdotal, the essential, the public, the private, the major, the minor, the individual, the collective, the personal, the political are all mixed, and if communities, themselves constructed in terms of this mix, are charged with the creation of public space and the public life it promotes, then on what terms should the recovery of public space be based? Having been captured by a culture industry that produces material culture in bad faith, the social world can only once again become a critical, deliberative formation of popular will if it is based on the exercise of rationality. This is the keystone of democratic legitimacy in a world in which there seems no longer to be a single and privileged stage specific to the political realm. Because he sees democracy as having lost its rationality (debates are turned into games; politics is entertainment), Habermas seeks to eliminate the affectual and imaginary from the public realm. For him, the problem of consumerism is ideological. If there is to be a social order the marketplace must be rejected from the public realm. Rawls also argues that rationality must be the basis of social order and, like Habermas, throws the baby out with the bathwater, by attempting to resolve the issue of consumerism by making public space achievable only if it is separated from the symbolic order. The symbolic (the aesthetic and the imaginary) must be confined to a private, nonpolitical realm, he says. But the role of the imaginary is not as a separate condition. Corner, Waldheim, Reed and Weller all see landscape architecture as having both political and affective agency. Mostafavi is explicit on this point. His vision of ecological urbanism achieves social transformation through the political practice of design (and design is an affective, aesthetic activity).

While it is an important consideration, the grounding in rationalism of people's lives in common, investigated by political philosophers, cannot give communities the basis for decision-making unless the affectual is part of the process. This does not mean the personal. "The inner connection between narcissism and the rhetoric of community has become increasingly apparent in identity politics, as community is ever more tightly defined in terms of those like oneself" (Villa 2001: 168). The re-rationalizing of the public sphere à la Habermas, itself an Enlightenment notion, is undercut both by its own elimination of the imaginary, and by the "ideology of intimacy," which assumes that social relationships are more real, believable and authentic the more they approach the inner psychological concerns of each person. This transmutes "political categories into psychological ones," giving us the personal political culture of contemporary society, where it becomes impossible to disagree with another's position without demonizing their person (Villa 2001: 164). We need to develop the landscape architectural project of public space design in a way that enriches the collective imaginary without recourse to tired formulations of community; in a way that enables *communitas* to evolve as an open system through the making of shared landscapes that are themselves open and evolutionary. This requires the conceiving of landscape architecture as performance, and the individual public design project as a "job for life" to which the landscape architect as public agent is committed and continually involved, guiding and regulating action.

In *The Human Condition* Hannah Arendt argued that politicization, as such, has no particular connection to the public sphere. She called for a performative approach to "world-making." Here, freedom of action, and the accomplishment of public space lies "in the performance itself" which she saw as an activity that brings it into existence (Arendt 1961: 153). A free and open public realm is constructed by the community through its free action, according to Arendt, and "is neither under the guidance of the intellect nor the dictates of the will." Arendt is not denying that intellect or will are necessary for the achievement of any action, nor is she claiming that goals or motives play no role in an agent's deliberations. Rather she is claiming that freedom of action does not reside in these categories but is manifest in the performance itself. Arendt's formulation is important because of the critical performative dimension to contemporary landscape architecture, which involves the framing of design itself as performance. In recent years, as we have noted, the emphasis has shifted from the designer or consultant as author to designer as performer, an actant whose work is an ongoing elaboration of the initial conditions that establish a project. The outcomes of such a performativity cannot already be known. Arendt regards contingency and indeterminacy as necessary conditions of political action, since the political future is something that is not susceptible to determination (Arendt 1958).

The potential of public space creation as a performative world-making is to be found in the emergent conditions of participatory democracy. The issue of communities' empowerment in the creation of their own public realm has occasioned the inception of liberatory community participation in the design process. This requires of the landscape architect the careful establishment of initial conditions for a playing out of project scenarios in ways that have not been envisaged, either by designer or by urban authorities. An "initial conditions" approach to the relationship between designer and the community he or she is serving involves a distancing of the author from cultural production and a discovery of accomplishment (and freedom of action) in the development of a cultural performance. If, as Mitchell and Van Deusen argue in their overview of the role of public space in the Downsview Park competition entries, the communities who will use – and continue to create – public space are to be involved in the decision-making process, then the designer must occupy a role that permits a "multiple, shared and combinatory" process. "It is not in design per se that publicness can be encouraged . . . but in how that design is administered and implemented" (Mitchell and Van Deusen 2001: 112). The designer's ongoing performance is as an orchestrator of the conditions for democracy. These conditions require the ability for communities to control their own public mileux. This means there must be a willingness on the part of urban administrators to let go, to have "faith in the democratic administration of space" and to "allow a space to host sometimes contentious publics" so that people can throw off their "limited role as consumers" and become "producers of their own contentious cultures, their own public spaces" (Mitchell and Van Deusen 2001: 112).

It is important in landscape architecture that we develop a working definition of community, since we so often consider ourselves as working with and for communities. The reader will perhaps discern the intention here, which is to argue that physical public open space is an assemblage, and that it is through the protocols of the assemblage that landscape architecture can assist with the building of community in the world. As suggested earlier, this requires a reformulation of the very idea of community. Instead of regarding community either as a simple collection of individuals, or as a hypostatized, deindividuated bloc similar to the "crowd" of Elias Canetti, philosopher Jean Luc Nancy defines community as a network of singularities that presents itself as being-in-common, a structure of shared life that is not "produced" but which evolves – emerges – in much the same way as the assemblages that species form in ecosystems – by drift (Nancy 1991a).

It is popularly thought that a community is a collective that makes up its own rules about how to live together. But this notion of community has long been problematized and now exists only as fabrication, or myth – an essentialist vestige of nineteenth-century social idealism. In social geography it is undermined by the symbolic power of the market economy (Adorno 1991; Zukin 1995; Fincher and Jacobs 1998), in science the controversial notion of a collective intelligence is studied in ways which provoke more questions than they answer (Woolley et al. 2010); in contemporary public art practice and theory, the concept of community is "highly ambiguous and problematic" (Kwon 2004: 6). It is particularly difficult to determine what can and cannot be tolerated by a community. Are the bounds, or limits to free speech to be policed by agencies external to the community, by the state, for instance? To what extent should freedom of action be curbed, and by whom? Don't communities work to silence voices they despise? Will a more democratic collective necessarily follow from a politics of prohibition? Can a community be truly multicultural? The identity or definition of (any) community remains open, like a situation, as a scene of political struggle. The common notion of the community as a coherent and unified social formation was put into question in the work of Nancy, followed by feminist social theorist Iris Marion Young (Young 1990). Nancy, in particular, proposes that the idea of community is necessarily unstable and "inoperative," enabling us to think beyond formulaic prescriptions of community and to open on to an altogether different model of collectivity and belonging. Communities share in and share with each other across a public realm in which certain physical actualizations occur that enable "communitarian associations" to present in specific or particular ways. In Nancy's and Young's work we come back to the primacy of the body, of embodied experience.

> Every day we can put our finger on the fact that in a certain sense, there is no sense left for us: no spoken, pronounced, enunciated, non-physical sense, that would make sense of all the rest. We come to touch a certain interruption of sense that has to do with the body, which is the body. And it is no accident, that with the body this interruption has to do/has to come to terms with the sense in the other sense, with the sense in the sense of feeling, in the sense of touching. To touch the interruption of sense, that is what for me is interesting in the affair of the body.
>
> (Nancy 2000: 112)

Nancy makes singularity and mortality the basis of community. We are alone, he says, we die and die alone. Community presents each of us with an openness in which we may understand that this is the case for all of us.

> A community is the presentation to its members of their mortal truth . . . It is the presentation of the finitude and the irredeemable excess that make up finite being: its

> death, but also its birth, and only the community can present me my birth, and along with it the impossibility of my reliving it, as well as the impossibility of my crossing over into my death.
>
> (Nancy 1991b: 15)

The individual lives in this mortality like an animal, like water in water. Other beings cross his or her path, and enter into relations, especially physical relations of propinquity, touchingness, visibility and scale. The landscapes in which such encounters occur are open and public, and enabling. This is public space as assemblage. It is not measurable. It is intensive rather than extensive. It is continually being remade. Beings enter and leave it and it reformulates. It is always in an act of recovery. Recall Delwyn Shepherd's Birdscapes project at Muriwai in New Zealand, outlined in Projects and Practices. What is critical in this project is that the landscape architect is assembling elements in such a way as to create a world in which it is possible for all components of the assemblage, human and nonhuman, to engender their own possibilities of existence. Dunes, trees, insects, birds, animals and humans are regarded as having in common a fit relation to an assemblage that continually evolves through its encounters with disturbance.

Public space is the making of a collective condition against a continual turbulence. Members of a community declare, Nancy says, their finitude, their mortal truth, their being-in-common with other finite beings, human and nonhuman. The formation of community is a condition of becoming in which sharing is what makes other beings other. Public space, then, is a fluid location of the collective encounter of humans and nonhumans in the constructed world, an encounter and negotiation between constituencies that not only speak different languages but for whom "life" is a variable that often includes the nonliving. It is only in their capacity as other that nonhumans share their being with ours, as a being-in-common. Landscape architecture does not create the stage for this community, but inscribes being-in-common within a territory that is occupiable, and enables the community of humans and nonhumans to share it in ways that only they can reveal through the interweaving of their different becomings. Community, according to Nancy, cannot be produced. No matter how much we work "with" a community we cannot create community – nor can we create a common world. The community of singularities is not limited to humankind; the idea of the public extends to animals, insects and plants. Our search for what makes up the common world requires a conception of this world as comprising interactive nonlinear systems, physical, social, cultural conditions that are always simultaneously local and global. We should think of landscapes as generated by, and therefore immanent to, the specific conditions they comprise. Finally, the task for the landscape architect is to know when and how to harness the energetic processes that drive the creation of new forms. If the landscape architect is attentive to what is already inherent in the situation being dealt with, then new forms can be developed that challenge both our functional and our aesthetic assumptions in the ongoing making of the world, even as they attempt to address age-old problems.

Returning to the hierarchically structured public spatialities referred to at the beginning of this section, it is very important that we sift out the symbolic and the imaginary from the market forces that now treat these as goods to be traded in. If one problem with the way Adorno, Corner, Waldheim and Weller conceive the urban imaginary is that they tend inadvertently to binarize it, with the symbolic on one side and the commercial ("the marketplace") on the other, this does not mean that the penetration of economic values into public space should be regarded as inevitable. It is a complex phenomenon, nevertheless. For we know now that the unintentional but ultimately high art/low art distinction the above writers portray is misleading. Art and social historians have shown, for instance, that so-called crafts practices of early centuries invested the "ordinary" with extraordinary symbolic capital. And in our own time Andy Warhol's career completely dissolved any vestige (however imaginary) of separation between art and marketplace long after the art world had been irradiated by capital. In landscape architecture we are watching the demise of community responsibility for the development of shared space as the terrains we perforce inhabit and use are increasingly formed according to the production and distribution of goods. Warhol's exposure of art's complicity with economics was part of a general resistance by artists to the commodification of the imaginary. Landscape architects, and all those who design the environments in which life is to be lived, should be alert to the expansion of markets and market values into spheres of life where they do not belong.[3] One way to oppose the marketization of the imaginary, if not completely eliminate it, is to initiate the self-organization of public space assemblages.

To do this we need first to consider how, with the advent of professional urban design, public *space* was elided as public *place*. For much of the twentieth century public urban open space was developed by urban designers, architects and planners. To a certain extent, urban design grew as a response to new requirements for the physical civic realm, transformed by the arrival of the personal motor vehicle, the central business district, entertainment precincts, and transit systems that carried urban workers out of the city at night and back in again in the morning. The social production of space changed irrevocably, transforming the urban field into the uncertain landscapes probed by film noir. Partially as a result of the opposing trends of the familiarization of politics described by Sennett and the defamiliarization of social space described by Lefebvre, and partially because of its rationalist/formalist tradition, the political objective of urban design was to hold the ambivalent and unpredictable reality of social production in check, to stop it from overflowing the whole field of urban reality. Urban design became a civic technique for the socialization of public space as *place*. In its drive to control the background open space in which human objects can appear, urban design developed an approach to spatial meaning by excluding what Žizek, after Lacan, calls the the Real – by transforming *what is* into an ongoing and persistent lack: a black hole which must continually be replenished (Žizek 1999: 42),[4] a perpetual loss which requires a perpetual recovery.

In urban design the spontaneous human production of collective affect is filled in by the professional production of place. Marc Augé's "non-place urban realm," a signifier of modernism, became the precursor of a place*ful*ness that was as empty and terrible as the conditions of place*less*ness. Place became the terrifying object that fills in the hole simply by means of its role in the structure of urban discourse that requires a – some, any – marker, a replacement for the absent condition. In the contemporary world place is, exactly, placelessness. The designed object, the Thing at the center – the designed urban landscape – is nothing but an effect of the structure. Its function is simply to fill the hole. Now the Thing itself – the public thing, *res publica* – is visible as an, as Žizek puts it, "incarnated, materialized emptiness" (1999: 42). This is how the imaginary is eliminated from public space – by the smooth functioning of a social machine that puts both urban design and public participation to work in the production of a placeless realm called place.[5] It is a supposedly open realm where humans interact in freedom – a realm from which, in principle, no one is excluded; where individuals participate in social discourse and make possible the "common good." The lack at the center, paradoxically, is precisely constitutive of this common good, the sense of community that Nancy repudiates because it constitutes the ground of a communion between the symbolic and the commercial that is distributed and shared out through immaterial and material space. It is in the name of the common good that public spaces fall into private hands. Granary Square, one of the biggest public squares in Europe, and a focal point for the regeneration of Kings Cross in London, is a managed private estate of ten plazas and parkland. The privatization of space is now the standard price of development, which is sold as bringing significant benefits to the populace in general (Vasagar 2012). This ambivalent space of disappearance and contingency, however, is open to the reappearance of surprising events: transformations, responses and articulations that are enabled by a collective that coheres in the very place that placelessness makes possible. It was asked earlier how landscape architects may contribute to public space formation. We must learn how to turn the situation to our advantage, politically and socially. Unless we are able to orchestrate the chaotic forces of expression and representation, rationality and the imaginary, we will be sucked into the black hole of urbanism – because the "instant environment machine" (Relph 1987) that produces the "soothing transmission of civic values" (Rowe 1999) is still the dominant force in urban landscape architecture.

If emergence theory is to offer some ways that the *urbus absconditas* – the problem of contemporary public space – can be considered, we must take seriously the concept of emergence. It is not a case simply of adding emergence to the mix as one-among-many, but of divesting landscape architecture of its substantives – nature, community, city, even public space – and rebuilding the urban landscape as, in Matless' terms, a "shuttle" which weaves together disparate realms of economic, political, scientific and aesthetic practice, seeing how its reverberations forge different ways of thinking and doing. For instance, the landscape architect may be thought of as not doing something *in* public space (such as creating a stage for events) but as developing an operation that produces a spatial publicness, a continual production of relationalities, holding in fusion the varying poles of spatial experience that public terrain encompasses or enables. This productivity may be considered as a feature of the kinds of social relationship and action that encourage the ongoing rehabilitation of the urban field by peoples, animals, processes and things. How do we facilitate the formation of milieux that enable bodies to intermingle in ways that are mutually beneficial, and through this intermingling create a common world? The efforts of the landscape architect are not constructed for consumption. They not only represent a practice (a practice of representation for instance), but enable life.

In order to delve into this further, let us postulate some urban terrain somewhere, as a stratum, a hierarchically ordered milieu whose patterns of production, connections to social and economic conditions, and inhabitational potentials are organized centrally in such a way as to preclude the actualization of other, alternative patterns. It is locked into a normative functioning determined by average values, without the possibility of – shall we say – developmental plasticity. Now, this is not the worst that can happen. Staying stratified – organized, signified, subjected – can be viewed as a resting point, a stasis, a condition in which experimentation does not occur, but could (Protevi 2009: 10–105). In any case let's say that, with respect to our thought experiment centered on an urban terrain, the potential for new conditions and patterns of production within this terrain has been overwritten by the social machine, which selects from the set of potential environmental and political connections, and consolidates a pattern of permissible aggregations. We are talking about a production of urban design here: a festival plaza, a revitalized waterfront. The thought experiment is motivated by this question: what remains when you take everything away – all the patterns of connection and production imposed on the terrain by the social machine? Perhaps the terrain will cease to work as part of a functional structure. Maybe it will enter a condition in which it is open to a field of new connections. Perhaps new patterns and triggers of behavior will emerge. Perhaps, in other words, it will move from equilibrium (a fixed set of habits) to an intensive realm where it may develop changeable patterns and habits. Emergence theory suggests that this new plane of existence can be reached by a systematic practice of disturbing the terrain's patterns in such a way as to realize a field in which any possible connection is equally probable. The key move is to construct a terrain in which patterning is flexible so that any one exclusive disjunction can be undone and an alternative patterning accessed.

An example is Savannah, Georgia, a city regarded as exhibiting a condition known as "place." But what if Savannah is not a place? What if it is a stratified terrain, a hierarchically ordered milieu that has resulted from a social machine that produces a narrative of place? This is not a production of difference, but a production of the same which has led to an ossification of production, and therefore of inhabitation and connection. Working together, these conditions are capable only of repetition, and repetition without difference is self-parody. What is required in order to move beyond the parodical production of place is a disruption that produces a mutation, a sport, in the form of a critical and emphatic production of anti-place. If peoples do not discover place but invent it, continual reinvention is necessary. If reinvention does not, or cannot occur, the collective is destined to a triviality of the same. The fetishization of place is produced by an illusion of history. The historical role of Savannah's twenty-four squares is to prevent the production of difference (and therefore of place). When one of these squares recently became available for redesign – a rescripting of the narrative of place – did mutation occur? The illusion is that by capturing and re-presenting historical circumstance, we can prevent what Augé calls anthropological place from becoming nonplace. But, is the movement to nonplace really an erasure of history? Does not this movement itself become historical? Is this movement itself not actually a *production of the historical*? (The historical being an interaction of the present with a version of the past.) To argue that we can prevent place from becoming nonplace by reference to historical conditions is to argue the modernist case – *pace* Fukuyama – that we are now beyond history. That history has ended and we are on a plateau of endless repetition; a repetition that can only be forestalled by a retrieval of historical conditions. Emergence theory provides an alternative to this argument.

How? The building of community may be advanced by thinking of public space as an assemblage that undergoes continual disturbance. To prepare the ground for this discussion it is necessary first to call on Badiou's notion of the situation, introduced in Chapter 03, Key Concepts, and how it relates to specificity. Landscape architecture has been influenced by the reification of public space, its physical instantiation as *this* space, *this* place, an idealized terrain. Alan How argues that this has narrowed the range of transformational procedures that public space can undergo (How 2003). Instead of eliding site/place/space as a permanent and immobile materiality, however, it may be regarded as transient and impermanent, even immaterial. When public space is conceived not simply as a physical arena, but as structured intermodally, intertextually and intersubjectively, the notion of site-specificity can be reframed. Seeing it as a situation challenges the idea of the site as a fixed location, and reconstructs it as no longer grounded but fluid, generated by the production of events and actions. A space-time condition that is always being reassembled, a situation is a field of possibilities that may include a physical terrain but, equally, may not. This is challenging for landscape architecture, but recall the trajectory of contemporary art. It is well known that in the twentieth century sculpture came down off its pedestals and spread itself, initially across the gallery floor, then across the deserts of New Mexico, and finally dematerialized into a performative relationality that incorporates global systems, local behaviors and endless series of transient, adventive events. Relational art is defined by Nicholas Bourriard as "a set of artistic practices which take as their theoretical and practical point of departure the whole of human relations and their social context, rather than an independent and private space" (Bourriard 2002: 113). Many artists whose work appears under this rubric are concerned with creating a social environment, an assemblage of shared encounters, rather than a physical piece at all. A 2008 artwork by Carsten Höller consisted of opening a restaurant nightclub, "The Double Club" in London, in collaboration with Germano Celant and Fondazione Prada for a six-month period. Fifty percent of its profits were donated to a charity that generates specialized projects to help abused women and children in the Democratic Republic of Congo. At Höller's 2010 show in the Hamburger Bahnhof, visitors could pay a thousand euros for a night on an exposed circular platform perched above twelve castrated reindeer, twenty-four canaries, eight mice and two flies. Gabriel Orozco's *Ping Pong Table* (1998) featured a lily pond surrounded by four ping-pong tables on which a four-person game of table tennis could be played. For *Home Run* (1993), an exhibition at the Museum of Modern Art, Orozco placed oranges in the windows of adjacent apartment buildings. Many artists have used travel, distance and international mail services as the media for their work, sending postcards to and from far-flung locations. Others construct works that appear simultaneously in different parts of the world, and then simply fade away. Similarly, surely, a landscape's relational engagement gives it a political valence that is as necessary and immediate as its status as aesthetic or ecological or eco-aesthetical artifact. The situation is a site of disturbance, of intertwining as much as of separation, and both occur at once, particularly the political and the aesthetic which are invented by the designers of landscapes as relation-events of what Massumi calls "lived abstraction" (Massumi 2012). No public square is apolitical; neither does any political act not incorporate aesthetic decisions.

Aligned with the evolving critique of a sculpture anchored in place, arts practice since the 1960s has conducted a thorough-going materialist investigation of institutional critique. The result is that site, once reified in the arts, has been reconfigured. Rather than a physical place, essentialist and authentic, the site has become "a relay or network of interrelated spaces and economies (studio, gallery, museum, art market, art criticism) which together frame and sustain art's ideological system" (Kwon 2004: 3). Arts activities have become dispersed across much broader cultural, social and discursive fields, organized intertextually through the nomadic movement of the artist. "The distinguishing characteristic of today's site-oriented art," writes Kwon, "is the way in which the artwork's relation to the actuality of a location and the social conditions of the institutional frame are both subordinate to a discursively determined . . . field of knowledge, intellectual exchange or cultural debate" (Kwon 2004: 26). This is quite different to how public space is understood in landscape architecture, where it is usually discussed as if it were a fixity instead of a set of outcomes of certain kinds of social relationship and human action (How 2003). For public spaces do not, as How puts it, "speak for themselves." It is the network of relations in which public spaces are embedded – their social matrix – that produces their significance as social explanations of community value.

A situation involves an adaptive and dynamic set of multiples. Consider a street corner where unemployed young people hang out or a parking lot where teens meet, as in Richard Linklater's film *Suburbia* (1996). The parking lot is connected inextricably, among other things, to a catalog of civic regulations and protocols that inhibit and extend its use. These protocols provide the opportunity for the mild transgression that ignites teen interest and guides their appropriation of the private/public space that parking lots instantiate. Such appropriation entangles spatial, economic and social trajectories through the application of compositional forces that can only be brought into being by drifting, deracinated teen-hood. The parking lot is not a site, but a condition. It is a modality composed of multiples – ambivalent spatiality, teen angst, darkness – all of which are composed of further multiples – laws, parents, school, unemployment, disenfranchisement, contingent edge conditions, time to burn – in short the lot is an open system. It is a situation that, despite its openness and complexity, can be recovered, gathered and collated in analysis and interpretation. The argument here is that unless these diverse situational affects are so gathered, the transformational potential of the parking lot cannot be effectively addressed through landscape architectural disturbance. How can the landscape architect who intervenes in this complex situation do so in such a way as to reinforce these young people's lived experience? How do we take seriously their empowerment? We do this through the extension of their encounters with the world that the lot enables. We create the city by discovering the lived materiality of the urban field and making this the means by which the urban is recovered from its dormancy. Physicality and subjectivity are critical to our experience of experience. This is the message of the situation. What urban landscape design can give us is the utter complicity of the situation in the construction of the subject's lived experience.

The development of contemporary public terrain, including parking lots, is a political act. Not a matter simply of the establishment of a central plaza or core site, public space is distributed along multiple broken continua: streets, parks, beach reserves, riparian edges, parking lots, set-backs, easements and so on. This has always been the case, but these supplementary ribbons and patches were usually supported by a singular magnetic condition that enabled a concentration of social and cultural capital at the intersection of significant urban flows. Now we find – in the towns, suburban precincts and urban fringes most people inhabit – that the spatial condition of civic freedom is a tangle of landscape structures whose role as public – political – milieux exceeds their primary function as highly controlled transit lines, water management systems and commercial geographies. In these heterogeneous, often unwilled, often ideological and planned, sometimes intensely contested territories, contemporary public space is forged as a collective condition of ambivalence and power. Public space is not only distributed and disconnected, but its potential as a collectively formulated civic terrain is regulated by the social machine. While sidewalks and streets, as Kohn says, "are practically the only remaining sites for unscripted political activity" nevertheless the management and policing of these serves to create a "carefully controlled, properly behaved, consuming public" (Mitchell and Van Deusen 2001; Kohn 2004: 4; Loukaitou-Sideris and Ehrenfeucht 2009).

In summary, throughout the twentieth century geographers, philosophers and urbanists on the one hand observed and decried the decline of authentic public space (a doctrine of loss), and on the other debated the conditions for its construction and renewal (a doctrine of recovery). Three factors stand out as common to all discussion. First, public space has become exclusionary and intolerant of difference. Second, it has been identified with capital and all its operations. Third, rather than being designed and administered by the people who use it, it is controlled either by the state or private corporations (Zukin 1995; Rowe 1999; Henaff and Strong 2001). In a socially saturated world where political activity is indistinguishable from general image-making and the traditional site of this activity – public space – has been given over to consumption, landscape architectural involvement can help create negotiated, democratic civic terrains.

Images are crucial to the imaginary. Landscape architecture is deeply involved in the production of images. Rather than banishing the scenographic to the suburbs and to new urbanist communities, ecological landscape urbanism has busied itself with creating the new scenographies of contemporary life. Image-production should be part of the very performativity that enables the emergence of affective, participatory design that is founded in the operations of the imaginary. Recalling Sartre, for whom the imaginary is a necessary component of human becoming, and relating his ontological freedom to Arendt's performative world-making freedom, it is just possible to see landscape architectural design as an enablement of free and open public space-making. An enactment of the social imaginary as the dimension through which humans create their ways of living together and of representing their collective life. We must always be alert to the whirring of the social machine, however. If the arguments of Habermas and Rawls for a rationalist formation of the public realm are extended to physical civic space, we can see that there is an opposition between such rationalist spatialities and the practice of performative design, which is open and indeterminate with respect to outcomes.

For the imaginary, understood as a production of emergent assemblages, cannot be grounded on democratic reason. Of course, reason calls continually on the resources of the imaginary, for instance in allegorical staging (floats covered in flowers) and national ceremony (presidential and military spectacle, for instance). If these forms of expression are instrumental props for rational power politics, then landscape architecture is implicated in the "impoverishment of the imaginary," as Corner calls it, not its inflorescence, when it unresistingly provides these resources to order. But if the imaginary, which is actually at the core of Nancy's being-in-common, and a critical feature of our forever becoming different, a site of continual transgression, is what emergence produces, then landscape architecture already inhabits that dynamic terrain. And if emergence is part of the continual reinvention of the whole symbolic order by means of which community is always becoming – "us" as memory, territory, language and custom – then the performance of landscape architecture is a crucial activity in open space construction. If not taken into account within the creative reassembly of the social, this whole layer of affect is liable to return and be glorified in forms that are, if not violent, then perverse (fascist display would be an extreme example). Georges Bataille has argued that it is society's "excess" that finds its outlet in the "glorious expenditure" of war, if it is not channeled towards a ream he calls sacred, where transgression is the key to life (as discussed in Chapter 05's Serpent of Pleasure and Sacred Groves).

Considered as an assemblage, public space cannot simply be spatial, in the sense of a physical domain. There is an important territorial component, but this must be considered carefully. While it activates an investment on the part of those who are involved in its becoming, the territorial milieu is not a privileged stage. Nor is it unmixed as to ownership and control. As argued above, public space is not necessarily a physically constructed or locatable "place." It is not necessarily site-specific. Nor is it merely terrain (though it includes terrain). It is a production of the interaction of transmissions and receptions that are material and spectral, intensive and metric, coagulant and dispersive, active and passive. It is not sufficient to think of civic landscape architecture as concerned only with the design of a physical place that backgrounds civic life. Signifiers have a symbolic power. We are in them as much as they are in us. In the Birdscapes example, the innumerable community meetings, the long discussions, the documents drawn up and signed, the squabbles, stories, histories and newspaper articles are all part of the assemblage, featuring as both expressive and processual components. Shepherd's own participation in these is critical, as are her plans, drawings, models. Her executive summaries, power point presentations and project budget lines are inseparable from the weekends the community is spending digging, planting, gathering seeds, mapping sightlines and flightlines and picnicking between times in the back dunes. Future states of the assemblage and its components may hinge on the outcome of discussions had while eating lunch and reading the local newspaper. Whole regimes of signs will affect this outcome. These signs are involved in the evaluation of the terrain according to different criteria: ease of access, economic inputs and outputs (maintenance, use as concession sites), inhabitation by vacationing surfers, mountain biker activities and so on.

This chapter has suggested strong connections between the idea of the imaginary and the notion of the ongoing construction of public space by the people who live within it. It mobilizes the overarching concept of assemblage as an interweaving of becomings that influence each other and the joint development of the space in which they find themselves. It proposes that the landscape architect is a performative designer whose continued involvement in communities catalyzes the making of the civic realm. It takes commitment to make a world.

07
CONCLUSION

The Introduction asked a number of questions about emergence theory. The Conclusion, also in interrogatory mode, asks quite rightly, if these have been addressed. The Introduction also said that the book is a voyage of discovery. Logically therefore, it did not have a specific landfall in mind. Putting these together we may consider whether anything of value has been returned to the home port from lands distant or near. One of the first queries, and really the motivation for the project, was how and why are ideas to do with emergence changing contemporary landscape discourse? Since the 1960s systems theory and the cascade of theoretical concepts that has tumbled out of it have spread through the sciences and the humanities. Three parallel developments have contributed to the growth of emergence theory in landscape architecture: the development of digital platforms for analysis, design and representation; conceptual shifts in ecology; and the exponential rise of the city as the primary foundation of life. These have been accompanied by another, deeper transition. This is the turn in social and cultural studies to a materialist conception of society. Currently, the two most quoted writers in the world are Michel Foucault and Pierre Bourdieu (Reuters 2007). Foucault advocated a materialist approach to history, and Bourdieu a materialist approach to society. Material objects and processes are now regarded not as the products of social processes, but as constitutive of them. In landscape architecture, environmental conditions are seen as the basis of social justice, quality of life and the health of the planet. Emergence theory provides a link between these realms, with a materialist explanatory framework that is adequate to social, ecological and urban phenomena. In emergence theory, the link is provided by the postulation that all systems are generated from abstract machines, or genetic diagrams, and that they gather richness and momentum through their interactions with their environments, permitting novel forms to emerge from these interactions. These ideas are changing landscape architecture in a number of ways:

- The designer is not a purveyor of top-down solutions but a discoverer of emergent processes.
- The designer works inside the system.
- Landscapes are seen as dynamic and unpredictable rather than static and biddable.

- All landscapes are open systems.
- Cities are seen as emergent systems rather than as geographies of function.
- There is no ontological difference between urban and natural process.
- Prosthetic landscapes are not seen as inferior to natural landscapes.
- Social capital may be built from within the system if appropriate initial conditions are established.
- A social ecology is an assemblage of human and nonhuman elements.
- The most effective designs are a mixture of bottom-up and top-down interactions.

We wanted to know at the beginning of the book what emergence theory really offers landscape architecture. The intervening pages have attempted to show that it suggests a reinvestment in the actual things of the world. It provides a means of actualizing close encounters with difference, in the sense both of continually becoming different, and of the radical difference of beings. It provides a rationale for organizing these encounters and maximizing difference. But why is this important? Seeing the world as a series of assemblages or overlapping, interactive systems enables us to develop ways of clarifying and deepening those relationships, of removing blockages and designing connections. Just as an automobile functions best when all its parts are moving together with the least friction, so planet Earth can function better when its parts are performing their functions with the greatest efficiency. Human beings are beginning to see that ecosystem and planetary functioning rely on the health and vitality of the many systems that connect to them. Designers are realizing that it is the connecting elements themselves that need attention, the places where systems meet, passages between geographies, edge conditions and components embedded in other components. These adjacencies, passageways and alembics are points of refinement and purification, where dissimilar forces disturb a gradient field and new conditions arise as a result. We think of men playing dominoes on a stoop, a deer grazing on the side of the interstate (Mom, look . . .!), a chameleon appearing on a windowsill, a rain garden channeling water from a parking lot. We think of a redesigned post-industrial riverfront transforming a city's sense of its self by opening its peoples up to all the cultures and histories a great river has to offer, or a regional approach to watershed design that takes account of systems that cannot easily be parceled into political or regulatory packages, and through this large scale landscape planning improves the resilience of myriad interlocking systems, habitats and assemblages.

We start to see how we actually need a new word to describe the metropolitan condition Mostafavi has called ecological urbanism. Our urban is more than urban and more than ecological. It is not a natural condition, nor is it an artificial one. These binaries will not do. Emergence can become critical to our understanding of the hybrid landscapes that have been fashioned from thousands of years of interaction with human and other species. Urban/natural systems are valorized by the movement of material and energy through them from source to sink. By means of their continuous exchange with the environment living systems elude thermodynamic equilibrium, staying as far as possible from it. These emergent landscapes – these hybridic citynatures, machinic ecologies, cyborg habitats – continually exchange flows of energy and matter with other systems that have different matter-energy conditions. The ongoing exchange of energetics takes the form of acquisition of energy input and dissipation of entropy output. In the cases both of natural river systems and prosthetic urban waterways, for instance, the development of order in the system is always matched by an increase in disorder in the adjacent, nested and overlapping systems to which it connects. The interaction of the hybrid urban nature with its hybrid rural nature is the condition of life of the system. These conditions are nearly merged. The whole landscape of the world is now a gradient field across which matter-energy intensities flourish and fade away as they transition from low entropy input (good energy) to high entropy output ("waste"). It's a dissipative three-dimensional continuum seeking its apotheosis in far-from-equilibrium life. The further the evolving planetary landscape is from equilibrium the more resilient and adaptive it is, as well as the more unpredictable and indeterminate, and the more likely its constituent landscapes are to flip from one condition to another in their reactions to ongoing and inevitable disturbance.

We see this in many situations. The Arab Spring is a classic example of a social system flipping into a new state as the result of a turbulent environmental condition. Occupy Wall Street could not have been predicted by an analysis of the political and social conditions to which it was a response. The earthquake that shattered Christchurch, New Zealand in February 2011 occurred within a geography where earthquakes are common (there have been more than 10,000 in the region since the big one in September 2010), but its location and magnitude combined to make it the biggest natural disaster the country has seen. The social and political fall-out from that event is transforming perceptions of the city – "It's a very bad place for us to be mentally – you can't build, innovate, be entrepreneurial. You lose motivation, capacity to get up and help ourselves. You can't make a city out of a depression." On the other hand, "The level of destruction that we've got there is such that we've actually got an opportunity to do something really fresh, and people are feeling optimistic now. What we had before was just mishmash from historical accident . . . Now we can think about it much more carefully and do something much, much better" (the *Guardian*, February 24, 2012: 9). A massive high entropy input has caused short-term chaos, but the long-term decrease in entropy as the city recovers could result in an increase in an ordering system not seen before in this country. Christchurch is going through a period where it is reducing the internal entropy brought about by the earthquake. It is doing this by developing self-organizing structures and information systems, social assets, economy and culture. Flows of resources from the rest of New Zealand are feeding the dissipative structure of Christchurch City, which is metabolizing them as if it were a set of organisms. They are being used to rebuild the city's ordered structure in time. The greater the size of the city, the greater its need to create the means to absorb energetic resources (low entropy inputs) and to get rid of entropic wastes – high entropy output, in this case of demolition waste, polluted water, inefficient local government regulations.

Is it not in this way that emergence theory connects the cultures of science and society to landscape architecture? The traditional separation of science from society has been critiqued across a range of disciplinary situations. With the development of social constructivism and the identification of radical historical discontinuities in scientific theory and practice by Kuhn, Foucault and others, what Jardine calls "the triumphal tales of scientific progress" were challenged. Scientific consensus was analyzed not as the product of disinterested inquiry, but of pursuit of social interests. "Appeals to reason and method no longer figured as routes to truth, but as *post facto* propaganda, means of confirming and celebrating scientific doctrines secured by less palatable means" (Jardine 2011: 3). The result has been a socialization of scientific activity. Actor network theorists have argued for investigation into the ways "centers of calculation" (laboratories, observatories, museums) identify, gather, analyze and distribute information. A more general recuperation of the persons and things once marginalized or framed as insignificant in the steady progress of science has occurred, and Latour has argued for a "republic of beings," an assembly of human and nonhuman "actants" within an ontological and epistemological democracy. This book has adopted Serres' term "regional episemologies" to describe the situatedness of knowledge and the dangers of universal forms of thought. With this downplaying of objective and dispassionate inquiry and the corresponding upsurge of diverse forms of emotional engagement in the sciences the mutual entanglement of the social and the scientific is now generally recognized. In landscape architecture, cartographic investigations try not to separate the social and political from the ecological, the biological, the geological and the meteorological, but instead to seek how each of these is informed, and constructed, by the others. Large-scale mining operations, hydroelectric power stations and agribusiness enterprises are inextricable from the landscapes and ecosystems within which they operate, and the social and cultural lives of the people and animals who live within them. The complex structures that interrelate these activities, transmitting information, bodies and mentalities from system to system, are precisely the subject matter of landscape architecture. Emergence theory enables designers to weave their performances through multiple assemblages, redefining, re-emphasizing and redistributing protocols and relationships in order to enhance the possibilities of existence of all the players in the systems. As emergentists, landscape architects acknowledge their embeddedness within these systems, and the motivated, contingent agendas that they pursue. There is a realization that theories are useful tools, but that the knowledges they generate for distribution and use are themselves effective only at certain times and in certain places for certain reasons. Not timeless but localized, not even ideational but embodied, not total but partial.

How is emergence theory informing practice? The first time this writer heard of emergence theory was in 1995 when a colleague came into the lunch room at Unitec School of Architecture and Landscape Architecture waving a copy of *AD* journal and said "The bastards, they've done it again!" The journal issue was called *Architecture After Geometry*. "Bloody architects," my friend said, "They've got to the new landscape theory before we have!" The issue included "From Object to Field" by Stan Allen. That edition of *AD* opened the flood gates and we were soon overwhelmed by articles and book sections discussing the roles of algorithms, self-organization, nonlinearity and feedback in architectural design. Influential pieces came in *ANY* magazine, another *AD* issue called "New Science = New Architecture" and then finally the landscape architecture explorations in Corner's *Recovering Landscape* and the various writings of Waldheim and other commentators on landscape urbanism.

The discussions we were having then were around the ideas of disturbance as a generator of design and the possibility of design being developed through the use of operations that removed the intentionality of the designer. From 2000 this author ran several landscape architecture studios that explored what he then called nonlinear design. The results were always interesting: novel disruptions of pre-existing terrain, provocative graphics and designs that seemed to access an ideational realm far removed from the exigencies of normal practice. As emergence and related concepts filtered into practice and studios in schools, several types of influence became discernible. Here, for the sake of brevity are eleven from a list that is much longer:

- Initial conditions are critical to later landscape performance.
- The designer can establish initial conditions and let other forces take over.
- Disturbance is not a remote or unlikely event – it is occurring all the time.

- Future states of a landscape are indiscernible.
- Natural systems and urban systems are open systems.
- Humans (designer, community and client) are part of the system.
- If people are part of the system they can be empowered by working with the system.
- Form is something that rises out of the designer's interaction with the landscape, not something that is imposed on it.
- Landscapes are continually finding new forms anyway.
- Being open ended means a landscape never reaches an achieved condition – there is no final form.
- There is no unchanging base against which landscape change occurs – everything is changing all the time.

Landscape architects trained to manage and control, to create form and space through the development of ideas derived from theories about place, identity, urbanism, even art history, found these, the conditions of emergence, difficult to deal with. To design with emergence is to see any situation as a set of complex adaptive systems that exhibit the primary characteristics of unpredictability, self-organization, autocatalysis and open-endedness. How does the designer engage with these conditions? Some approaches include:

Event-based scenario planning
Establishing initial conditions
Operational design (topographical, hydrological, ecological)

One of the issues discovered with nonlinear design early on was the complete lack of control associated with repetitive disturbance-based operations. The designer literally does not know what he or she is doing, and must become comfortable with that. Another challenge is that it is difficult to assign value to processes that occur through their own energetic self-organizational capacities. If this outcome has been produced by means of a response to some kind of disturbance, and that one has too – how do we decide which is the better? It is like trying to value one kind of natural process over another. Is the shedding of leaves by a liquidambar a more or less valuable process than the licking of fur off a wolf cub's face by its mother?

One of the practical challenges these issues are incubating is sometimes put like this: if there is no final product, how can a design proposition depict a desired outcome? If the landscape itself selects its future states through a process of interaction with indeterminable environmental fluctuations, how do we know what to depict – and how does the client know what they are getting? These questions are based on past epistemologies of design. They are not actually landscape architectural questions. The garden always defies the gardener. The client might know what he or she is getting from the landscape architect: a package of documents that describes a proposition and provides an analytical rationale for it, based on a thorough investigation of the site, its physical context and all the social and economic forces that influence it. But the client will never know what the final outcome will be, or even when an evaluation is best conducted. The best professional firms will map alternative futures for the client, showing how elements may change according to the influence of speculated contextual dynamics, but no outcome can ever be certain and unintended consequences are more plentiful in landscape architecture than in any other design discipline.

How does emergence theory enable us to reformulate relationships between humans and nonhumans, between humans and the world? Humans and nonhumans are seen as inhabiting the same ontological field. There are no superior and inferior beings. Humans and nonhumans encounter each other as members of a collective that works together to create a common world in which each species can articulate its potential. Creatures are no longer seen as having essences with fixed and indisputable boundaries but as exhibiting habits that change and evolve through negotiation with other creatures and with the world. Humans are in the system, whatever the system is. They are not subjects to the world's objects.

This author realized quite early in his career as a landscape architect that what was standing between him and a profound appreciation of landscape was landscape architecture.

NOTES

01 Introduction

1. The term is Bruno Latour's, from *The Politics of Nature* (2004: 246).

02 Narratives

1. Systems theory was founded by Ludwig von Bertalanffy, who worked on the development of biological systems theory in the 1930s and 1940s. His magnum opus, *General Systems Theory: Foundations, Development, Applications*, was published in 1976.
2. See www.red3d.com/cwr/index.html for Craig Reynolds homepage and www.red3d.com/cwr/boids/ for the procedural model of Boids.
3. The philosopher Michel Serres popularized the centrality of relationships in his five-volume series of books, *Hermes*, which began publication in 1969.
4. Interestingly, Gaston begins by saying that in the twenty-first century "'urban ecology came of age."' This is echoed in Lister (2010) who states in her first sentence "'Ecology has come of age'."
5. See, for instance, most of *New Science = New Architecture?*, issue 67(9/10) of *AD* (1997); Bettum and Hensel (2000); Batty (2000); Salingaros (1998); Davidson et al. (1997).
6. Clearly this is not good for the people who look after cities. Stability and order are much easier to administrate than unpredictability and flux. Most twentieth-century urban design has supported a symbiotic relationship between cities and the elected bodies that administer them. The ideal of the masterplan, for instance, demonstrates how predictability and investment go hand in hand. It is no coincidence that the primary visual tool of New Urbanism is the masterplan.

03 Key concepts

1. Respectively: Mostafavi and Nalje 2003; Corner 1999c; Schroder 2001; Waldheim 2006a.
2. See Vista's website: www.landscape-architects.eu/about-vista.html.
3. Hayles advises by way of counter-argument that the introduction of disorder into a system does not *necessarily* imply a movement to greater complexity. See Hayles 1990:196–216.

04 Projects and practices

1. See Ted Hughes' translation *Tales From Ovid* (1997) for vivid renderings of the poet's vision.
2. The Daphne myth has been the subject of numerous paintings and sculptures. Tiepolo (fifteenth century, Florence) and Pollaiuolo (eighteenth century, Venice) both capture Daphne at the moment of her *cri de coeur*, as she calls to her father the river god Peneus to save her and she begins to change. Bernini (seventeenth century, Rome) also takes the moment of transformation as his subject.

3. Deleuze has distinguished between the movement from the possible to the real on the one hand, and the transition from the virtual to the actual on the other. Unlike the movement from possible to real, the transition from virtual to actual does not assume a set of predefined non-physical but existent forms which achieve physical reality as material forms that resemble them. Deleuze's distinction does not involve resemblance of any kind. Structures as different as spheres and cubes emerge from the same topological point.
4. Tom Carey (2007), www.asla.org/lamag/lam07/february/feature 2.html. Last accessed September 20, 2011.
5. Matthew Urbanski (2007), www.asla.org/lamag/lam07/february/feature 2.html. Last accessed September 20, 2011.
6. Robin Moore (2007), www.asla.org/lamag/lam07/february/feature 2.html. Last accessed September 20, 2011.

05 Encounters

1. Latin for beautiful place.
2. To think of one and many as a binary condition is itself problematic. For Bergson, for instance, it is not a question of opposing the multiple to the one, but of seeing these as two types of multiplicity. Deleuze follows Bergson on this point: "The notion of multiplicity saves us from thinking in terms of one and multiple" (Deleuze 1991: 43).
3. The Greek Theocritus (born c.300 BCE) and the Latin poet Virgil (70–19 BCE) provide the pastoral origin of the *locus amoenus*, in the former's *Idylls* and the latter's *Eclogues*.
4. While in prison the Roman statesman Boethius (c.480–524 CE) consoles himself with the idea of the mutability of all earthly fortune (Boethius 1969).
5. St. Anselm (1033–1109) was a scholastic philosopher and clergyman. Quoted in Pearsall and Salter (1973: 25).
6. See Gurevich (1985) and Bartlett (2001), both of whom argue that the widespread idea that early medieval people were indifferent to nature is demonstrably false. Harvey's (1990) research also supports this argument.
7. In the last thirty years a considerable body of research has focused on the centrality of the landscape in Greek cultures. This has been conducted on the one hand by French structuralist classical scholars such as Detienne, Gernet, Vernant and Vidal-Naquet, and on the other by British and American archeologists of whom Cole, Osbourne, Rackham and Shipley are representative examples.
8. The rich repertoire of texts and images that comprise the literature on Arkadia is brought together in Harry Levin's *The Myth of The Golden Age in the Renaissance* (1969). For secondary literature see also Edgar Wind's *Pagan Mysteries in the Renaissance* (1967) and Micea Eliade's *The Myth of the Eternal Return* (1954).
9. Even when Peter Levi, the Penguin translator of Pausanius, visited the site of Phygalia in 1963 the road up to it was "not practicable except on foot; the mountain track is easier but requires a guide." Today there is a sealed road to Bassae but it is still a tortuous and demanding drive, particularly from the Arkadian side of the mountain.
10. Pausanius visited the Argive Heraion and the Temple of Poseidon outside Mantinea. The remains of these sanctuaries are still clearly visible today.
11. Polignac argues that Athens was the only exception to the bipolarity of ancient Greek cities.
12. A table of Homeric references can be found in Tom Turner's recent *Garden History* (2005).
13. By the use of the term declination I intend to denote a movement away from a primary organizational prescription. Like the term declension, also employed here, it has its etymological roots in the Latin *declinatio*, defined by Websters as "turning aside" (*Webster's New Collegiate Dictionary*, 1974. London: Bell and Sons).
14. "In the classical period there were three gymnasia, all outside the city walls and all associated with a hero or god. The Academy was dedicated to the hero Akodemos, the Lyceum was associated with Apollo, and Kynosarges with Heraklites" (Webster 1969: 49).
15. Classics scholar I.M. Finley develops an extended treatment of the role of the gift in archaic Greek society in his *The World of Odysseus* where he declares that "[n]o single detail in the life of the heroes receives so much attention in the *Iliad* and the *Odyssey* as gift-giving, and always there is frank reference to adequacy, appropriateness, recompense" (2002: 54).

16. Girard similarly links violence with the sacred. He sees violence as something exterior to human being, and part of all the other outside forces that threaten human being. "Violence is the heart and secret soul of the sacred" (1977).
17. See Walter Otto's *Dionysus: Myth and Cult* for a description of the Dionysian rituals where "everything that has been locked up is released" (1965).
18. Frazer (1925) still gives the most complete account (in English at least) of patterns of tree-worship. But see also Burkert (1985: 114ff.) and Cook (1914: Vol. 2, Bk 1, 392–416).
19. Both Diana and Artemis were also conceived as oak-goddesses. See Cook (1914: 417).
20. History of Garden Design in Ancient Greece (www.gardenvisit.com/got/3/greek_garden_history.htm).
21. But see Cook (1914: 394–422) for a discussion which shows the difficulties involved in relating legend (such as Zeus being associated with oaks) to the archeological record.
22. The display at the museum of Nemea includes quotations from many nineteenth-century visitors, including Christopher Wordsworth, *Greece* (1844); Richard Farrer, *A Tour in Greece* (1882); and Charles Henry Hanson, *The Land of Greece* (1886).
23. The author visited Nemea and most of the other sites discussed in this article on field trips in 2004 and 2005.
24. This bowl is on display in the museum at Delphi.
25. www.gardenvisit.com/got/3/greek_garden_history.htm. Last accessed December 15, 2012.
26. "According to the traditional local observance," Pausanius writes, "I slaughtered nothing to the goddess; the sacred law for her sacrifice dictates that private individuals and once a year the whole Phygalian community should take the fruit of cultivated trees, particularly the grape, and the honeycomb, and greasy unspun wool, and lay them on the altar in front of the cave, with oil poured over them."
27. Nearly all the groves mentioned by Pausanius are dedicated to one of these gods.
28. This is not to say that frenzy and madness can no longer occur here. Dionysian rites and "enthusiasm," prophetic possession and oracular visions were still possible in a landscape not specifically marked as sacred. See E.R. Dodds, *The Greeks and the Irrational* (1951).
29. Throughout his *Idylls*, poems about amorous young shepherds and goatherds who play truth-or-dare games, the third century BCE Greek poet Theocritus investigates the imaginative possibilities inherent in the association of transgressive behavior with a sensuous landscape setting.
30. One is reminded of Bataille's figure of the garden path that, once abandoned, is soon overgrown, "swarming with plant and animal life . . . If the path is paved with asphalt, it is for a long time sheltered from the pressure. This means that the volume of life possible . . . will not be realized . . . but the pressure is there . . . Without exploding, its extreme exuberance pours out in a movement always bordering on explosion" (Bataille 1988: 30).
31. Meyer (2008) highlights the continuities between these writers' framing of cities as emergent and dynamic and the systems-based theorizing of landscape urbanists.
32. See Strathausen (2003: 25), who locates the uncanny in "the destruction and the re-invention of visual coherence."
33. The term is Deleuze's (see 1992: 23).
34. The *nouvelle roman* was a French literary phenomenon of the 1950s through the 1980s. Robbe-Grillet, Natalie Sarraulte and Marguerite Duras were its primary exponents.
35. The term *mathexis universalis* is introduced by Alan Weiss in his memorable analysis of baroque gardens (Weiss 1998).
36. The phrase is Kathryn A. Hoffman's (1997).
37. The City of Auburn is in Lee County.
38. Pers. Comm. Holly Jantz, Auburn University, January 2011.
39. Megafauna are animals whose weight lies between 100lbs and 200lbs.
40. Pers. Comm. Richard Thorp, Alabama Department of Conservation, October 19, 2011.
41. Pers. Comm. Dana Johnson, USDA Wildlife Services Officer, October 19, 2011.
42. The identification of ecologies by reference to dominant species, practiced by Ethridge and many nonecologists, has been put into question, with the rise of ecosystem ecology in which it is considered to be misleading. This is because, on the one hand, the "dominance" of species within ecosystems changes through time and under different conditions and, on the other, dominance is regarded as a factor of the conceptual frame used by the observer.

06 Propositions

1. www.wjhooddesign.com
2. www.jamesrosecenter.org
3. See Sandel (2012), who refutes the idea that markets have no moral impact, and shows with many examples how things that are supposed to be a communal good get marketized and turned into cash.
4. Žizek does not discuss urban design, public space or place, but "reality" in general. I have adapted his argument.
5. Social machine is Deleuze and Guattari's term. It is adapted here to signify the process of capitalism that produces the contemporary civic realm as a function of economic objectives, and patterns the affective involvement of people and public space.

BIBLIOGRAPHY

Aben, R. and de Witt, S. (1999) *The Enclosed Garden: History and Development of the Hortus Conclusus and its Reintroduction into the Present-day Landscape*, Rotterdam: 010 Publishers.

Adorno, T.W. (1991) *The Culture Industry: Selected essays on mass culture*, ed. J.M. Bernstein, London: Routledge.

Agamben, G. (2003) *The Open: Man and Animal*, Palo Alto, CA: Stanford University Press.

Ahl, V. and Allen, T.F.H. (1996) *Hierarchy theory, a vision, vocabulary and epistemology*, New York: Columbia University Press.

Aldrich, R. (1955) *Kiss Me Deadly*, United Artists.

Allen, R.E. (1966) *Greek Philosophy: Thales to Aristotle*, New York: The Free Press.

Allen, S. (1997) "From Object to Field," *AD Architecture After Geometry*, 76(5/6): 24–31.

Allen, S. (1998) "Diagrams Matter," *ANY*, 23: 16–19.

Allen, S. (2004) "Field Conditions," *Quaderns*, 241 Q1.0: 22–35.

Allen, T.F.H. and Hoekstra, T.W. (1992) *Toward a Unified Ecology*, New York: Columbia University Press.

Allen, T.F.H. and Starr, T.B. (1982) *Hierarchy: perspectives for ecological complexity*, Chicago: University of Chicago Press.

Almy, D. (ed.) (2007) *On Landscape Urbanism*, Austin, TX: Center for American Architecture and Design.

Aquinas, T. (1963) *Summa Theologiae*, vol. 8, trans. T. Gilbey, London and New York: Blackfriars.

Arendt, H. (1958) *The Human Condition*, Chicago: The University of Chicago Press.

Arendt, H. (1961) *Between Past and Future*, New York: The Viking Press.

Aristotle (1952a) *De Anima*, vol. 8, *Great Books of the Western World*, London: Encyclopedia Brittanica.

Aristotle (1952b) *On the Heavens*, vol. 8, *Great Books of the Western World*, ed. R.M. Hutchins, London: Encyclopedia Brittanica.

Aristotle (1952c) *Physics*, vol. 8, *Great Books of the Western World*, London: Encyclopedia Brittanica.

Arnheim, R. (1958) *Film as Art*, London: Faber and Faber.

Arnheim, R. (1977) *The Dynamics of Architectural Form*, Berkeley and Los Angeles: University of California Press.

Assad, A. and Packard, N.H. (2008 [1992]) "Emergence," in *Emergence: Contemporary Readings in Philosophy and Science*, M.A. Bedau and P. Humphreys (eds.), Cambridge, MA: MIT Press.

Augustine, Saint (1931) *The City of God*, trans. John Healey, London and Toronto: J.M. Dent and Sons.

Bacon, E.N. (1967) *Design of Cities*, London: Thames and Hudson.

Badiou, A. (2000) *Deleuze: The Clamour of Being*, Minneapolis: University of Minnesota Press.

Banham, R. (1971) *Los Angeles: The Architecture of Four Ecologies*, London: Penguin.

Barabasi, A.L. (2002) *Linked: The New Science of Networks,* Cambridge, MA: Perseus.
Baridon, M. (1998) "The Scientific Imagination and the Baroque Garden," *Studies in the History of Gardens and Designed Landscapes*, 18(1): 5–19.
Barnett, R. (1997) "Gardens Without Meaning," *Landscape Review,* 3(2): 22–42.
Barnett, R. (2007) "Sacred Groves: Sacrifice and the Order of Nature in Ancient Greek Landscapes," *Landscape Journal*, 26(2): 252–269.
Barnett, R. (2012) *Emergence in Landscape Architecture*, New York: Routledge.
Bartlett, R. (2001) *Medieval Panorama*, London: Thames and Hudson.
Bataille, G. (1929) "L'Informe," *Documents*, 1 (December 1929).
Bataille, G. (1985) 'The formless' in *Visions of Excess: Selected Writings 1927–1939*, ed. and trans. Alan Stoekl, Minneapolis: University of Minnesota Press.
Bataille, G. (1988) *The Accursed Share*, New York: Zone Books.
Bataille, G. (1992) *Theory of Religion*, New York: Zone Books.
Batty, M. (2000) "Less is More, More is Different: Complexity, Morphology, Cities and Emergence," *Environmental and Planning B: Planning and Design*, 27(2): 167–168.
Batty, M. (2005) "Complexity in Regional Science Alonso Prize Lecture." www.complexcity.info/2011/11/12/complexity-in-regional-science-the-alonso-lecture/ (last accessed December 15, 2012).
Baudelaire, C. (1961) *Selected Verse*, London: Penguin.
Baudelaire, C. (1965) *The Painter of Modern Life and Other Essays*, London: Phaidon.
Baudrillard, J. (1990) *Seduction,* London: Macmillan.
Bedau, M.A. and Humphreys, P. (eds.) (2008) *Emergence: Contemporary Readings in Philosophy and Science*, Cambridge, MA: MIT Press.
Bell, M. and Leong, S.T. (eds.) (1998) *Slow Space*, New York: The Monacelli Press.
Benjamin, W. (1973a) *Charles Baudelaire: A Lyric Poet in the High Era of Capitalism*, London: New Left Books.
Benjamin, W. (1973b) *Illuminations*, Glasgow: Fontana.
Benson, J. and Rowe, M. (2000) *Landscape and Sustainability*, London: Routledge.
Berger, A. (2001) "Learning From Downsview," *Landscape Architecture* (March): 132–137.
Berger, J. (1980) "Why Look at Animals?" *About Looking*, London: Writers and Readers Publishing Cooperative.
Bergson, H. (1998) *Creative Evolution*, New York: Dover Publications.
Berman, M. (1988) *All That is Solid Melts into Air*, New York: Penguin.
Berrizbeitia, A. (2007) "Re-Placing Process," in *Large Parks*, J. Czerniak and G. Hargreaves (eds.), New York: Princeton Architectural Press.
Bettum, J. and Hensel, M. (2000) "Channeling Systems: dynamic processes and digital time-based methods in urban design," *Contemporary Processes in Architecture AD* 70(3): 36–43.
Birge, D. (1982) *Sacred Groves in the Ancient Greek World*, Berkeley: University of California, PhD thesis.
Blitz, J.H. (2008) *Moundville*, Tuscaloosa: The University of Alabama Press.
Boethius (1969) *The Consolation of Philosophy*, trans. V.E. Watts, Harmondsworth: Penguin.
Bogdanovich, P. (1967) *Fritz Lang in America*, New York: Praeger.
Bogue, R. (2003) *Deleuze on Cinema,* New York: Routledge.
Bois, Y.-A. and Kraus R. (2000) *Formless: A User's Guide*, New York: Zone Books.
Bourdieu, Pierre (1972) *Outline of a Theory of Practice*, Cambridge: Cambridge University Press.
Bourgain, P. (2002) "La paradis perdu," in *Sur la Terre Comme au Ciel: Jardins d'Occident a la Fin du Moyen Age*, E. Antoine (ed.), Paris: Reunion des Musees Nationaux.
Bourriard, N. (2002) *Relational Aesthetics*, Paris: Les Presse du Reel.
Borsch-Supan, H. (2000) *Antoine Watteau,* Cologne: Konemann.
Braley, C. (1994) "Historical Period Indian Archeology of the Georgia Coastal Plain," in *Georgia Archeological Context Reports*, Atlanta: Georgia Department of Natural Resources.
Braudel, F. (1995) *The Mediterranean and the Mediterranean World in the Age of Philip II*, Berkeley: University of California Press.

Braund, K.E.H. (1993) *Deerskins and Duffels: The Creek Indian Trade with Anglo-America, 1685–1815*, Lincoln: University of Nebraska Press.

Brennan, A. (2010) "Green Building and a Critique of Biomimicry," *JustMeans*. www.justmeans.com/Green-Building-a-critique-of-biomimicry-part-1/15922.html (last accessed November 4, 2010).

Buchanan, M. (2002) *Nexus: Small Worlds and the Ground-breaking Science of Networks*, London: W.W. Norton and Company.

Burkert, W. (1983) *Homo Necans: The Anthropology of Ancient Greek Sacrificial Ritual and Myth*, Berkeley: University of California Press.

Burkert, W. (1985) *Greek Religion*, Oxford: Blackwell.

Caillois, R. (1959) *Man and the Sacred*, Glencoe, IL: The Free Press.

Caillois, R. (1984) "Mimicry and Legendary Psychaesthenia," trans J. Shepley, *October* 31 (Winter): 17–32.

Camille, M. (1998) *The Medieval Art of Love: Objects and Subjects of Desire*, London: Lawrence King.

Capra, F. (1996) *The Web of Life*, London: Flamingo.

Carroll, M. (2003) *Earthly Paradises: Ancient Gardens in History and Archeology*, London: The British Museum Press.

Carroll-Spillecke, M. (1992) "The Gardens of Greece From Homeric to Roman Times," *Journal of Garden History*, 12(2): 84–101.

Casey, E.S. (1997) *The Fate of Place: A Philosophical History*, Berkeley and London: University of California Press.

Chaucer, G. (1987) *The Riverside Chaucer*, 3rd edn, ed. L.D. Benson, Boston: Houghton Miffin.

Cherchi, P. (1994) *Andreas and the Ambiguity of Courtly Love*, Toronto: University of Toronto Press.

Cilliers, P. (1998) *Complexity and Postmodernism: Understanding Complex Systems*, London and New York: Routledge.

Clayton, P. (2004) *Mind and Emergence: From Quantum to Consciousness*, Oxford: Oxford University Press.

Clayton, P. and Davies, P. (eds.) (2006) *The Re-Emergence of Emergence: The Emergentist Hypothesis from Science to Religion*, Oxford: Oxford University Press.

Cole, S.G. (1994) "Demeter in the Ancient Greek City and its Countryside," in *Placing the Gods: Sanctuaries and Sacred Space in Ancient Greece*, S.E. Alcock and R. Osborne (eds.), Oxford: Clarendon Press.

Cole, S.G. (2000) "The Landscape of Artemis," *Classical World*, 93: 471–482.

Collinge, S.K. (2009) *Ecology of Fragmented Landscapes*, Baltimore: The Johns Hopkins University Press.

Collingwood, R.G. (1954) *The Idea of Nature*, London: Oxford University Press.

Colman, Felicity J. (2010) "Affect," in *The Deleuze Dictionary*, Adrian Parr (ed.), Edinburgh: Edinburgh University Press.

Colonna, F. (1999) *Hypnerotomachia Poliphili*, trans. J. Godwin, London: Thames and Hudson.

Conan, M. (ed.) (2007) "Middle Eastern Garden Traditions: Unity and Diversity," *Dumbarton Oaks Colloquium on the History of Landscape Architecture*, Washington, DC: Dumbarton Oaks.

Connolly, P. (2006) "Making Landscape Urbanism Landscape Architectural," in *The Mesh Book*, J. Blood and J. Raxworthy (eds.), Melbourne: RMIT Press.

Cook, A.B. (1914) *Zeus: A Study in Ancient Religion*, Cambridge: Cambridge University Press.

Cook, R.E. (2000) "Do Landscapes Learn? Ecology's 'New Paradigm' and Design in Landscape Architecture," in *Environmentalism in Landscape Architecture*, M. Conan (ed.), Washington, DC: Dumbarton Oaks.

Coombes, D. (2007) *Changing the Channel on TV*, Auckland: Unitec New Zealand. Master of Landscape Architecture Thesis.

Copleston, F. (1962) *A History of Philosophy: Greece and Rome*, New York: Image Books.

Le Corbusier (1929) *The City of Tomorrow and Its Planning*, trans. Frederick Etchells, London: John Rodher.

Le Corbusier (1967) *The Radiant City*, London: Faber.

Corner, J. (1990) "A Discourse on Theory I: Sounding the Depths – Origins, Theory and Representation," *Landscape Journal*, 9(2): 61–78.

Corner, J. (1991) "A Discourse on Theory 2: Three Tyrannies of Contemporary Theory and the Alternative of Hermeneutics," *Landscape Journal*, 10(2): 115–133.

Corner, J. (1997) "Ecology and Landscape as Agents of Creativity," in *Ecological Design and Planning*, G.F. Thompson and F.R. Steiner (eds.), New York: John Wiley and Sons.

Corner, J. (1999a) "Field Operations," *AD Architecture of the Borderlands*, 69(7/8): 52–55.

Corner, J. (1999b) "The Agency of Mapping," in *Mappings*, D. Cosgrove (ed.), London: Reaktion Books.

Corner, J. (1999c) *Recovering Landscape: Essays in Contemporary Landscape Architecture*, New York: Princeton University Press.

Corner, J. (1999d) "Eidetic Operations and New Landscapes," in *Recovering Landscape: Essays in Contemporary Landscape Architecture*, New York: Princeton University Press.

Corner, J. (2003) "Landscape Urbanism," in *Landscape Urbanism: A Manual for the Machinic Landscape*, Mohsen Mostafavi and Ciro Nalje (eds.), London: Architectural Association.

Corner, J. (2006) "Terra Fluxus," in *The Landscape Urbanism Reader*, C. Waldheim (ed.), New York: Princeton Architectural Press.

Cornwall, A. and Schatten, V. (2007) *Spaces of Change*, London: Zed Books.

Cosgrove, D., (ed.) (1999). *Mappings*, London: Unwin Hyman.

Craft, C. (2011) "Urban Floodplain Design: Utilizing the Regenerative Disturbance of Flooding to Create Form and Habitat Within the Landscape," *School of Architecture and Landscape Architecture*, Auburn: Auburn University, Master of Landscape Architecture.

Cresswell, T. (2006) *On the Move: Mobility in the Modern Western World*, London and New York: Routledge.

Cresswell, T. (2010) "New cultural geography – an unfinished project," *Cultural Geographies*, 17(2): 169–174.

Crosby, A.W. (1986) *Ecological Imperialism: The Biological Expansion of Europe 900–1900*, Cambridge: Cambridge University Press.

Crosby, A.W. (1994) *Germs, Seeds and Animals: Studies in Ecological History*, Armonk and London: M.E. Sharpe.

Cullen, G. (1961) *Townscape*, London: Architectural Press.

Cuthbert, A.R. (2003) *Designing Cities: Critical Readings in Urban Design*, Oxford: Blackwell.

Czerniak, J. (ed.) (2002) "Introduction: Appearance, Performance: Landscape at Downsview," in *CASE: Downsview Park, Toronto*, J. Czerniak (ed.), New York: Prestel Publishing.

Daniels, S. (2010) "Maps of Making," *Cultural Geographies*, 17(2): 181–184.

Dante (1814) *The Divine Comedy*, trans. H.F. Carey, London: Frederick Warne.

Daston, L. and Lunbeck, E. (eds.) (2011) *Histories of Scientific Observation*, Chicago: University of Chicago Press.

David, J. and Hammond, R. (2011) *High Line*, New York: Farrar, Straus and Giroux.

Davidson, P. and Bates, D. with Kipnis, J. (1997) "Future Generations University," *Architecture After Geometry AD* 67(5/6): 32–37.

Davis, M. (1990) *City of Quartz: Excavating the Future in Los Angeles*, London: Verso.

Davis, M. (2001) "Bunker Hill: Hollywood's Dark Shadow," in *Cinema and the City: Film and Urban Societies in a Global context*, M.S.a.T. Fitzmaurice (ed.), Oxford: Blackwell.

De Burgh, W.G. (1953) *The Legacy of the Ancient World*, Harmondsworth: Penguin.

De Certeau, M. (1984) *The Practice of Everyday Life*, Berkeley: University of California Press.

De Landa, M. (1998) "Deleuze, Diagrams and the Genesis of Form," *Any*, 23: 30–33.

De Landa, M. (1999a) *A Thousand Years of Nonlinear History*, New York: Swerve Editions.

De Landa, M. (1999b) "Immanence and Transcendence in the Genesis of Form," in *A Deleuzian Century?* I. Buchanan (ed.), Durham, NC: Duke University Press.

De Landa, M. (1999c) "The Nonlinear Development of Cities," in *Eco-Tec: Architecture of the In-between*, A. Marras (ed.), New York: Princeton Architectural Press.

De Landa, M. (1999d) "Extensive Borderlines and Intensive Borderlines," *Architectural Design*, 69 (7–8): 78–79.

De Landa, M. (2006) *A New Philosophy of Society: Assemblage Theory and Social Complexity*, London and New York: Continuum.
De Lorris, G. and de Meun, J. (1983) *The Romance of the Rose*, trans. C. Dahlberg, New York: Princeton University Press.
Dear, M.J. (2000) *The Postmodern Urban Condition*, Oxford: Basil Blackwell.
Debus, A.G. (1977) *The Chemical Philosophy: Paracelsian Science and Medicine in the 16th and 17th Centuries*, New York: Science History Publications.
Debus, A.G. (1978) *Man and Nature in the Renaissance,* Cambridge: Cambridge University Press.
Deleuze, G. (1991) *Bergsonism*, New York: Zone Books.
Deleuze, G. (1992) *Cinema 1: The Movement Image*, London: The Athlone Press.
Deleuze, G. (1994) *Difference and Repetition*, New York: Columbia University Press.
Deleuze, G. and Guattari, F. (1987) *A Thousand Plateaus: Capitalism and Schizophrenia,* Minneapolis and London: University of Minnesota Press.
Deleuze, G. and Guattari, F. (1994) *What Is Philosophy?,* New York: Columbia University Press.
Deng, Q. (2012) "Designing Open Systems," *School of Architecture, Planning and Landscape Architecture*, Auburn: Auburn University, Master of Landscape Architecture Thesis.
DeStefano, S. (2010) *Coyote at the Kitchen Door: Living with Wildlife in Suburbia*, Cambridge, MA: Harvard University Press.
Detienne, M. (1979) *Dionysus Slain*, Baltimore: The Johns Hopkins University Press.
Deutsche, R. (1998) *Evictions: Art and Spatial Politics*, Cambridge, MA: MIT Press.
Dimendberg, E. (2004) *Film Noir and the Spaces of Modernity*, Cambridge, MA: Harvard University Press.
Dodds, E.R. (1951) *The Greeks and the Irrational*, Berkeley: University of California Press.
Duncan, J. and Ley, D. (eds.) (1993) *Place/Culture/Representation*, London: Routledge.
Dupuy-Vachey, M.-A. (2006) *Fragonard*, Paris: Editions Terrail/Edigroup.
Durkheim, E. (1915) *The Elementary Forms of the Religious Life*, London: Allen and Unwin.
Easterling, K. (1999) *Organization Space: Landscapes, Highways and Houses in America*, Cambridge, MA: MIT Press.
Easthope, A. (1997) "Cinecities in the Sixties," in *The Cinematic City*, D.B. Clarke (ed.), London: Routledge.
Eco, U. (1986) *Art and Beauty in the Middle Ages*, New Haven and London: Yale University Press.
Edson, Michael M., Foin, Theodore C. and Knapp, Charles M. (1981) "Emergent Properties and Ecological Research," *American Naturalist*, 118: 593–596.
Edwards, P., Popkin, R.H. and Allen, R.E. (1966) *Greek Philosophy: Thales to Aristotle*, New York: Free Press.
Eliade, M. (1954) *The Myth of the Eternal Return*, Princeton: Princeton University Press.
Eliade, M. (1959) *Cosmos and History*, New York: Harper and Row.
Eliot, T.S. (ed.) (1932) "Dante," in *Selected Essays*, London: Faber and Faber.
Ellis, G.F.R. (2006) "On the Nature of Emergent Reality," in *The Re-Emergence of Emergence: The Emergentist Hypothesis from Science to Religion*, P. Clayton and P. Davies (eds.), Oxford: Oxford University Press.
Erlich, P.R. and Erlich, A.H. (1981) *Extinction: The Causes and Consequences of the Disappearance of Species*, New York: Random House.
Ethridge, R. (2003) *Creek Country: The Creek Indians and Their World,* Chapel Hill and London: The University of North Carolina Press.
Euripides (1958) *Three Great Plays of Euripides*, trans. R. Warner, New York: Mentor Books.
Euripides (1960) *Hippolytus*, in *The Complete Greek Tragedies*, ed. D.G. Lattimore, trans. D. Greve, Chicago: University of Chicago Press.
Euripides (1994) *Hippolytus*, in *The Greek Tragedies in New Translations*, ed. W. Arrowsmith, trans. R. Bagg, London: Oxford University Press.
Farr, D. (2008) *Sustainable Urbanism: urban design with nature*, Hoboken, NJ: John Wiley and Sons.
Felson, A.J. and Pickett, S.T.A. (2005) "Designed Experiments: New Approaches to Studying Ecosystems," *Frontiers in Ecology and Environment*, 3(10): 549–556.

Feltham, O. and Clemens, J. (2003) "An Introduction to Alain Badiou's Philosophy," in *Infinite Thought: Truth and the Return of Philosophy*, O. Feltham and J. Clemens (eds.), London: Continuum.
Fincher, R. and Jacobs, J.M. (eds.) (1998) *Cities of Difference*, New York and London: The Guilford Press.
Finley, I.M. (2002) *The World of Odysseus*. Cambridge: The Folio Society.
Fleming, J.V. (1969) *The Roman de la Rose: A Study in Allegory and Iconography*, Princeton: Princeton University Press.
Fox, C.H. (2006) "Coyotes and Humans: Can We Coexist?" in *Proceedings of the 22nd Vertebrate Pest Conference,* Davis: University of California, Davis.
Frampton, K. (2008) "Manuel de Sola-Morales," in *Manual De Sola-Morales: A Matter of Things,* A. Geuze, M. de Sola-Morales and K. Frampton (eds.), Rotterdam: NAi Publishers.
Frank, N. (1996) "The Crime Adventure Story: A New Kind of Detective Film," in *Perspectives on Film Noir*, R.B. Palmer (ed.), New York: GK Hall.
Frankfort, H., Frankfort, H.A., Wilson, J.A. and Jacobsen, T. (1971) *Before Philosophy: The Intellectual Adventure of Ancient Man*, Baltimore: Penguin.
Frazer, J. (1925) *The Golden Bough: A Study in Magic and Religion*, London: Macmillan.
Freeman, K. (1950) *Greek City States*, London: MacDonald.
Frow, J. (1997) *Time and Commodity Culture: Essays in Cultural Theory and Postmodernity*, Oxford: Clarendon Press.
Gaston, K.J. (ed.) (2010) *Urban Ecology*, Cambridge: Cambridge University Press.
Gehrt, S.D. (2007) "Ecology of Coyotes in Urban Landscapes," in *Proceedings of Conference on Wildlife Damage Management*, Lincoln: University of Nebraska Press.
Giamatti, A.B. (1966) *The Earthly Paradise and the Renaissance Epic*, Princeton: Princeton University Press.
Giedion, S. (1967) *Space, Time and Architecture*, Cambridge, MA: Harvard University Press.
Girard, R. (1977) *Violence and the Sacred*, Baltimore: The Johns Hopkins University Press.
Glacken, C. (1967) *Traces on the Rhodian Shore: Nature and Culture in Western Thought from Ancient Times to the End of the Eighteenth Century*, Berkeley: University of California Press.
Gleick, J. (1998) *Chaos: The Amazing Science of the Unpredictable*, London: Vintage.
Golley, F.B. (1993) *A History of the Ecosystem Concept in Ecology: More Than the Sum of the Parts*, New Haven: Yale University Press.
Gombrich, E.H. (1972) *The Story of Art*, Oxford: Phaidon.
Goodman, P. and Goodman, P. (1947) *Communitas: Means of Livelihood and Ways of Life*, Chicago: Chicago University Press.
Gordon, I.L. (ed. and trans.) (1960) *The Seafarer*, London: Penguin.
Gousset, M.-T. (2002) "Le jardin d'amour: La quete du paradis sur terre," in *Sur la Terre Comme au Ciel: Jardins d'Occident a la Fin du Moyen Age*, E. Antoine (ed.), Paris: Reunion des Musées Nationaux.
Grady, W. (1994) *The Nature of Coyotes: Voice of the Wilderness,* Vancouver and Toronto: Greystone Books.
Graham, S. (2011) "Global Homelands Security and Extraterritorialization," in *New Geographies: Scales of the Earth*, Vol. 4, Cambridge: Harvard University Press.
Grant, E. (1996) *The Foundations of Modern Science in the Middle Ages*, Cambridge: Cambridge University Press.
Gray, C. (2007) "Turning the Field: Contradictions in Landscape Urbanism," *Kerb* 15: 94–100.
Greenblatt, S. (2011) *The Swerve: How the World Became Modern*, New York and London: W.W. Norton and Company.
Grimm, N.B., Faeth, S.H., Golubiewski, N.E., Redman, C.L., Wu, J., Bai, X. and Briggs, J.M. (2008) "Global Change and the Ecology of Cities," *Science*, 319: 756–760.
Gunn, A.M.F. (1952) *The Mirror of Love: A Reinterpretation of the Roman de la Rose*, Lubbock: Texas Tech Press.
Gurevich, A.J. (1985) *Categories of Medieval Culture*, trans. G.L. Campbell, London: Routledge and Kegan Paul.
Habermas, J. (1989) *The Structural Transformation of the Public Sphere: An Enquiry into a Category of Bourgoise Society*, Cambridge: Polity Press.

Haila, Y. and Taylor, P. (2001) "The Philosophical Dullness of Classical Ecology and a Levinsian Alternative," *Biology and Philosophy*, 16(1): 93–102.
Hamilton, R. (1982) *Euripides' Hippolytus: Commentary*, Bryn Mawr, PA: Thomas Library.
Harai, J.V. and Bell, D.F. (eds.) (1982) *Hermes: Literature, Science, Philosophy*, Baltimore: The Johns Hopkins University Press.
Haraway, D. (2004) *The Haraway Reader*, New York and London: Routledge.
Harris, M. (1979) *Cultural Materialism: The Struggle for a Science of Culture*, Walnut Creek, CA: AltaMira Press.
Harrison, R.P. (1992) *Forests: The Shadow of Civilization*, Chicago: University of Chicago Press.
Harrison, R.P. (2008) *Gardens: An Essay on the Human Condition*, Chicago and London: The University of Chicago Press.
Harvey, D. (1990) *The Condition of Postmodernity*, Cambridge: Blackwell.
Harvey, D. (2006) "The Political Economy of Public Space," in *The Politics of Public Space*, S. Low and N. Smith (eds.), New York: Routledge.
Harvey, J. (1990) *Medieval Gardens*, London: Batsford.
Hawks, H. (1946) *The Big Sleep*, Warner Brothers.
Hayden, D. (1995) *The Power of Place: Urban Landscapes as Public History*, Cambridge, MA and London: MIT Press.
Hayden, D. (1996) "Urban Landscape or Public History," in *The Power of Place: Urban Landscapes as Public History*, Cambridge, MA: MIT Press.
Hayden, D. (2004a) *A Field Guide to Sprawl*, New York and London: W.W. Norton and Company.
Hayden, D. (2004b) *Building Suburbia: Green Fields and Urban Growth, 1820–2000*, New York: Vintage Books.
Hayden, D. (2006) "Building the American Way: Public Subsidy, Private Space," in *The Politics of Public Space*, S. Low and N. Smith (eds.), New York: Routledge.
Hayles, N.K. (1990) *Chaos Bound: Orderly Disorder in Contemporary Literature and Science*, New York: Cornell University Press.
Heatwole, H.F. and Taylor, J. (1987) *Ecology of Reptiles*, Chipping Norton: Surrey Beatty.
Hegarty, P. (2003) "Formal Insistence," *The Semiotic Review of Books* 13(2): 6–9.
Hegel, G.W.F. (1977) *Phenomenology of Spirit*, Oxford: Oxford University Press.
Heidegger, M. (1962) *Being and Time*, Oxford: Blackwell.
Henaff, M. and Strong, T.B. (eds.) (2001) *Public Space and Democracy*, Minneappolis: University of Minnesota Press.
Hensel, M., Menges, A. and Weinstock, M. (eds.) (2004) *Emergence: Morphogenetic Design Strategies*, special issue of *Architectural Design*, 74(3), Chichester: Wiley-Academy.
Hensel, M., Menges, A. and Weinstock, M. (eds.) (2006) *Techniques and Technologies in Morphogenetic Design*, special issue of *Architectural Design*, 76(2), London: Wiley Academy.
Heppenstall, A.J., Crooks, A.T., Lee, S.M. and Batty, M. (eds.) (2012) *Agent-Based Models of Geographical Systems*, New York: Springer.
Hess, D.J. (2009) *Localist Movements in a Global Economy: Sustainability, Justice, and Urban Development in the United States*, Cambridge, MA and London: MIT Press.
Hester, R.T. (2006) *Design for Ecological Democracy*, Cambridge, MA and London: MIT Press.
Hill, K. (2002) "Urban Ecologies: Biodiversity + Urban Design," in *CASE: Downsview, Park Toronto*, J. Czerniak (ed.), New York: Prestel Publishing.
Hill, K. (2005) "Shifting Sites," in *Site Matters: Design Concepts, Histories, and Strategies*, Carol J. Burns and Andrea Kahn (eds.), Routledge: New York.
Hinman, K. (2011) "Kudzu: How a Wonder Vine Unveiled by Japan at the 1876 Centennial Turned into a Stealth Invader and Began Eating America," *American History Illustrated*, 46(2): 38.
Hoffmann, K.A. (1997) *Society of Pleasures: Interdisciplinary Readings in Pleasure and Power During the Reign of Louis XIV*, New York: St Martin's Press.
Holland, J.H. (1998) *Emergence: From Chaos to Order*, New York: Basic Books.

Hollier, D. (1989) *Against Architecture: The Writings of George Bataille*, Cambridge, MA: MIT Press.
Holling, C.S. (1978) *Adaptive Environmental Assessment and Management*, London: John Wiley and Sons.
Homer (1965) *The Odyssey*, trans. R. Fitzgerald, London: Panther Books.
Honderich, T. (ed.) (2005) *The Oxford Companion to Philosophy*, Oxford: Oxford University Press.
Hood, W. (2012) "Hood Studio." http://www.wjhooddesign.com/ (last accessed December 15, 2012).
Hough, M. (1990) *Out of Place: Restoring Identity to the Regional Landscape*, New Haven and London: Yale University Press.
How, A. (2003) "Traditions in Social Theory," in *Critical Theory*, I. Craib (ed.), New York: Palgrave.
Howes, L.L. (1997) *Chaucer's Gardens and the Language of Convention*, Gainesville: University Press of Florida.
Howland, M. (2002) "The Legacy of Contamination and the Redevelopment of Inner City Industrial Districts," Lincoln Institute of Land Policy working paper, University of Maryland.
Hughes, T. (1997) *Tales from Ovid*, London: Faber and Faber.
Hull, D.L. (2000) "The Metaphysics of Evolution," in *The Philosophy of Ecology: From Science to Synthesis*, D.R. Keller and F.B. Golley (eds.), Athens and London: The University of Georgia Press.
Hulten, K.G.P. (1975) *Jean Tinguely: Meta*, London: Thames and Hudson.
Hunt, J.D. (1989). *The Figure in the Landscape: Poetry, Painting and Gardening During the Eighteenth Century*, Baltimore: The Johns Hopkins University Press.
Hunt, J.D. (2000) *Greater Perfections: The Practice of Garden Theory*, Philadelphia: The University of Pennsylvania Press.
Jackson, P. (1989) *Maps of Meaning*, London: Unwin Hyman.
Jameson, F. (1992) *Postmodernism, or the Cultural Logic of Late Capitalism*, Durham, NC: Duke University Press.
Jardine, N. (2011) "Chalk to Cheese," *The Times Literary Supplement*, 5671 (December 16).
Johnson, S. (2002) *Emergence: The Connected Lives of Ants, Cities and Software*, London: Penguin.
Kauffman, S.A. (1993) *The Origin of Order: Self-Organization and Selection in Evolution*, New York: Oxford University Press.
Kay, J.J. (2008) "Framing the Situation: Developing a System Description," in *The Ecosystem Approach: Complexity, Uncertainty, and Managing for Sustainability*, D. Waltner-Toews, J.J. Kay, and N.-M. Lister (eds.), New York: Columbia University Press.
Keller, D.R. and Golley, F.B. (eds.) (2000) *The Philosophy of Ecology: From Science to Synthesis*, Athens and London: University of Georgia Press.
Koestler, A. (1967) *The Ghost in the Machine*, New York: Macmillan.
Kohn, M. (2004) *Brave New Neighborhoods: The Privatization of Public Space*, New York and London: Routledge.
Koolhaas, R. (1995) "The Generic City," in *S,M,L,XL*, R. Koolhaas and B. Mau (eds.), New York: Monacelli Press.
Krutnik, F. (1997) "Something More Than Night: Tales of the Noir City," in *The Cinematic City*, D.B. Clarke (ed.), London: Routledge.
Kwinter, S. (1992) "Emergence: or the Artificial Life of Space," *Anywhere*, 2: 162–171.
Kwinter, S. (1994) "The Complex and the Singular," *Anywhere*, 3: 186–197.
Kwinter, S. (1998) "Leap in the Void: A New Organon?" *Anyhow*, 7: 22–27.
Kwinter, S. (2001) *Architecture of Time: Toward a Theory of the Event in Modernist Culture*, Cambridge, MA: MIT Press.
Kwon, M. (2004) *One Place After Another: Site-Specific Art and Locational Identity*, Cambridge, MA: MIT Press.
Landsberg, S. (1995) *The Medieval Garden*, London: British Museum Press.
Lapsey, R. (1997) "Mainly in Cities and at Night," in *The Cinematic City*, D.B. Clarke (ed.), London: Routledge.
Latour, B. (1993) *Why We Have Never Been Modern*, London: Prentice-Hall.
Latour, B. (1999) "The Trouble with Actor Network Theory." www8.open.ac.uk/researchprojects/iccm/library/23 (last accessed September 23, 2012).

Latour, B. (2004) *The Politics of Nature: How to bring the Sciences into Democracy*, Cambridge, MA: Harvard University Press.
Latour, B. (2005) *Reassembling the Social: An Introduction to Actor-Networking Theory*, Oxford and New York: Oxford University Press.
Latour, B. and Weibel, P. (eds.) (2005) *Making Things Public: Atmospheres of Democracy*, Cambridge, MA and London: MIT Press.
Leach, N. (ed.) (2009) "Digital Morphogenesis," *Architectural Design*, 79(1), Chichester: Wiley-Academy.
Lee, C. (2005) "Collaborative Models to Achieve Environmental Justice and Healthy Communities," in *Power, Justice, and the Environment*, D.N. Pellow and R.J. Brulle (eds.), Cambridge, MA: MIT Press.
Lefebvre, H. (1991) *The Production of Space*, Oxford: Blackwell.
Lefebvre, H. (1995) *Introduction to Modernity: Twelve Preludes, September 1959–May 1961*, London: Verso.
Lefebvre, H. (1996) *Writings on Cities*, Oxford: Blackwell.
Lefebvre, H. (2003) *The Urban Revolution*, Minneapolis: University of Minnesota Press.
Leibniz, G.W. (1973) "Monadology," in *Leibniz: Philosophical Writings*, London: J.M. Dent & Sons.
Levin, H. (1969) *The Myth of the Golden Age in the Renaissance*, New York: Oxford University Press.
Levins, R. and Lewontin, R.C. (2000) "Dialectics and Reductionism in Ecology," in D.R. Keller and F.B. Holley (eds.) *The Philosophy of Ecology*, Athens and London: The University of Georgia Press.
Lindenmayer, D.B. and Fischer, J. (2006) *Habitat Fragmentation and Landscape Change: An Ecological and Conservation Synthesis*, London: Island Press.
Linklater, R. (1996) *SubUrbia*, Castle Rock Entertainment.
Lister, N.-M. (1997) "A Systems Approach to Biodiversity Conservation Planning," http://www/ryerson.ca/~surp/content/faculty_administration/biographies/faculty/nml/Lister_EnvMonitAsmt.pdf (accessed December 18, 2012).
Lister, N.-M. (1998) "A Systems Approach to Biodiversity Conservation Planning," *Environmental Monitoring and Assessment*, 49: 123–155.
Lister, N.-M. (2007) "Sustainable Large Parks: Ecological Design or Designer Ecology?" in *Large Parks*, J. Czerniak and G. Hargreaves (eds.), New York: Princeton Architectural Press.
Lister, N.-M. (2008) "Bridging Science and Values: The Challenge of Biodiversity Conservation," in *The Ecosystem Approach: Complexity, Uncertainty, and Managing for Sustainability*, D. Waltner-Towes, J.J. Kay and N.-M. Lister (eds.), New York: Columbia University Press.
Lister, N.-M. (2010) "Insurgent Ecologies: (Re)Claiming Ground in Landscape and Urbanism," in *Ecological Urbanism*, M. Mostafavi and G. Doherty (eds.), Baden: Lars-Mueller Publishers.
Lloyd, G.E.R. (1992) "Greek Antiquity: The Invention of Nature," in *The Concept of Nature*, J. Torrance (ed.), Oxford: Oxford University Press.
Longus (1989) *Daphnis and Chloe*, trans. P. Turner, London: Penguin.
Lorimer, Hayden (2010a) "Forces of nature, forms of life: calibrating ethology and phenomenology," in *Taking Place: Non-Representational Geographies*, Anderson, B. and Harrison P. (eds.), London: Ashgate.
Lorimer, Hayden (2010b) "Walking: New forms and spaces for studies of pedestrianism," in *Geographies of Mobilities: Practices, Spaces and Subjects*, Cresswell, T. and Merriman, P. (eds.), London: Ashgate.
Loukaitou-Sideris, A. and Ehrenfeucht, R. (2009) *Sidewalks: Conflict and Negotiation over Public Space*, Cambridge, MA: MIT press.
Lovejoy, A.O. (1964) *The Great Chain of Being*, Cambridge, MA: Harvard University Press.
Lucretius (1928) *De Rerum Natura*, trans. W.H.D. Rouse, London: Heinemann.
Lucretius (1999) *On the Nature of the Universe*, trans. Robert Melville, Oxford: Oxford University Press.
Lynch, K. (1960) *The Image of the City*, Cambridge, MA: MIT Press.
Massumi, B. (2012) *Semblance and Event*, Cambridge, MA: MIT Press.
Mate, R. (1950) *D.O.A.*, United Artists.
Matless, D. (1998) *Landscape and Englishness*, London: Reaktion.
Marcuse, H. (1972) *Counterrevolution and Revolt*, Boston: Beacon Press.
Maturana, H.R. and Varela, F.J. (1980) *Autopoeisis and Cognition: The Realization of the Living*, Dordrecht / Boston: D. Reidel Publishing Company.
McGrane, B. (1989) *Beyond Anthropology*, New York: Columbia University Press.

Merleau-Ponty, M. (1969) *The Essential Writings of Merleau-Ponty*, New York: Harcourt, Brace and World.

Meurk, C.D. and Swaffield, S.R. (2007) "Cities as Complex Landscapes: Biodiversity opportunities, landscape configurations and design directions," *New Zealand Garden Journal*, 10(1): 10–20.

Meyer, E. (2001) "The Post-Earth Day Conundrum: Translating Environmental Values into Landscape Design," in *Environmentalism in Landscape Architecture*, M. Conan (ed.), Washington, DC: Dumbarton Oaks.

Meyer, E. (2008) "Introduction: Extending Landscape Urbanism's History," in *Lunch: Territory*, S.F. Levy and D. Malda (eds.), Charlottesville: University of Virginia, 3. www.arch.virginia.edu/lunch/print/territory/rooting.html (last accessed September 25, 2012).

Meyer, J. (1996) "The Functional Site: or the Transformation of Site Specificity," *Documents*, 7 (Fall): 20–27.

Meyer, J. (1997) "Nomads," *Parkett*, 35 (May): 205–214.

Michel, John (1973) *The Earth Spirit*, London: Thames and Hudson.

Miller, A., Martin, C. and Kerr, S. (eds.) (2007) *Kerb Journal of Landscape Architecture*, 15, Melbourne: RMIT University Press.

Miller, S.G. (1995) *Nemea*, San Francisco: University of California Press.

Mirzoeff, N. (2006) "The Flickers of Seduction: The Ambivalent and Surprising Painting of Watteau," in *Antoine Watteau: Perspectives on the Artist and the Culture of his Time*, M.D. Sheriff (ed.), Newark: University of Delaware Press.

Mitchell, M. (2003) *The Right to the City: Justice and the Fight for Public Space*, New York: Guildford Press.

Mitchell, M. (2009) *Complexity: A Guided Tour*, Oxford: Oxford University Press.

Mitchell, D. and Van Deusen, R. (2001) "Downsview Park: Open Space or Public Place?" in *CASE: Downsview Park, Toronto*, J. Czerniak (ed.), New York: Prestel GSD.

Moore, Kathryn (2012) *Overlooking the Visual: Demystifying the Art of Design*, New York: Routledge.

Morton, T. (2007) *Ecology without Nature: Rethinking Environmental Aesthetics*, Cambridge, MA: Harvard University Press.

Morton, T. (2010) *The Ecological Thought*, Cambridge, MA: Harvard University Press.

Mostafavi, M. and Doherty, G. (eds.) (2010) *Ecological Urbanism*, Baden: Lars-Muller Publishers.

Mostafavi, M. and Nalje, C. (eds.) (2003) *Landscape Urbanism: A Manual for the Machinic Landscape*, London: Architecture Association.

Mozingo, L.A. (1997) "The Aesthetics of Ecological Design: Seeing Science as Culture," *Landscape Journal*, 16(1): 46–59.

Mumford, L. (1938) *The Culture of Cities*, New York: Harcourt Brace.

Mumford, L. (1961) *The City in History*, London: Secker and Warburg.

Murray, A.S. (1988) *Who's Who in Mythology*, London: Bracken Books.

Nancy, J.L (1991a) "Of Being-In-Common," in *Community at Loose Ends*, Miami Theory Collective (ed.), Minneapolis: University of Minnesota Press.

Nancy, J.L. (1991b) *The Inoperative Community*, Minneapolis: University of Minnesota Press.

Nancy J.L (2000) *Being Singular Plural*, Stanford: Stanford University Press.

Newton, N.T. (1971) *Design on the Land: The Development of Landscape Architecture*, Cambridge, MA: Belknap Press.

Niemelä, J., Breuste, J.H., Guntenspergen, G., McIntyre, N.E., Elmqvist, T. and James, P. (eds.) (2011) *Urban Ecology: Patterns, Processes and Applications*, Oxford: Oxford University Press.

Nietschze, F. (1962) *Philosophy in the Tragic Age of the Greeks*, Chicago: Chicago University Press.

Norton, B.G. (1992) "A New Paradigm for Environmental Management," in *Ecosystem Health: New Goals for Environmental Management*, R. Constanza, B.G. Norton and B.D. Haskells (eds.), Washington, DC: Island Press.

Norton, D. (ed.) (2008) *The Bible*, London: The Folio Society.

Odum, E.P. (1971) *Fundamental of Ecology*, Philadelphia: WT Saunders and Co.

Odum, E.P. (1997) *Ecology: A Bridge Between Science and Society*, Sunderland: Sinauer and Associates.

Odum, H.T. (1983) *Systems Ecology: An Introduction*, New York: John Wiley.

Olmsted (Jr.), F.L. (1928) *Forty Years of Landscape Architecture*, New York: GP Putnam's Sons.

O'Neill, R.V., DeAngelis, D.L., Waide, J.B. and Allen, T.H.F. (1986) *A Hierarchical Concept of Ecosytems*, Princeton: Princeton University Press.
Osbourne, R. (1987) *Classical Landscape with Figures: The Ancient Greek City and its Countryside*, London: George Philip.
Otto, W. (1965) *Dionysus: Myth and Cult*, Dallas, TX: Spring Publications.
Otto, W. (1978) *The Homeric Gods*, New York: Octagon Books.
Ovid (1957) *The Art of Love*, trans. R. Humphries, Bloomington: Indiana University Press.
Ovid (1986) *Metamorphoses*, trans. A.D. Melville, Oxford: Oxford University Press.
Oxford English Dictionary (1996) Oxford: Oxford University Press.
Parr, A. (2005) *The Deleuze Dictionary*, Edinburgh: Edinburgh University Press.
Patton, P (1994) "Anti-Platonism and Art," in *Gilles Deleuze and the Theater of Philosophy*, C.V. Boundas and D. Olkowski (eds.), London: Routledge.
Pausanius (1971) *Guide to Greece*, London: Penguin.
Pavel, M.P. (2009) *Breakthrough Communities: Sustainability and Justice in the Next American Metropolis*, Cambridge, MA: MIT Press.
Pearsall, D. and Salter, E. (1973) *Landscapes and Seasons of the Medieval World*, Toronto: University of Toronto Press.
Pellow, D.N. and Brulle, R.J. (eds.) (2005) *Power, Justice, and the Environment: A Critical Appraisal of the Environmental Justice Movement*, Cambridge, MA and London: MIT Press.
Pickett, S.T.A., Kolasa, J. and Jones, C.G. (1994) *Ecological Understanding: The Nature of Theory and the Theory of Nature*, San Diego: Academic Press.
Pilkey, O. and Hume, T.M (2001) "The Shoreline Erosion Problem: Lessons from the Past," *Water and Atmosphere*, 9(2): 22–23.
Plato (1952a) *Parmenides*, in *The Dialogues of Plato*, ed. B. Jowett, Chicago: Chicago University Press.
Plato (1952b) *Phaedrus*, in *The Dialogues of Plato*, ed. B. Jowett, Chicago: Chicago University Press.
Plato (1952c) *Symposium*, in *The Dialogues of Plato*, ed. B. Jowett, Chicago: University of Chicago Press.
Plato (1952d) *Timaeus*, in *Great Books of the Western World*, ed. M. Maynard, London: Encyclopaedia Brittanica.
Plato (1952e) *The Republic*, in *The Dialogues of Plato*, ed. B. Jowett, Chicago: Chicago University Press.
Plotinus (1884) *The Six Enneads*, in *Great Books of the Western World*, ed. S. McKenna and B.S. Page, Chicago: University of Chicago Press.
Polignac, F.D. (1995) *Cults, Territory, and the Origins of the Greek City-State*, Chicago: University of Chicago Press.
Poole, K. (2004) "Potentials for Landscape as Infrastructure," in *The Mesh Book: Landscape / Infrastructure*, J. Blood and J. Raxworthy (eds.), Melbourne: RMIT Press.
Popov, N. (2011) *Generative Urban Design with Cellular Automata and Agent Based Modelling*. Unpublished paper, Unitec, New Zealand.
Prest, J. (1981) *The Garden of Eden: The Botanic Garden and the Re-Creation of Paradise*, London: Yale University Press.
Pretor-Pinney, G. (2010) "Carried along on the crest of a wave," in the *Guardian*, April 4, pp. 32–33.
Prigogine, I. (1996) *The End of Certainty: Time, Chaos and the New Laws of Nature*, New York: The Free Press.
Prigogine, I. and Stengers, I. (1982) "Postface: Dynamics From Leibniz to Lucretius," in *Hermes: Literature, Science, Philosophy*, J.V. Harari and D. Bell (eds.), Baltimore: The Johns Hopkins University Press.
Prigogine, I. and Stengers, I. (1984) *Order Out of Chaos: Man's New Dialogue With Nature*, London: Flamingo.
Protevi, J. (2009) *Political Affect: Connecting the Social and the Somatic*, Minneapolis: University of Minnesota Press.
Pulliam, H.R. and Johnson, B.R. (2002) "Ecology's New Paradigm: What Does It Offer Designers and Planners?" in *Ecology and Design: Frameworks For Learning*, B.R. Johnson and K. Hill (eds.), Washington, DC: Island Press.

Pulselli, R.M and Tiezzi, E. (2009) *City Out of Chaos: Urban Self-Organization and Sustainability*, Southampton: WIT Press.

Putnam, R.D. (2000) *Bowling Alone: The Collapse and Revival of American Community*, New York: Simon and Schuster.

Raban, J. (1974) *Soft City*, London: Hamish Hamilton.

Rackham, O. (1990) "Ancient Landscapes," in *The Greek City From Homer to Alexander*, O. and S.P. Murray (eds.), Oxford: Clarendon Press.

Rawls, J. (1971) *A Theory of Justice*, New York: Basic Books.

Raxworthy, J. and Blood, J. (eds.) (2004) *The Mesh Book*, Melbourne: RMIT Press.

Reed, C. (2007) "Landscape Urbanism Practices: Precis and Case," in *On Landscape Urbanism*, D. Almy (ed.), Austin, TX: Center for American Architecture and Design.

Reed, P. (ed.) (2005) *Groundswell: Constructing the Contemporary Landscape*, New York: The Museum of Modern Art.

Rehmann, A. and Anbrine, S. (2007) "Unity and Diversity of Mughal Garden Experiences," in *Middle Eastern Garden Traditions: Unity and Diversity*, M. Conan (ed.), Washington, DC: Dumbarton Oaks.

Reinhardt, H. (2000) "German Gardens in the Eighteenth Century: Classicism, Rococo and Neo-Classicism, in *The History of Garden Design: The Western Tradition from the Renaissance to the Present Day,* M. Mosser and G. Teysott (eds.), London: Thames and Hudson and MIT Press.

Relph, E. (1987) *The Modern Urban Landscape*, Baltimore: The Johns Hopkins University Press.

Renfrew, C. and Bahn, P. (2008) *Archeology: Theories, Methods and Practices*, London: Thames and Hudson.

Reuters (2007) Reuters' ISI Web of Science. www.timeshighereducation.co.uk (last accessed November 18, 2012).

Reynolds, C. (2012) *Boids: Background and Update.* www.red3d.com/cwr/boids/ (last accessed December 15, 2012).

Richardson, N. (ed.) (2003) *The Homeric Hymns*, London: Penguin.

Richter, H. (1965) *Dada: Art and Anti-Art,* London: Thames and Hudson.

Robbe-Grillet, A. (1961) *L'Année Dernière à Marienbad*, Paris: Les Editions Minuit.

Robbe-Grillet, A. (1965) *For a New Novel: Essays on Fiction,* New York: Grove Press.

Rorty, R. (1979) *Philosophy and the Mirror of Nature*, Princeton: Princeton University Press.

Rose, M. (2006) "Gathering 'dream of presence': a project for the cultural landscape," *Environment and Planning D: Society and Space* (24): 537–544.

Rowe, P.G. (1999) *Civic Realism,* Cambridge, MA: MIT Press.

Rudd, G. (2007) *Greenery: Ecocritical Readings of Late Medieval English Literature*, Manchester and New York: Manchester University Press.

Ruggles, D.F. (2007) *Islamic Gardens and Landscapes,* Philadelphia: University of Pennsylvania Press.

Russell, B. (1961) *History of Western Philosophy*, London: George Allen and Unwin.

Salingaros, N. (1998) "Mathematical Theory of the Urban Web," *Journal of Urban Design*, 3: 53–71.

Salingaros, N.A. (2000) "Complexity and Urban Coherence," *Journal of Urban Design*, 5: 291–316.

Sandel, M. (2012) *What Money Can't Buy: The Moral Limits of Markets*, New York: Farrar, Strauss and Giroux.

Santana, E.M. (2010) "Food Habits and Anthropogenic Supplementation in the Diet of Coyotes (Canis latrans) Along an Urban-Rural Gradient," *Wildlife and Forestry*, Auburn, Auburn University, Master of Science.

Sartre, J.P. (2004) *The Imaginary: A Phenomenological Psychology of the Imagination*, London and New York: Routledge.

Sasaki, H. (2002) *Theory in Landscape Architecture*, ed. S. Swaffield, Philadelphia: University of Pennsylvania Press.

Saunders, P.T. (1997) "Nonlinearity: What it is and Why it Matters," *AD New Science – New Architecture*, 67(9/10): 52–57.

Schmitz, O.J. (2010) *Resolving Ecosystem Complexity*, Princeton and Oxford: Princeton University Press.

Schroder, T. (2001) *Changes in Scenery: Landscape Architecture in Europe*, Basel: Birkhauser.

Schroder, T. and Joosten, H. (2005) *Buro Kiefer: Recombinations*, Oostkamp: Verlage.

Scully, V. (1979) *The Earth, the Temple and the Gods: Greek Sacred Architecture*, New Haven: Yale University Press.
Sebald, W.G. (1998) *The Rings of Saturn*, London: The Harvill Press.
Sennett, R. (1976) *The Fall of Public Man*, New York: Vintage Books.
Sennett, R. (1994) *The Uses of Disorder*, New York: W.W. Norton.
Serres, M. (1977) *La Naissance de la Physique dans la Texte de Lacrece: Fleuves et Turbulences*, Paris: Minuit.
Serres, M. (1980) *Le Parasite*, Paris: Grasset.
Serres, M. (1982a) *Hermes: Literature, Science, Philosophy*, ed. J.V. Harari and D.F. Bell, Baltimore: The Johns Hopkins University Press.
Serres, M. (1982b) "The Origin of Geometry," in *Hermes: Literature, Science, Philosophy*, ed. J.V. Harari and D.F. Bell, Baltimore: The Johns Hopkins University Press.
Serres, M. (1982c) "Lucretius, Science and Religion," in *Hermes: Literature, Science, Philosophy*, ed. J.V. Harari and D.F. Bell, Baltimore: The Johns Hopkins University Press.
Serres, M. (1982d) "Language and Space: From Oedipus to Zola," in *Hermes: Literature, Science, Philosophy*, ed. J.V. Harari and D.F. Bell, Baltimore: The Johns Hopkins University Press.
Serres, M. (1982e) "Turner Translates Carnot," in *Hermes: Literature, Science, Philosophy*, ed. J.V. Harari and D.F. Bell, Baltimore: The Johns Hopkins University Press.
Serres, M. (1982f) *The Parasite*, Baltimore: The Johns Hopkins University Press.
Serres, M. (1991) *Rome: The Book of Foundations*, Baltimore: The Johns Hopkins University Press.
Serres, M. (1992) *The Natural Contract*, Ann Arbour: The University of Michigan Press.
Serres, M. with B. Latour (1995) *Conversations on Science, Culture, and Time*, Ann Arbor: The University of Michigan Press.
Seward, B. (1960) *The Symbolic Rose*, Dallas: Spring Publications.
Sheffield, J. (ed.) (2009) *Systemic Development*, Litchfield Park, AZ: ISCE Publishers.
Shepherd, D. (2009) "Redefining Coastal Erosion," *School of Architecture and Landscape Architecture*, Auckland, Unitec New Zealand, Master of Landscape Architecture.
Sheriff, M.D. (1990) *Fragonard: Art and Erotism*, Chicago and London: University of Chicago Press.
Sheriff, M.D. (ed.) (2006) *Antoine Watteau: Perspectives on the Artist and the Culture of his Time*, Newark: University of Delaware Press.
Siewers, A.K. (2009) *Strange Beauty: Ecocritical Approaches to Early Medieval Landscape*, New York: Palgrave Macmillan.
Simpson, P. (2005) *Pohutukawa and Rata: New Zealand's Iron-Hearted Trees*, Wellington: Te Papa Press.
Smith, V. (2007) "Seize the Moment in the Strategy: Downsview Park, Toronto, Canada," *Kerb*, 15: 34–39.
Spinoza, B. (2002) *Ethics*, in *Spinoza: Complete Works*, ed. M.L. Morgan, Indianapolis and Cambridge, MA: Hackett Publishing Company.
Soja, E. (1980) "The Socio-spatial Dialectic," *Annals, Association of American Geographers*, 70: 207–225.
Soja, E. (1989) *Postmodern Geographies: The Reassertion of Space in Critical Social Theory*, London: Verso.
Soja, E. (1996) *Thirdspace: Journeys to Los Angeles and Other Real-and-Imagined Places*, Cambridge, MA: Blackwell.
Sola-Morales, I. (1996) "Terrain Vague," in *Anyplace*, C. Davidson (ed.), Cambridge, MA: MIT Press.
Sola-Morales Rubio, I. (2002) *Terrain Vague*, http://www.scribd.com/doc/48643515/de-Sola-Morales-I-Terrain-Vague (last accessed October 4, 2012).
Spicer, A. (2002) *Film Noir*, Harlow: Pearson Education.
Stagoll, C. (2005) "Difference," in *The Deleuze Dictionary*, A. Parr (ed.), New York: Columbia University Press.
Steele, J. (2005) *Ecological Architecture: A Critical History*, New York: Thames and Hudson.
Stengers, I. (1997) *Power and Invention: Situating Science*, Minneapolis: University of Minnesota Press.
Stengers, I. (2000) *The Invention of Modern Science*, Minneapolis: The University of Minnesota Press.
Stivale, C. (ed.) (2005) *Gilles Deleuze: Key Concepts*, Montreal and Kingston: McGill-Queen University Press.
Strathausen, C. (2003) "Uncanny Spaces: The City in Ruttmann and Vertov," in *Screening the City*, M. Shiel and T. Fitzmaurice (eds.), London: Verso.

Strogatz, S. (2003) *Sync: The Emerging Science of Spontaneous Order*, London: Penguin.

Tafuri, M. (1990) *The Sphere and the Labyrinth*, Cambridge, MA: MIT Press.

Taylor, C. (2004) *Modern Social Imaginaries*, Durham, NC: Duke University Press.

Taylor, M. (1993) "The Phobic Object: Abjection in Contemporary Art," in *Abject Art: Repulsion and Desire in American Art*, J.B. Levi, C. Rouser, L.C. Jones and S. Taylor (eds.), New York: Whitney Museum of American Art.

Taylor, M. (2001) *The Moment of Complexity: Emergent Network Culture*, Chicago: The University of Chicago Press.

Taylor, M. (2005) *Unruly Complexity: Ecology, Interpretation, Engagement*, Chicago and London: University of Chicago Press.

Theocritus (1978) *The Poems of Theocritus*, trans. A. Rist, Chapel Hill: University of North Carolina Press.

Thompson, J.B. (1984) *Studies in the Theory of Ideology*, Berkeley: University of California Press.

Torrance, J. (1992) *The Concept of Nature*, Oxford: Oxford University Press.

Trosper, R.L. (2005) "Emergence Unites Ecology and Society," *Ecology and Society*, 10(1): article 14.

Turner, T. (2005). *Garden History: Philosophy and Design 2000 bc–2000 ad*. London: Spon Press.

Turner, W. (1910) "Honorius of Autun," in *The Catholic Encyclopaedia*, New York: Robert Appleton Company.

Ulanowicz, R.E. (1997) *Ecology: The Ascendent Perspective*, New York: Columbia University Press.

Ulmer, E.G. (1945) *Detour*, Producers Releasing Company.

van Gerwen, R. (2004) "Force Fields in the Daily Practice of a Dutch Landscape Architect," in *The Mesh Book*, J. Blood and J. Raxworthy (eds.), Melbourne: RMIT Press.

Vasagar, J. (2012) "Public Spaces Fall Into Private Hands," the *Guardian*, June 11. www.guardian.co.uk/uk/2012/jun/11/granary-square-privately-owned-public-space (last accessed September 23, 2012).

Vidler, A. (1992) *The Architectural Uncanny: Essays in the Modern Unhomely*, Cambridge, MA: MIT Press.

Villa, D.R. (2001) "Theatricality in the Public Realm of Hannah Arendt," in *Public Space and Democracy*, M. Henaff and T.B. Strong (eds.), Minneapolis: University of Minnesota Press.

Virgil (1907) *Eclogues and Georgics*, trans. T.F. Royds, London: Everyman's Library.

Waldheim, C. (2001) "Park = City? The Downsview Park Competition," *Landscape Architecture*, 91(3): 80–85, 98–99.

Waldheim, C. (2002) "Landscape Urbanism: A Genealogy," *Praxis: Journal of Writing and Building*, 4: 10–17.

Waldheim, C. (2006a) *The Landscape Urbanism Reader*, New York: Princeton Architectural Press.

Waldheim, C. (2006b) "Landscape as Urbanism," *The Landscape Urbanism Reader*, C. Waldheim (ed.), New York: Princeton Architectural Press.

Waldheim, C. (2007) "Indeterminate Emergence," *Kerb*, 15: 14–17.

Waldrop, M.M. (1992) *Complexity: The Emerging Science at the Edge of Order and Chaos*, New York: Simon and Schuster Paperbacks.

Wall, A (1999) "Programming the Urban Surface," in *Recovering Landscape*, J. Corner (ed.), New York: Princeton University Press.

Waltner-Toews, D., Kay, J.J. and Lister, N.-M. (eds.) (2008) *The Ecosystem Approach: Complexity, Uncertainty, and Managing for Sustainability*, New York: Columbia University Press.

Watson, P.F. (1979) *The Garden of Love in Tuscan Art of the Early Renaissance*, Philadelphia: Art Alliance Press.

Watson, R.N. (2011) "The Ecology of Self in A Midsummer Night's Dream," in *Ecocritical Shakespeare*, L. Bruckner and D. Brayton (eds.), London: Ashgate.

Webber, M. (1964) "The Urban Place and Non-Place Urban Realm," *Exploration into Urban Structure*, Philadelphia: University of Pennsylvania Press.

Weinstock, Michael (2004) "Morphogenesis and the Mathematics of Emergence," in Hensel, M., Menges, A. and Weinstock, M. (eds.), *Emergence: Morphogenetic Design Strategies*, special issue of *Architectural Design*, 74(3). Chichester: Wiley-Academy.

Weiss, A.S. (1998) *Unnatural Horizons: Paradox and Contradiction in Landscape Architecture*, Princeton: Princeton Architectural Press.

Weller, R. (2006/2007) "Global Theory/Local Practice," *Kerb*, 15: 66–71.
Whitehead, A.N. (1961) *Adventures in Ideas*, New York: The Free Press.
Wickman, P.W. (1999) *The Tree That Bends: Discourse, Power and the Survival of the Maskoki People*, Tuscaloosa and London: The University of Alabama Press.
Wind, E. (1967) *Pagan Mysteries in the Renaissance*, Harmondsworth: Penguin.
Wise, J.M. (2005) "Assemblage," in *Gilles Deleuze: Key Concepts*, C.J. Stivale (ed.), Montreal and Kinsgston: McGill-Queens University Press.
Woodbury, M. (2012) "Reframing the Collective," *School of Architecture, Planning and Landscape Architecture,* Auburn, Auburn University, Master of Landscape Architecture Thesis.
Woolf, H.B. (1974) *Webster's New Collegiate Dictionary* (Gen. Ed.), London: G. Bell and Sons, Ltd.
Woolley, A.W., Chabris, C.F., Pentland, A., Hashmi, N. and Malone, T.W. (2010) "Evidence for a Collective Intelligence Factor in the Performance of Human Groups," *Science* 330 (6004): 686–688.
Wu, J. (2008) "Toward a Landscape Ecology of Cities: Beyond Buildings, Trees, and Urban Forest," in *Ecology, Planning and Management of Urban Forests: International Perspectives*, M.M. Carriero, Y.C. Song, and J.G. Wu (eds.), New York: Springer.
Wycherley, X. (1949) *How the Greeks Built Cities*, London: Macmillan.
Wylie, J. (2007) *Landscape*, Oxford and New York: Routledge.
Young, I.M. (1990) "The Ideal of Community and the Politics of Difference," in *Feminism/Postmodernism*, L.J. Nicholson (ed.), New York: Routledge.
Young, I.M. (2000) *Inclusion and Democracy*, Oxford: Oxford University Press.
Zaera-Polo, A. (2005) *Groundswell: Constructing the Contemporary Landscape*, P. Reed (ed.), New York: The Museum of Modern Art.
Žizek, S. (1993) "'The Thing That Thinks': The Kantian Background of the Noir Subject," in *Shades of Noir*, J. Kopjec (ed.), London: Verso.
Žizek, S. (1999) "The Spectre of Ideology," in *The Žizek Reader*, E. Wright and E. Wright (eds.), Blackwell: Oxford.
Žizek, S. (2008) *In Defence of Lost Causes*, London: Verso.
Zukin, S. (1995) *The Cultures of Cities*, Oxford: Blackwell Publishers.

INDEX

For Product Safety Concerns and Information please contact our EU representative GPSR@taylorandfrancis.com Taylor & Francis Verlag GmbH, Kaufingerstraße 24, 80331 München, Germany